PLF

DATE r

Fluid Motions in Volcanic Conduits:
A Source of Seismic and Acoustic Signals

Geological Society books refereeing procedures

The Society makes every effort to ensure that the scientific and production quality of its books matches that of its journals. Since 1997, all book proposals have been refereed by specialist reviewers as well as by the Society's Books Editorial Committee. If the referees identify weaknesses in the proposal, these must be addressed before the proposal is accepted.

Once the book is accepted, the Society Book Editors ensure that the volume editors follow strict guidelines on refereeing and quality control. We insist that individual papers can only be accepted after satisfactory review by two independent referees. The questions on the review forms are similar to those for *Journal of the Geological Society*. The referees' forms and comments must be available to the Society's Book Editors on request.

Although many of the books result from meetings, the editors are expected to commission papers that were not presented at the meeting to ensure that the book provides a balanced coverage of the subject. Being accepted for presentation at the meeting does not guarantee inclusion in the book.

More information about submitting a proposal and producing a book for the Society can be found on its web site: www.geolsoc.org.uk.

It is recommended that reference to all or part of this book should be made in one of the following ways:

LANE, S. J. & GILBERT, J. S. (eds) 2008. *Fluid Motions in Volcanic Conduits: A Source of Seismic and Acoustic Signals*. Geological Society, London, Special Publications, **307**.

CHOUET, B., DAWSON, P. & MARTINI, M. 2008. Shallow-conduit dynamics at Stromboli Volcano, Italy, imaged from waveform inversions. *In*: LANE, S. J. & GILBERT, J. S. (eds) 2008. *Fluid Motions in Volcanic Conduits: A Source of Seismic and Acoustic Signals*. Geological Society, London, Special Publications, **307**, 57–84.

GEOLOGICAL SOCIETY SPECIAL PUBLICATION NO. 307

Fluid Motions in Volcanic Conduits:
A Source of Seismic and Acoustic Signals

EDITED BY

S. J. LANE and J. S. GILBERT
University of Lancaster, UK

2008
Published by
The Geological Society
London

THE GEOLOGICAL SOCIETY

The Geological Society of London (GSL) was founded in 1807. It is the oldest national geological society in the world and the largest in Europe. It was incorporated under Royal Charter in 1825 and is Registered Charity 210161.

The Society is the UK national learned and professional society for geology with a worldwide Fellowship (FGS) of over 9000. The Society has the power to confer Chartered status on suitably qualified Fellows, and about 2000 of the Fellowship carry the title (CGeol). Chartered Geologists may also obtain the equivalent European title, European Geologist (EurGeol). One fifth of the Society's fellowship resides outside the UK. To find out more about the Society, log on to www.geolsoc.org.uk.

The Geological Society Publishing House (Bath, UK) produces the Society's international journals and books, and acts as European distributor for selected publications of the American Association of Petroleum Geologists (AAPG), the Indonesian Petroleum Association (IPA), the Geological Society of America (GSA), the Society for Sedimentary Geology (SEPM) and the Geologists' Association (GA). Joint marketing agreements ensure that GSL Fellows may purchase these societies' publications at a discount. The Society's online bookshop (accessible from www.geolsoc.org.uk) offers secure book purchasing with your credit or debit card.

To find out about joining the Society and benefiting from substantial discounts on publications of GSL and other societies worldwide, consult www.geolsoc.org.uk, or contact the Fellowship Department at: The Geological Society, Burlington House, Piccadilly, London W1J 0BG: Tel. +44 (0)20 7434 9944; Fax +44 (0)20 7439 8975; E-mail: enquiries@geolsoc.org.uk.

For information about the Society's meetings, consult *Events* on www.geolsoc.org.uk. To find out more about the Society's Corporate Affiliates Scheme, write to enquiries@geolsoc.org.uk.

Published by The Geological Society from:
The Geological Society Publishing House, Unit 7, Brassmill Enterprise Centre, Brassmill Lane, Bath BA1 3JN, UK

(*Orders*:   Tel +44 (0)1225 445046, Fax +44 (0)1225 442836)
Online bookshop: www.geolsoc.org.uk/bookshop

The publishers make no representation, express or implied, with regard to the accuracy of the information contained in this book and cannot accept any legal responsibility for any errors or omissions that may be made.

British Library Cataloguing in Publication Data

A catalogue record for this book is available from the British Library.

ISBN 978-1-86239-262-5

Typeset by Techset Composition Ltd, Salisbury, UK

Printed by MPG Books Ltd, Bodmin, UK

Distributors

North America
For trade and institutional orders:
The Geological Society, c/o AIDC, 82 Winter Sport Lane, Williston, VT 05495, USA
Orders: Tel +1 800-972-9892
            Fax +1 802-864-7626
            E-mail: gsl.orders@aidcvt.com

For individual and corporate orders:
AAPG Bookstore, PO Box 979, Tulsa, OK 74101-0979, USA
Orders: Tel +1 918-584-2555
            Fax +1 918-560-2652
            E-mail: bookstore@aapg.org
            Website: http://bookstore.aapg.org
India
Affiliated East-West Press Private Ltd, Marketing Division, G-1/16 Ansari Road, Darya Ganj, New Delhi 110 002, India
Orders: Tel +91 11 2327-9113/2326-4180
            Fax +91 11 2326-0538
            E-mail: affiliat@vsnl.com

Contents

Acknowledgements

Steve Lane and Jennie Gilbert thank Angharad Hills and staff of the Geological Society Publishing House for their help with the production of this book. Jennie and Steve also thank the following people who gave their precious time to assist in helping the authors improve the scientific and editorial quality of the papers submitted to this Special Publication.

Phil Leat
British Antarctic Survey, High Cross, Madingley Road, Cambridge, CB3 0ET, UK

Jenni Barclay
School of Environmental Sciences, University of East Anglia, Norwich, NR4 7TJ, UK

Rick Aster
Department of Earth and Environmental Science and Geophysical Research Center, New Mexico Institute of Mining and Technology, 801 Leroy Place, Socorro, NM 87801, USA

Minoru Takeo
Earthquake Research Institute, 1-1-1 Yayoi, Bunkyo-ku, Tokyo, Japan

Luca D'Auria
Istituto Nazionale di Geofisica e Vulcanologia, Osservatorio Vesuviano, UF Centro di Monitoraggio, Via Diocleziano, 328 – 80124, Napoli, Italy

Jeremy Phillips
Department of Earth Sciences, University of Bristol, Wills Memorial Building, Queens Road, Bristol BS8 1RJ, UK

Dork Sahagian
Earth and Environmental Sciences, Lehigh University, 31 Williams Dr, Bethlehem, PA 18015-3126, USA

Mie Ishihara
Earthquake Research Institute, 1-1-1 Yayoi, Bunkyo-ku, Tokyo, Japan

Jacopo Taddeucci
Istituto Nazionale di Geofisica e Vulcanologia, Dept. of Seismology and Tectonophysics, Via di Vigna Murata 605, 00143, Roma, Italy

Amanda Clarke
School of Earth and Space Exploration, Arizona State University, Tempe, AZ 85287-1404, 480-965-6590, USA

Hugh Tuffen
Department of Environmental Science and Lancaster Environment Centre, Lancaster University, Lancaster LA1 4YQ, UK

Antonella Longo
Istituto Nazionale di Geofisica e Vulcanologia, Sezione di Pisa, via della Faggiola 32, 56126 Pisa, Italy

Oleg Melnik
Institute of Mechanics, Moscow State University, 1-Michurinskii prospect, 119192, Moscow, Russia

Alison Rust
Department of Earth Sciences, University of Bristol, Wills Memorial Building, Queens Road, Bristol BS8 1RJ, UK

Sonia Calvari
Istituto Nazionale di Geofisica e Vulcanologia, Piazza Roma 2, 95123 Catania, Italy

Mike James
Department of Environmental Science and Lancaster Environment Centre, Lancaster University, Lancaster LA1 4YQ, UK

Bernard Chouet
US Geological Survey, 345 Middlefield Road, MS 910, Menlo Park, California, CA 94025, USA

Jeff Johnson
Department of Earth Sciences, University of New Hampshire, Durham, NH 03824, USA

Silvio De Angelis
Montserrat Volcano Observatory, Flemmings, Montserrat, West Indies

Bruce Julian
US Geological Survey, 345 Middlefield Rd, MS977, Menlo Park, CA 94025, USA

Lionel Wilson
Department of Environmental Science and Lancaster Environment Centre, Lancaster University, Lancaster LA1 4YQ, UK

Liz Parfitt
Cynlais, Conwy Old Road, Dwygyfylchi, Conwy LL34 6RB, UK

Harry Pinkerton
Department of Environmental Science and Lancaster Environment Centre, Lancaster University, Lancaster LA1 4YQ, UK

The consequences of fluid motion in volcanic conduits

JENNIE S. GILBERT & STEPHEN J. LANE

[1]*Department of Environmental Science, Lancaster Environment Centre, Lancaster University,
LA1 4YQ, UK (e-mail: j.s.gilbert@lancaster.ac.uk)*

Abstract: When volcanoes are active, there are characteristic signs such as ground movement, sounds, heat and ejected material. Each of these signs is a result of, and hence an information source for, fluid motion in volcanic conduits. Here we briefly review some of the links between these signs and fluid flow processes and suggest future directions that should allow advancement of eruption forecasting as these links become understood more fully. Cross-fertilization between increasingly realistic numerical and experimental models, diverse geophysical data sources, and chemical and physical evidence in eruptive products can be achieved by simultaneously applying these approaches to well-studied volcanoes.

Volcanic activity comprises a wide range of phenomena that result from complex non-linear interactions and feedback mechanisms. Processes that start in the volcanic plumbing system determine the nature of subsequent events and control eruption styles. On emerging from the vent, volcanic material enters the atmosphere and the ensuing interactions are key in determining the consequent transport of volcanic debris that defines the impact on the environment and on human lives and infrastructure. Our capability to mitigate volcanic hazards relies in large part on forecasting eruptive events, and this, in turn, requires a high degree of understanding about the physical and chemical processes operating during volcanism. The ability to interpret surface observations and measurements in terms of subsurface processes is a key step in eruption forecasting.

The flow of magma in volcanic plumbing systems is a prerequisite for surface eruptive activity. For magma flow to take place, force must be applied and pressure gradients must exist. Forces and pressures in magma systems will change with time on a range of scales, resulting in motion of conduit walls that, if sufficiently large, will be measurable. Fluids in motion also exert drag on the walls of a conduit, variations of which will create ground motion (e.g. Green *et al.* 2006). Changes in force, pressure and drag may also couple into a range of resonant or other cyclical processes. This suggests that a major source mechanism of ground motion at active volcanoes is magma flow, and that this ground motion may, therefore, be interpreted in terms of fluid flow within the active volcanic conduit.

Volcanoes are largely monitored by measurement of the consequences of fluid flow within conduit systems, including motion of the ground over a wide range of timescales (e.g. Chouet

2003; Chadwick *et al.* 2006), motion of the atmosphere in response to eruption (e.g. Matoza *et al.* 2007), thermal signatures (e.g. Ball & Pinkerton 2006), and the physicochemical nature of ejected material, for example gas chemistry (Burton *et al.* 2007), crystal zonation (Blundy *et al.* 2007), and pyroclast morphology (Lautze & Houghton 2007). Many of these natural signals have the same underlying source process, namely the motion of gases, liquids and solids within and beneath volcanic edifices. The linking of passive geo-physicochemical signals to the dynamics of fluids formed the focus of a Workshop entitled 'The Physics of Fluid Oscillations in Volcanic Systems', held at Lancaster University on 7 and 8 September 2006, and seeded the production of this Special Publication.

Fluid flow in volcanoes is a difficult phenomenon to observe in action, especially with increasing volcanic explosivity index (VEI, Newhall & Self 1982; Pyle 2000). High-VEI eruptions do not happen very often, and the larger the explosion the less frequent it is, with, for example, repeat timescales of order 10^3 years for VEI 6 events such as the Krakatau 1883 eruption. This makes detailed syn-event measurements of large explosive episodes infrequent on the timescale of contemporary science; an event like Krakatau in 1883 has never been monitored by modern methods. Medium-VEI events such as vulcanian explosions frequently destroy near-field observation equipment (e.g. Voight *et al.* 1998), suggesting that proximal measurements during high-VEI eruptions are very difficult to obtain even when an event does occur. Processes that contribute to an explosive event are difficult to access directly; a prime example of this is how to obtain data about the nature of flow in the volcanic conduit system during explosive activity. Direct measurements are not possible,

From: LANE, S. J. & GILBERT, J. S. (eds) *Fluid Motions in Volcanic Conduits: A Source of Seismic and Acoustic Signals*. Geological Society, London, Special Publications, **307**, 1–10.
DOI: 10.1144/SP307.1   0305-8719/08/$15.00 © The Geological Society of London 2008.

and our main source of information is the deformation of the conduit wall created by flow that is unsteady on a wide range of spatial and temporal scales. The current level of understanding of high-VEI events is based on combining information from different volcanoes, which introduces further complexities. However, a small population of accessible, persistently active, low-VEI volcanoes with relatively uncomplicated magma rheology, such as Erebus (Antarctica) and Stromboli (Italy), produce weakly explosive and highly repeatable eruptions on timescales of hours and, therefore, lend themselves to detailed study. Kilauea Volcano in Hawaii is heavily instrumented and continuously active. Dome-building and vulcanian eruptions, such as the Soufrière Hills, West Indies, and Sakurajima, Japan, are currently undergoing intensive study. It is from such lower-VEI volcanoes that the linking of magma flow to measurable effects is most likely. In future, a unified process-based interpretation of passive geo-physicochemical signals honed on low-VEI volcanoes should be applied in order to increase understanding and forecasting of high-VEI events.

The approach taken to gain insight into flow in volcanic conduits relies on field observations and measurements of events, their products and consequences, combined with theoretical analysis and models of processes, and laboratory experimental modelling of materials and mechanisms. Laboratory and numerical approaches may predict currently uninterrupted and undetected volcanic signals, as well as providing knowledge of system behaviour (e.g. Longo *et al.*; Kurzon *et al.*; Rust *et al.*; James *et al.*, 2008). A range of types of field observations of both volcanic activity and post-eruption volcanic rocks provides the 'ground truth' for numerical and experimental modelling, as well as providing vital historical data. However, using one or two types of field observation in isolation makes it difficult to attribute those observations to a specific fluid-flow source process in a volcanic conduit. A combined interpretation of many types of observation and measurement at individual volcanoes is likely to be a powerful way forward.

Flow of low-viscosity magma

Eruptions that typically occur at Stromboli and Hawaii involve low-viscosity basalt magma. Strombolian eruptions are discrete, gas-rich events that result from the large-scale separation of water vapour from magma. Hawaiian eruptions are longer-lived and relatively continuous events where water vapour and magma are erupted with less separation. See Parfitt & Wilson (2008) for

further details and the relationship between eruptive styles and magma viscosity.

Ohminato *et al.* (1998) linked a particular ground deformation signal measured at Kilauea Volcano, Hawaii, to a conceptual fluid-flow source process (Fig. 1). The key components of this source process were conduit geometry and the presence of two phases of significantly different compressibility, viscosity and density, namely water vapour and silicate melt. The interaction of conduit geometry with gas-liquid flows, where gas bubbles are of similar dimension to the conduit radius, provides a rich vein of source processes with which to generate seismic and acoustic signals.

Inversion of highly repeatable seismic measurements at Stromboli (e.g. Auger *et al.* 2006; Chouet *et al.* 2003) suggests that magma motion generates a downward-directed force of about 10^8 N, together with an expansion of the source region (Fig. 2). Inversion also reveals that the conduit is an inclined dyke. This has important consequences for low-VEI systems where gravity often plays a major role in magma flow. Magma generally contains 1–5 wt% water; however, Strombolian eruptions at Stromboli are much more water rich (Chouet *et al.* 1974; Blackburn *et al.* 1976). This strongly suggests that water vapour is separating from the parent magma to form the erupted material. The chemistry of gases erupted during the Strombolian activity also suggests rapid separation of magma and volatile (on diffusion timescales) from depths between 900 m and 2.7 km (Burton *et al.* 2007), supporting the previous observations.

Intermittent gas-rich eruptions at Stromboli suggest the periodic rise and burst of large overpressured bubbles of water vapour separated by relatively bubble-poor magma. In order to understand how Strombolian eruptions are created, it is critical to study the processes that transform a bubbly magma deep in the volcano to the burst of one large bubble at the surface. Water-gas systems, extensively studied for industrial purposes (e.g. Mudde 2005), show that large bubbles form by coalescence of many small bubbles when the overall gas volume fraction exceeds about 0.25 (Clift *et al.* 1978). However, geometric considerations indicate that the time intervals between large bubble bursts would be similar to burst duration with such a high gas volume fraction. Stromboli erupts for tens of seconds, separated by gaps of tens of minutes (Chouet *et al.* 2003), suggesting that other mechanisms operate to promote the formation of gas-rich regions some 1 km or more below the top of the magma column.

One approach to gain understanding of volcanic processes is to carry out small-scale laboratory experiments using liquids and gases analogous to magma. Jaupart & Vergniolle (1988, 1989) injected

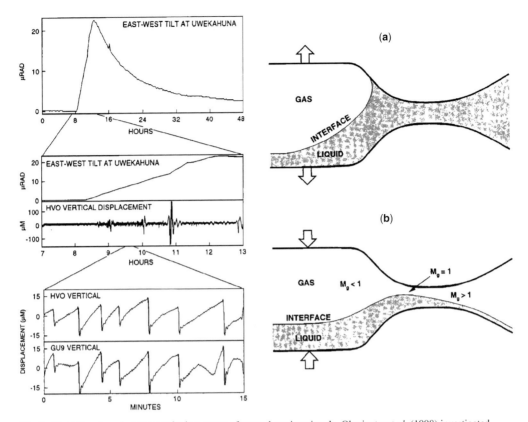

Fig. 1. Fluid flow in volcanoes results in a range of ground motion signals. Ohminato *et al.* (1998) investigated repeatable, small-amplitude, saw-tooth displacement waveforms of period 1–3 minutes (lower left panel) that were superimposed on the rising limb of a large-amplitude asymmetric waveform (top left) of approximate period 3 days occurring at Kilauea Volcano, Hawaii, on February 1, 1996. The large waveform was attributed to perturbation of magma flow feeding an eruption at the East Rift. The source of the saw-tooth displacement waveform was constrained to a small volume 1 km below the northeastern edge of the Halemaumau pit crater. The seismic source is interpreted as the slow inflation then rapid deflation of a sub-horizontal crack, but the underlying fluid-flow process is not identifiable from seismic data. One hypothetical source mechanism involves the interaction of conduit geometry with a separated gas-liquid flow. In a conduit of constant dimension, this flow may be completely separated, wavy, or slug flow; however, a constriction limits flow rate when full of viscous liquid increasing the pressure gradient in the constriction (**a** in right panel). In this condition, pressure upstream of the constriction increases causing the dyke to inflate and generate upward ground motion. Downstream, pressure is likely to decrease at this time. Eventually, the liquid supply into the constriction is exhausted and low-viscosity gas can pass through (**b** in right panel). The escape of gas, possibly limited by choked flow (M_g is Mach number of gas), results in relatively rapid reduction of the pressure gradient in the constriction. Deflation of the upstream dyke section generates the downward ground motion. The process then repeats. (Reprinted with kind permission of AGU.)

small air bubbles into the base of a flat-roofed, metre-sized tank with a narrow, vertical outlet tube, and filled with a water-based liquid. The small bubbles collected under the tank roof, which acted as a single large trap. This is not unexpected in itself, but the intermittent escape of gas up the vertical outlet tube, as the accumulating bubble layer collapsed, provided a powerful mechanism for creating one large bubble from many small ones when the average gas volume fraction was much less than 0.25. Gas-rich regions erupting at

Stromboli are a direct consequence of conduit geometry, and not flow pattern change (i.e., from bubbly to slug (Mudde 2005)) in a pipe.

Similar laboratory experiments carried out by Ripepe *et al.* (2001) also showed the presence of pressure fluctuations both within and above the liquid in the tank (Fig. 3) as the foam layer collapsed and gas escaped up the outlet pipe. In volcanoes, pressure changes in the plumbing system result in motion of the rock walls. This motion is detectable as seismic activity; therefore,

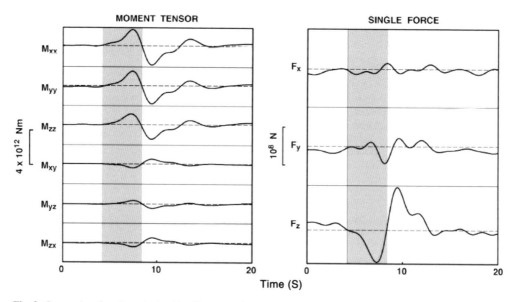

Fig. 2. Source time functions obtained by Chouet *et al.* (2003) from repeatable seismic waveforms at Stromboli Volcano, Italy. The moment tensor components (left panel) indicate that the source expands (shaded region) at the same time as a downward force (right panel, shaded region) of about 10^8 N is applied to the edifice. Analysis of seismic signals can potentially yield the geometry and motion history of the conduit, but does not 'see' into the conduit to reveal the underlying fluid accelerations. Chouet *et al.* (2008) provide further interpretation of these data. (Reprinted with kind permission of AGU.)

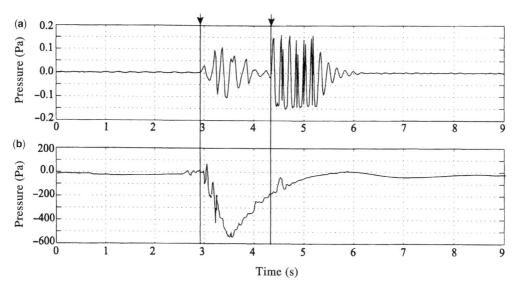

Fig. 3. The trapping of bubbles at a geometric boundary, and their consequent escape to generate a pulsatory process from a steady input, was demonstrated by Jaupart & Vergniolle (1988, 1989). Such a mechanism may be volcanically relevant where the timescale for gas-liquid separation is short compared to magma residence time in the conduit. Similar laboratory experiments by Ripepe *et al.* (2001) showed pressure fluctuations above the liquid surface (**a**) and within the experimental tank (**b**); the left-hand arrow indicates the start of slug ascent, and the right-hand arrow the time of slug burst. These experimental fluid-flow mechanisms may be likened to seismic and acoustic sources at volcanoes, and illustrate the ubiquity of seismic and acoustic source processes wherever liquid is accelerated, especially in the presence of changes in tube geometry. (Reprinted with kind permission of AGU.)

these experiments demonstrated the link between a fluid-dynamic process and a potential seismic source in volcanic systems.

At Stromboli, the trapping geometry is suggested as being between 900 m and 2.7 km below the vent (Burton *et al.* 2007). Chouet *et al.* (2008) identified a potential trap in the form of a dyke more horizontal than vertical, up which escaping gas travels, which intersects the vent feeder dyke some 900 m below the surface.

Following scaling arguments, Seyfried & Freundt (2000) injected gas into a water-filled column with a larger-diameter reservoir above. It was noted that gas-slug bursting in the column was a quiescent process, but when the water surface was in the reservoir the slug burst was more vigorous. These experiments again emphasized the importance of conduit geometry in determining fluid flow and, therefore, volcano-seismic, acoustic and thermal behaviour. Quiescent slug burst was also observed in vertical and inclined tubes of constant cross-section (James *et al.* 2004), but detailed pressure measurements revealed the presence of pressure oscillations due to bouncing of liquid on the rising gas slug (Fig. 4). The pressures measured, and the resultant forces generated, could not,

however, account for those revealed by analysis of seismic data from Stromboli (Chouet *et al.* 2003). Tube geometry, in the form of an upward widening, was found to generate forces and pressures that scale approximately to those measured at Stromboli (James *et al.* 2006). Figure 5 illustrates the flow of liquid linked to seismic signals generated at Stromboli. The main physical process is the deceleration of a volume of liquid falling under gravity, which explains why the observed force is downward. James *et al.* (2006) also found that the force does not have to couple into the conduit in the source region of fluid motion, but into any surface with a horizontal component below the source. This experimental insight indicated that seismic signals do not come from point sources, but instead from regions of the conduit wall. Analysis of seismic data to investigate extended rather than point sources has been developed by Nakano *et al.* (2007), and application of this to Stromboli is likely to expand knowledge of the conduit geometry as well as the fluid-dynamic source process. These experiments also demonstrated that conduit geometry is at the root of both the fluid and coupling processes, and explains why the seismic source was invariant in space. Any movement in source position

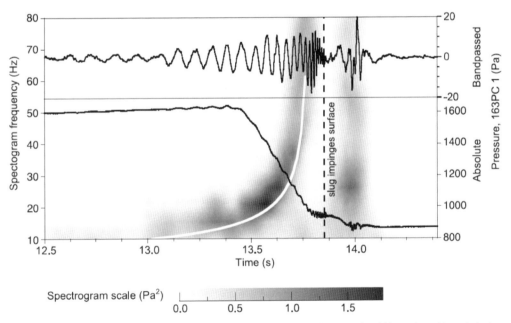

Fig. 4. Pressure changes occur within liquid (lower black trace) as a gas slug ascends within a tube, with a relatively large pressure drop as the slug passes the pressure sensor (163PC 1). Filtering reveals a lower-amplitude gliding oscillation (upper black trace) that can be modelled as bouncing of liquid on the gas slug during ascent (white model line on monochrome spectrum). This illustrates the close linkage between a fluid flow process (slug ascent) and stimulated characteristic pressure fluctuations (bouncing liquid plug). It highlights the potential for understanding seismic and acoustic source mechanisms in terms of their fluid dynamics. (Reprinted from James *et al.* (2004) with the kind permission of Elsevier.)

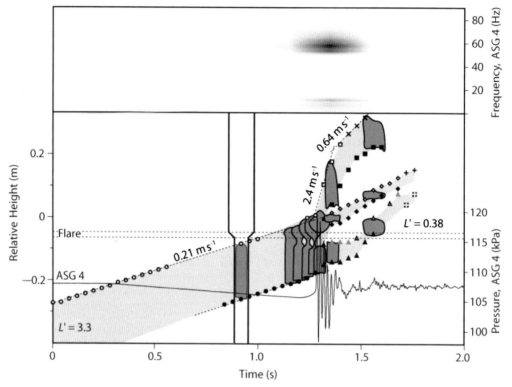

Fig. 5. The subvertical ascent of a gas slug (initial slug length to diameter ratio (L') is 3.3 in the lower tube, going to 0.38 in the upper tube) through a tube flare is a possible candidate for the fluid-dynamic source mechanism of seismic signals inverted by Chouet *et al.* (2003). The shaded region in the lower panel indicates the space–time history of the gas slug and its daughter bubbles after break-up above the tube flare. Slug break-up is caused by the formation of a closing liquid annulus forming a piston. Downward deceleration of this liquid piston produces increasing pressure and force acting downward on the apparatus. Pressure changes in excess of 10% of the static value occur (labelled ASG 4) and show characteristic frequency spectra resulting from multiple stimulated mechanisms (upper panel). The force scales to a value close to 10^8 N (Fig. 2). Contrast this mechanism with that proposed for a subhorizontal conduit (Fig. 1). (Reprinted from James *et al.* (2006) with kind permission of AGU.)

would indicate conduit evolution and heighten expectations of changing eruption behaviour.

In this Special Publication, the analysis of seismic signals from Stromboli (Chouet *et al.*) is extended to show the presence of a second, deeper source location that further constrains both the geometry of conduits feeding the surface vents and the flow processes operating. This deeper source falls in the depth range for slug formation indicated by gas chemistry (Burton *et al.* 2007). Also presented are numerical modelling results of gas slug expansion (James *et al.*) on approach to the magma surface that controls the explosivity of Strombolian eruptions, and interpretations of acoustic signals (Vergniolle & Ripepe; Vergniolle) measured during Strombolian and Hawaiian activity at Etna Volcano, Sicily, in terms of flow processes in the conduit. Thermal emission during a range of low-VEI events (Marchetti & Harris) at Pu'u 'O'o

crater, Kilauea, Hawaii gives insight into near-surface conduit geometry as well as degassing processes. Fluid motion resulting from convective overturn at several kilometres depth may theoretically produce detectable ground deformation (Longo *et al.*) and extend the timescale of forecasting larger-scale eruptions.

Flow of high-viscosity magma

The richness of oscillatory processes in volcanic systems intensifies as magma viscosity increases and rheology becomes more complex. Non-linear dependence of magma viscosity on water content and temperature (e.g. Hess & Dingwell 1996), combined with a yield strength (e.g. Pinkerton & Norton 1995), strain-rate-dependent viscosity, and transition to brittle solid behaviour at high strain rates (e.g. Dingwell & Webb 1989), makes these systems less accessible to the use of numerical and analogue models in the interpretation of oscillatory processes. The number density of measurement devices in the study of the Soufrière Hills Volcano (SHV), Montserrat, West Indies (Voight *et al.* 2006), is lower than that at Stromboli because of the greater degree of hazard for both personnel and equipment, and because the repeat timescales of events are longer. However, detailed analysis of data from the SHV displays magma flow behaviour different to, and perhaps more complex than, that of Stromboli. One major difference between Stromboli and the SHV is formation of a rheologically stiffened andesite plug in the top few hundred metres of the Montserrat conduit (Sparks 1997). This plug forms as water escapes from magma within the upper conduit, increasing melt viscosity and promoting crystal nucleation and growth. Basaltic magma at Stromboli has a lower viscosity and allows gas bubbles to separate readily from the liquid and promote magma circulation; a process that is damped out at higher magma viscosity.

The advantage of more complex ground deformation is that there is more information with which to constrain flow processes. Tuffen *et al.* (2003) provided rare evidence of a postulated seismic source preserved in rock; namely, brittle fracture and healing during flow of viscous magma. This adds solid-state processes to the list of flow-generated seismic sources. For the SHV, Green & Neuberg (2006) correlated ground deformation over a wide range of frequencies, specifically seismic and tilt signals, to reveal families of waveforms. The source of seismic signals appeared stationary in the conduit, and the tilt signal resulted from overpressure in the conduit as well as wall drag on the upward-flowing magma (Green *et al.* 2006). Neuberg *et al.* (2006) proposed a flow model to explain the measured ground deformation (Fig. 6) that incorporates the known rheology of viscoelastic magma at SHV. This work illustrates the importance of identifying the physical processes responsible for particular signals in giving clues to subterranean flow that can be used to constrain experimental and theoretical approaches to understanding flow in volcanic conduits.

Iverson *et al.* (2006) modelled the low-VEI flow of magma into the growing 2004–05 lava dome at Mount St. Helens, USA, based on multiparametric data from petrological, gas chemistry, seismic (drumbeat earthquakes), thermal, mechanical property and geodetic sources. These data are consistent with flow in a near-equilibrium stick-slip mode with gouge material mainly sourced from the magma plug. The 2004–05 eruption can be considered as a continuation of the eruptive activity of the 1980s. Inversion of data from a temporary seismic

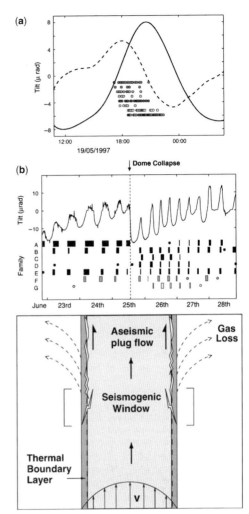

Fig. 6. The importance of measuring ground motion over a wide bandwidth is demonstrated by the correlation of tilt and families of seismic data, (**a**), measured at Soufrière Hills Volcano, Montserrat. The solid line shows tilt and the dashed line the first time-derivative of tilt. Symbols show different families of seismic signal. Dome collapse caused rapid major structural changes to the conduit, significantly altering the nature of the seismic and tilt signals (**b**). The canonical flow mechanism proposed to explain the relationship between seismic and tilt signals (lower panel) involves the transition from ductile to brittle behaviour as water escapes, causing increases in magma viscosity. The seismic source mechanism is the onset of brittle fracture and plug flow operating on the timescale of the tilt cycles. There is also the possibility that the seismogenic window is geometrically constrained (Chouet 2006, comment at Workshop); for instance a conduit narrowing would increase the rate of shear and promote brittle fracture at the flow margin. (Reprinted from Neuberg *et al.* (2006), with kind permission of Elsevier).

network (Waite *et al.* 2008) indicates two seismic sources operating. Waite *et al.* (2008) interpretation of the drumbeat earthquakes at MSH is not a stick-slip process, but the resonance of a steam-filled crack. The crack is likely supplied with water degassing from the magma and vaporized ground-water. These two contrasting interpretations based on large amounts of data demonstrate how difficult complex systems can be to understand.

In this Special Publication, a numerical model of stick-slip processes (Lensky *et al.*) explores the phenomena underpinning tilt cycles at the SHV. Lane *et al.* take a complementary analogue exper-imental approach to understanding flow processes generating the same tilt cycles, and reach a conclusion different to that of Lensky *et al.*, thus illustrating the complexities of this system. Conduit geometry is considered to play a major role in the separation of gas from liquid in low-viscosity magmas, but geometry may also influence the behaviour of more viscous dome-building magmas as shown by seismic evidence (Ohminato, 2008) from Asama Volcano, Japan.

Future directions

Difficulty in both collection and interpretation of volcanic ejecta in the context of flow processes in volcanic conduits is reflected in the paucity of research in this area. Chemical and physical evi-dence from eruptive products of unsteady processes during flow remains the ultimate test for many pro-posed source mechanisms. Such evidence could include the physical effects of brittle fracture and viscous healing, gas separation by chemical species, physical structure of pyroclasts, and chemical heterogeneity in phenocrysts and micro-lites. Collection, analysis and flow-based interpret-ation of gaseous, liquid and solid ejecta, as well as searching for evidence of flow processes in volcanic rocks, are key future directions in linking the evidence for past flow to specific flow behaviour.

The geometry of the volcanic conduit system appears to play a major role in the source mechan-ism of a number of different eruptive and oscillatory processes. Knowledge of the influence of conduit geometry on magma flow may provide a means of monitoring changes in geometry and discriminating between different flow behaviours using field measurements. Direct observation of conduit geo-metry can be made in the field from exhumed dykes, sills and other conduits. However, the force and pressure fields these radiated when active are unknown. Research at Stromboli has shown that monitoring ground motion resulting from fluid flow in active conduits with a network of wide-band

sensors has great potential for revealing the size, shape and position of geometrically important zones. Continuation of conduit-imaging work at Stromboli is crucial to understanding the way in which active conduits evolve in time and space, and respond to changes in magma flow rate or nearby slope failure, and how geometry changes can be linked to hazard forecasting.

Oscillatory processes over a wide range of periods are ubiquitous in magma flow. The sources of these processes lie in the physicochem-ical behaviour of volcanic systems from the surface downwards. In order to gain an understand-ing of these processes, and therefore improve forecast quality, the forward-modelling link between source mechanism and measurable effect first needs to be strengthened and extended to more volcanoes. This requires cross-fertilization between increasingly realistic numerical and exper-imental models of fluid- and elastodynamics, spatially and temporally dense field measurements of diverse geophysical signals at all frequencies, and chemical and physical evidence in eruptive products, new and old. Some of these opportunities are in place, and are under rapid continued develop-ment, but they will require consistent linking and multigroup collaboration to provide more comprehensive understanding.

We thank all the authors and reviewers who have made this Special Publication possible. The Geological Society of London, in particular Angharad Hills, are thanked for their assistance with the Workshop on 'The Physics of Fluid Oscillations in Volcanic Systems', held at Lancaster University in 2006, and for agreeing to publish this volume. IAVCEI, Steve McNutt, and the Faculty of Science and Technology at Lancaster University are thanked for facilitating financial assistance to developing world scientists and students, and running costs for the Workshop underpinning this volume (http://www.es.lancs.ac.uk/seismicflow/). We are grateful to Jenni Barclay and Phil Leat for constructive and helpful reviews of this paper.

References

AUGER, E., D'AURIA, L., MARTINI, M., CHOUET, B. & DAWSON, P. 2006. Real-time monitoring and massive inversion of source parameters of very long period seismic signals: An application to Stromboli Volcano, Italy. *Geophysical Research Letters*, **33**, Art. No. L04301.

BALL, M. & PINKERTON, H. 2006. Factors affecting the accuracy of thermal imaging cameras in volcanology. *Journal of Geophysical Research – Solid Earth*, **111**, B11203, doi:10.1029/2005JB003829.

BLACKBURN, E. A., WILSON, L. & SPARKS, R. S. J. 1976. Mechanisms and dynamics of Strombolian activity. *Journal of the Geological Society, London*, **132**, 429–440.

BLUNDY, J., CASHMAN, K. & BERLO, K. 2007. Volatile fluxing and magma storage at Mount St. Helens volcano. *Geochimica et Cosmochimica Acta*, **71**, A101.

BURTON, M., ALLARD, P., MURÉ, F. & LA SPINA, A. 2007. Magmatic gas composition reveals the source depth of slug-driven Strombolian explosive activity. *Science*, **317**, 227–230.

CHADWICK, W. W., JR., GEIST, D. J., JÓNSSON, S., POLAND, M., JOHNSON, D. J. & MEERTENS, C. M. 2006. A volcano bursting at the seams: Inflation, faulting, and eruption at Sierra Negra volcano, Galápagos. *Geology*, **34**, 1025–1028.

CHOUET, B. 2003. Volcano seismology. *Pure and Applied Geophysics*, **160**, 739–788.

CHOUET, B., HAMISEVICZ, B. & MCGETCHIN, T. R. 1974. Photoballistics of volcanic jet activity at Stromboli, Italy. *Journal of Geophysical Research*, **79**, 4961–4976.

CHOUET, B., DAWSON, P., OHMINATO, T., ET AL. 2003. Source mechanisms of explosions at Stromboli Volcano, Italy, determined from moment-tensor inversions of very-long-period data. *Journal of Geophysical Research*, **108**, 2019, doi:10.1029/2002JB001919.

CLIFT, R., GRACE, J. R. & WEBER, M. E. 1978. *Bubbles, Drops and Particles*. Academic Press.

DINGWELL, D. B. & WEBB, S. L. 1989. Structural relaxation in silicate melts and non-Newtonian melt rheology in geologic processes. *Physics and Chemistry of Minerals*, **16**, 508–516.

GREEN, D. N. & NEUBERG, J. 2006. Waveform classification of volcanic low-frequency earthquake swarms and its implication at Soufrière Hill Volcano, Montserrat. *Journal of Volcanology and Geothermal Research*, **153**, 51–63.

GREEN, D. N., NEUBERG, J. & CAYOL, V. 2006. Shear stress along a conduit wall as a plausible source of tilt at Soufrière Hills volcano, Montserrat. *Geophysical Research Letters*, **33**, L10306, doi:10.1029/2006GL-25890.

HESS, K.-U. & DINGWELL, D. B. 1996. Viscosities of hydrous leucogranitic melts: A non-Arrhenian model. *American Mineralogist*, **81**, 1297–1300.

IVERSON, R. M., DZURISIN, D., GARDNER, C. A. ET AL. 2006. Dynamics of seismogenic volcanic extrusion at Mount St Helens in 2004–05. *Nature*, **444**, 439–443.

JAMES, M. R., LANE, S. J., CHOUET, B. & GILBERT, J. S. 2004. Pressure changes associated with the ascent and bursting of gas slugs in liquid-filled vertical and inclined conduits. *Journal of Volcanology and Geothermal Research*, **129**, 61–82.

JAMES, M. R., LANE, S. J. & CHOUET, B. 2006. Gas slug ascent through changes in conduit diameter: laboratory insights into a volcano-seismic source process in low-viscosity magmas. *Journal of Geophysical Research (Solid Earth)*, **111**, B05201, doi:10.1029/2005JB003718.

JAUPART, C. & VERGNIOLLE, S. 1988. Laboratory models of Hawaiian and Strombolian eruptions. *Nature*, **331**, 58–60.

JAUPART, C. & VERGNIOLLE, S. 1989. The generation and collapse of a foam layer at the roof of a basaltic magma chamber. *Journal of Fluid Mechanics*, **203**, 347–380.

LAUTZE, N. C. & HOUGHTON, B. F. 2007. Linking variable explosion style and magma textures during 2002 at Stromboli volcano, Italy. *Bulletin of Volcanology*, **69**, 445–460.

MATOZA, R. S., HEDLIN, M. A. H. & GARCÉS, M. 2007. An infrasound array study of Mount St. Helens. *Journal of Volcanology and Geothermal Research*, **160**, 249–262.

MUDDE, R. F. 2005. Gravity-driven bubbly flows. *Annual Review of Fluid Mechanics*, **37**, 393–423.

NAKANO, M., KUMAGAI, H., CHOUET, B. & DAWSON, P. 2007. Waveform inversion of volcano-seismic signals for an extended source. *Journal of Geophysical Research – Solid Earth*, **112**, Art. No. B02306.

NEUBERG, J., TUFFEN, H., COLLIER, L., GREEN, D., POWELL, T. & DINGWELL, D. 2006. The trigger mechanism of low-frequency earthquakes on Montserrat. *Journal of Volcanology and Geothermal Research*, **153**, 37–50.

NEWHALL, C. G. & SELF, S. 1982. The explosivity index (VEI) – an estimate of explosive magnitude for historical volcanism. *Journal of Geophysical Research*, **87**, 1231–1238.

OHMINATO, T., CHOUET, B., DAWSON, P. & KEDAR, S. 1998. Waveform inversion of very long period impulsive signals associated with magmatic injection beneath Kilauea Volcano, Hawaii. *Journal of Geophysical Research (Solid Earth)*, **103**, 23839–23862.

PARFITT, E. A. & WILSON, L. 2008. *Fundamentals of Physical Volcanology*. Blackwell Publishing.

PINKERTON, H. & NORTON, G. 1995. Rheological properties of basaltic lavas at sub-liquidus temperatures – laboratory and field-measurements on lavas from Mount Etna. *Journal of Volcanology and Geothermal Research*, **68**, 307–323.

PYLE, D. M. 2000. Sizes of volcanic eruptions. *In*: SIGURDSSON, H., HOUGHTON, B., MCNUTT, S. R., RYMER, H. & STIX, J. (eds) 2000. *Encyclopaedia of Volcanoes*, Academic Press, 263–269.

RIPEPE, M., CILIBERTO, S. & DELLA SCHIAVA, M. 2001. Time constraints for modeling source dynamics of volcanic explosions at Stromboli. *Journal of Geophysical Research (Solid Earth)*, **106**, 8713–8727.

SEYFRIED, R. & FREUNDT, A. 2000. Experiments on conduit flow and eruption behaviour of basaltic volcanic eruptions. *Journal of Geophysical Research (Solid Earth)*, **105**, 23727–23740.

SPARKS, R. S. J. 1997. Causes and consequences of pressurisation in lava dome eruptions. *Earth and Planetary Science Letters*, **150**, 177–189.

TUFFEN, H., DINGWELL, D. B. & PINKERTON, H. 2003. Repeated fracture and healing of silicic magma generate flow banding and earthquakes? *Geology*, **31**, 1089–1092.

VOIGHT, B., HOBLITT, R. P., CLARKE, A. B., LOCKHART, A. B., MOLLER, A. B., LYNCH, L. & MCMAHON, J. 1998. Remarkable cyclic ground deformation monitored in real-time on Montserrat, and its use in eruption forecasting. *Geophysical Research Letters*, **25**, 3405–3408.

VOIGHT, B., LINDE, A. T., SACKS, I. S., *ET AL*. 2006. Unprecedented pressure increase in deep magma reservoir triggered by lava-dome collapse. *Geophysical Research Letters*, **33**, L03312, doi:10.1029/2005GL024870.

WAITE, G. P., CHOUET, B. & DAWSON, P. B. 2008. Eruption dynamics at Mount St. Helens imaged from broadband seismic waveforms: Interaction of the shallow magmatic and hydrothermal systems. *Journal of Geophysical Research*, **113**, B02305, doi:10.1029/2007JB005259.

Damping of pressure waves in visco-elastic, saturated bubbly magma

I. KURZON[1,2], V. LYAKHOVSKY[2], O. NAVON[1] & N. G. LENSKY[2]

[1]*Institute of Earth Science, The Hebrew University, Jerusalem 91904, Israel*
(e-mail: ittai.kruzon@mail.huji.ac.il)

[2]*The Geological Survey of Israel, 30 Malkhe Israel St., Jerusalem 95501, Israel*

Abstract: The attenuation of pressure waves in a saturated bubbly magma is examined in a model, coupling seismic wave-propagation with bubble growth dynamics. This model is solved analytically and numerically, including effects of diffusion of volatiles, visco-elasticity and bubble number density. We show that wave attenuation is controlled mainly by the *Peclet* and *Deborah* numbers. The *Peclet* number is a measure of the relative importance of advection to diffusion. The *Deborah* number is a visco-elastic measure, describing the importance of elasticity in comparison to viscous melt deformation. We solve numerically for wave attenuation for various magma properties corresponding to a wide range of *Peclet* and *Deborah* numbers. We show that the numerical solution can be approximated quite well for frequencies above 1 Hz, by an analytical end-member solution, obtained for high *Peclet* and low *Deborah* numbers. For lower frequencies, volatile transport should be accounted for, leading to higher attenuation with respect to the analytical solution. However, if the *Deborah* number is increased, either by longer relaxation time or by higher frequencies, then attenuation decreases with respect to the analytical solution. Therefore, visco-elasticity leads to a significant improvement of the resonating qualities of a magma-filled conduit and widens the depth and frequency ranges where pressure waves will propagate efficiently through the conduit.

The interaction of pressure waves with bubbly liquids has been discussed in the literature. It was confronted either by studying non-linear radial oscillations of a single bubble (Keller & Kolodner 1956; Keller & Miksis 1980; accounting also for liquid compressibility — Prosperetti & Lezzi 1986; Lezzi & Prosperetti 1987; and including the effect of heat transfer between bubble and liquid — Prosperetti *et al.* 1988) or by exploring linear pressure waves in bubbly liquids (Caflisch *et al.* 1985a,b present a rigorous mathematical approach while Commander & Prosperetti 1989 present a more heuristic approach following the suspension-model of Van Wijngaarden 1968). Commander & Prosperetti (1989) focused on linear pressure waves propagating in liquids containing small concentrations of gas bubbles. They constrained their treatment to conditions of no volatiles flux between bubble and melt, and to wavelengths that are much longer than the typical radius of bubbles.

Ichihara *et al.* (2004) presented a model accounting for effects of bubble resonance and scattering of pressure waves propagating through a visco-elastic bubbly liquid. This model was further developed by Ichihara & Kameda (2004) to include also thermal effects and non-ideality of the gas in the bubble on the acoustic bulk properties of a bubbly magma.

The problem of mass transfer between bubbles and melt in volcanic systems was confronted by Collier *et al.* (2006) in order to obtain the attenuation of pressure waves propagating through a magma-filled conduit, assuming an incompressible bubbly magma. They concluded that there is a very limited range in depth which allows waves to propagate with minimum damping.

In this study, we integrate the theory of pressure waves propagating in bubbly liquids (Commander & Prosperetti 1989) with the theory of bubble growth in magmas (Navon & Lyakhovsky 1998). Analytical solutions are presented (Table 1) for four bubble-growth end-member regimes: viscous, diffusive, equilibrium and no mass flux. Visco-elasticity and melt compressibility are introduced by a new visco-elastic bubble growth model. This model is used to generate a visco-elastic, numerical solution for the *Q* factor, and to understand better the physics of seismic-wave attenuation in a saturated, visco-elastic bubbly magma.

Theory

The wave equation for bubbly liquids

Commander & Prosperetti (1989) studied the propagation of pressure waves through a mixture

From: LANE, S. J. & GILBERT, J. S. (eds) *Fluid Motions in Volcanic Conduits: A Source of Seismic and Acoustic Signals.* Geological Society, London, Special Publications, **307**, 11–31.
DOI: 10.1144/SP307.2   0305-8719/08/$15.00 © The Geological Society of London 2008.

Table 1. *The general structure of this paper*

	Analytical solutions for the viscous model			Numerical solutions
No mass flux (NMF) regime	Viscous regime	Diffusive regime	Equilibrium regime	
Low-viscosity Compressible (equation 25)				
Incompressible (equation 26)		Incompressible (equation 36)	$Q \to \infty$ (equation 38)	Incompressible, Viscous (Fig. 4)
Compressible (equation 24)	Compressible (equation 30)	Compressible (equation 35)	$Q \to \infty$ (equation 38)	Visco-elastic (Fig. 5)

Generalization to the finite shell model (equations 39–42)
The relations between the numerical and analytical solutions (Figs 4, 5)
The dependency of damping on the *Peclet* number (Figs 5, 7)
The dependency of damping on n_d, f_E and *De* (Fig. 6)
Analytical approximations for the visco-elastic numerical solution (Fig. 8)
The visco-elastic $Q(f, depth)$ profile of a magma-filled conduit (Fig. 9)

of liquid and gas bubbles. They simplified the expressions for the continuity and momentum equations of a mixture and derived from them the following wave equation (see Appendix A for a detailed description):

$$\frac{1}{c^2}\ddot{\delta P} - \nabla^2 \delta P = \rho_m \ddot{\phi}, \qquad (1)$$

where $\delta P(x, t)$ is the amplitude of the pressure wave (which is small in comparison to the ambient pressure); ρ_m and c are the density and speed of sound in the melt. The variable ϕ stands for the

gas volume fraction and is defined as

$$\phi = \frac{4}{3}\pi R^3 n_d, \qquad (2)$$

where R is the bubble radius and n_d is the number density (number of bubbles in a unit volume of the bubbly magma) which is assumed constant for $\phi \ll 1$.

We search for a solution to (1) assuming the following expression for a propagating wave:

$$\delta P(x, t) = A \exp(ikx - i\omega t), \qquad (3)$$

Table 2. *Constants and parameters used in the model*

Constants/parameters	Symbol	Value
P-wave velocity in melt	c	2300 m/s
Melt density	ρ_m	2300 kg/m^3
Bulk modulus	K	12 GPa
Solubility constant	K_H	4.11×10^{-6} Pa$^{-1/2}$
Molecular weight of water	M	0.018 kg/mol
Ideal gas constant	G	8.314 J/(mol · K)
Melt temperature	T	1123 K
Wave amplitude	δP	0.1 MPa
Applied frequency	f	**0.0001–1000 Hz**
Number density	n_d	**10^{12} m^{-3}–10^{15} m^{-3}**
Initial pressure	P_0	**20 MPa–80 MPa**
Initial radius	R_0	**0.1 μm–10 μm**
Diffusivity	D	**10^{-12} m^2/s–10^{-10} m^2/s**
Viscosity	η	**10^5 Pa · s–10^8 Pa · s**
Shear modulus	μ	**0.1 GPa–1 GPa**

The values are based mainly on Collier *et al.* (2006) and Hurwitz & Navon (1994). The viscosity range accounts for the effective viscosity, considering the possibility of a high-viscosity skin around growing bubbles (Lensky *et al.* 2001; Mourtada-Bonnefoi & Mader 2001). The shear modulus range is based on Mungall *et al.* (1996), Romano *et al.* (1996) and Dingwell (1998).

where ω is the angular frequency of the propagating pressure wave, k is the related wave number and A is the slowly changing wave-amplitude. Understanding the attenuation of such waves is the main aim of this paper.

The Q factor

Commonly, attenuation is expressed by the quality factor, Q, describing how well a system could maintain its oscillatory motion. Of several expressions suggested in the literature, the most frequently used has been defined by Aki & Richards (2002):

$$Q = \frac{2\pi E(\omega)_{max}}{\Delta E(\omega)}, \quad (4)$$

where ω is the angular frequency of the wave, E_{max} is the maximum energy stored in one cycle at a specific frequency and ΔE is the energy dissipated in one cycle. O'Connell & Budiansky (1978) obtained Q through the constitutive equation of a visco-elastic medium:

$$Q = \frac{|\text{Re}M(\omega)|}{|\text{Im}M(\omega)|}, \quad (5)$$

where $M(\omega)$ is the complex elastic modulus of the material. In this definition, Q refers to the ratio between the real and imaginary parts of the complex modulus.

Collier et al. (2006) modified the definition given by (5) and obtained the following definition for Q:

$$Q(\omega) = \frac{|\text{Re}V_{bm}^2|}{|\text{Im}V_{bm}^2|} = \frac{1}{\tan(\psi)}. \quad (6)$$

This expression indicates that the Q factor could be obtained by two methods: either by the ratio between the real and imaginary parts of V_{bm}, the complex wave velocity in the bubbly magma, or by ψ, the phase lag between the applied stress and the resulting strain rate.

Saturated bubbly magma

We assume that for a saturated bubbly magma, bubble radius oscillates around its initial saturation value, R_0, and that the amplitude of the oscillations is small ($\delta R(x, t) \ll R_0$). For constant R_0 we may write that:

$$R = R_0 + \delta R(x, t)$$

$$\dot{R} = \dot{\delta R}(x, t) \quad (7a-c)$$

$$\ddot{R} = \ddot{\delta R}(x, t).$$

Furthermore, solving for small oscillations we assume that

$$R^3 \cong R_0^3 + 3R_0^2 \delta R. \quad (8)$$

We substitute (8) into (2), differentiate ϕ twice with respect to time (assuming constant n_d and ρ_m) and substitute the resulting expression into (1) leading to

$$\frac{1}{c^2}\ddot{\delta P} - \nabla^2 \delta P = \rho_m 4\pi n_d R_0^2 \ddot{\delta R}. \quad (9)$$

Equation 9 is the fundamental wave equation of our model, describing pressure waves in a saturated bubbly magma. This equation should be combined with the pressure-dependent δR, derived from the bubble growth model, in order to obtain a full solution for wave propagation and attenuation.

Bubble growth dynamics

The growth of bubbles in a liquid is controlled by two main processes. The first process is the diffusion of dissolved volatiles from a supersaturated melt into bubbles after decompression or from bubbles into an undersaturated melt under compression. The second process is the visco-elastic deformation of the melt. Proussevitch et al. (1993) suggested that the bubbly magma can be modelled as a grid of closely packed spherical cells, where each cell includes a spherical bubble enclosed by a finite shell of melt. Assuming that there is no interaction between the cells, modelling of one cell provides a general solution for the bubble growth process. The cell model was further developed by Proussevitch & Sahagian (1996), Lyakhovsky et al. (1996), Navon & Lyakhovsky (1998) and Lensky et al. (2004), showing good agreement with experimental results.

Since bubble growth dynamics is essential for the present study, we present below a brief description of the model. The model assumes a Newtonian incompressible fluid, with spherical symmetry and using spherical coordinates.

Consider a bubble of radius R_0 in equilibrium with a melt shell of radius S_0 at pressure P_0. Water mass balance dictates that

$$R_0^3 = S_i^3 (C_i - C_0) \frac{\rho_m}{\rho_g}. \quad (10)$$

where ρ_g is the gas density, C_i is the water concentration in the melt when the bubble dissolves and shrinks, leaving a melt sphere of radius S_i ($S_i^3 = S^3 - R^3$) and C_0 is the equilibrium water

14 I. KURZON *ET AL.*

concentration. C_0 (and any other equilibrium concentration) is related to the pressure through the solubility law of water, that for pressures up to 200 MPa, is commonly approximated as a square-root relation:

$$C_0 \approx K_H \sqrt{P_0}, \qquad (11)$$

where K_H is Henry's constant.

If pressure is changed to a new pressure $P_a \neq P_0$, the system is no longer in equilibrium and volatiles diffuse through the melt and transfer to or from the bubble according to:

$$\frac{d\left(\frac{4}{3}\pi R^3 \rho_g\right)}{dt} = 4\pi D \rho_m R^2 \left(\frac{\partial C}{\partial r}\right)_R, \qquad (12)$$

where D is the diffusivity of water in the melt and C is the water concentration in the melt.

The concentration gradient at the interface is obtained by solving the diffusion-advection equation:

$$\frac{\partial C}{\partial t} + v(r)\frac{\partial C}{\partial r} = \frac{1}{r^2}\frac{\partial}{\partial r}\left(Dr^2\frac{\partial C}{\partial r}\right), \qquad (13)$$

where $v(r)$ is the radial melt velocity. For an infinite shell and quasi-static concentration profile, the concentration gradient on the bubble-melt interface is (Lyakhovsky *et al.* 1996):

$$\left(\frac{\partial C}{\partial r}\right)_R = \frac{(C_0 - C_R)}{R}, \qquad (14)$$

where C_R is the concentration of volatiles on the bubble-melt interface, and C_0 is the equilibrium pressure on the outer boundary of the shell.

The gas density in the bubble is related to pressure by the ideal gas law:

$$\rho_g = \frac{M}{GT}P_g, \qquad (15)$$

where M is the molecular weight of water, G is the gas constant and T is the absolute temperature, which is assumed constant in this model. For an infinite shell, P_a, P_g and $v(r)$ are related by the momentum and continuity equations:

$$\frac{\dot{R}}{R} = \frac{1}{4\eta}(P_g(t) - P_a(t)), \qquad (16)$$

where η is the viscosity of the melt and

$$v(r) = \dot{R}\frac{r^2}{R^2} \qquad (17)$$

Navon & Lyakhovsky (1998) showed that following an instantaneous pressure drop, three regimes can be identified (Fig. 1): (i) a viscous regime where diffusion is efficient, gas pressure is maintained close to P_0 and bubble growth is controlled mainly by viscous deformation; (ii) a diffusive regime where gas pressure falls so that P_g is close to P_a, and growth is controlled mainly by diffusion; and (iii) an equilibrium regime where melt and bubbles approach chemical equilibrium and R approaches equilibrium at $P = P_a$.

The different growth regimes can be characterized with the following timescales (Fig. 1): $\tau_V = 4\eta/(P_0 - P_a)$ for the viscous growth; $\tau_D = R_0^2/D$ for the diffusive growth and $\tau_E = S_i^2/D$ for reaching equilibrium. In cases where diffusive mass transfer is negligible, an additional timescale is defined: $\tau_{NMF} = 4\eta/3P_0$ (Barclay *et al.* 1995), where NMF stands for 'No Mass Flux'. These timescales, and especially the relations between them, provide the key for understanding the dependency of damping on the various processes occurring in the bubbly magma. Furthermore, the inverse of these time scales can be related to the frequency of pressure waves propagating through a bubbly magma.

The model discussed above accounts for viscous deformation of the melt neglecting the elastic component of melt deformation; this assumption is allowed for low Deborah numbers: $De \equiv \eta\dot{\tau}/\mu\tau < 1$, where μ is the shear modulus of the melt, τ is the applied stress and $\dot{\tau}$ is the rate of stress-loading. This number corresponds to the ratio between the Maxwell relaxation time $\tau_m = \eta/\mu$ and the characteristic time of loading $\tau/\dot{\tau}$, and reflects the nature of deformation in the material. For propagating waves (according to (3) $\tau/\dot{\tau} = 1/\omega$ and the Deborah number is expressed as $De = \omega\eta/\mu$. For a magma with constant η and μ, De is proportional to the frequency of the applied pressure wave. For low frequencies, $De < 1$, deformation is viscous and elastic strains are negligible. At very high frequencies, $De \gg 1$, deformation is mostly elastic and viscous components could be ignored. For $De \approx 1$, elastic and viscous components should be considered, and for an infinite shell we obtain two additional terms in the momentum equation:

$$\frac{\dot{R}}{R} = \frac{1}{4\eta}\left(P_g(t) - P_a(t)\right)$$
$$+ \frac{1}{4\mu}\frac{d(P_g(t) - P_a(t))}{dt} - \frac{1}{3K}\frac{dP_a(t)}{dt}. \qquad (18)$$

(See Appendix B for the full derivation.) The first term in the right-hand side of the equation stands for viscous resistance (as in equation 16), the

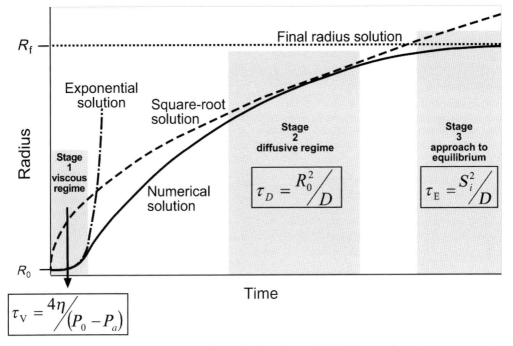

Fig. 1. Bubble growth solutions as a function of time (after Lensky *et al.* 2002). The numerical solution corresponds to the analytical end-member growth regimes. Also seen are the characteristic timescales for growth (see text for further explanation).

second term is the elastic component and the third term accounts for melt compressibility.

Analytical solutions

The Q Factor for No Mass Flux (NMF) conditions: a benchmarking study

The 'No Mass Flux' approximation describes bubble growth dynamics when diffusion of volatiles between melt and bubbles is negligible; this is actually the case solved by Commander & Prosperetti (1989). Since we are considering small vesicularities, we will focus our treatment on the infinite shell model, when the bubble does not 'see' the outer boundary of the shell. A generalization to a finite shell is given in Appendix C.

The no mass flux condition states that the mass of gas in the bubble is constant:

$$P_g R^3 = P_0 R_0^3. \tag{19}$$

Substitution of (8) in (19) yields

$$P_g \approx \frac{P_0 R_0^3}{R_0^3 + 3R_0^2 \delta R} \approx P_0\left(1 - 3\frac{\delta R}{R_0}\right). \tag{20}$$

Substitution of (20), (7a) and (8) in (16) leads to the following expression:

$$\frac{\dot{\delta R}(t)}{R_0} = \frac{1}{4\eta}\left[P_0\left(1 - 3\frac{\delta R(t)}{R_0}\right) - P_a(t)\right]. \tag{21}$$

Defining $P_a(t) = P_0 + \delta P(t)$, the solution for this equation becomes:

$$\delta R(t) = \exp\left[-\frac{t}{\tau_{NMF}}\right]$$
$$\times \left\{\int -\frac{R_0}{4\eta}\exp\left[\frac{t}{\tau_{NMF}}\right]\delta P(t)dt\right\}. \tag{22}$$

Taking the second derivative of (22) in respect to time, substituting the result into (9) and applying (3) leads to the following equation for the inverse of the complex phase velocity:

$$\frac{1}{V_{bm}^2} = \frac{1}{c^2} + \frac{\rho_m \pi n_d R_0^3}{\eta}\frac{(1 + i\omega\tau_{NMF})\tau_{NMF}}{(1 + \omega^2\tau_{NMF}^2)}. \tag{23}$$

Following the Quality factor definition (6), we obtain the following expression for $Q(\omega)$:

$$Q = \frac{\eta\omega}{\rho_m \pi n_d R_0^3 c^2} + \frac{1}{\omega} \cdot \left[\frac{9P_0^2}{\rho_m 16\eta\pi n_d R_0^3 c^2} + \frac{3P_0}{4\eta} \right].$$
(24)

Commander & Prosperetti (1989) obtained a similar solution. However, since in their treatment they accounted only for fluids of low viscosity (e.g. water), the first term was naturally neglected, resulting in:

$$Q = \frac{1}{\omega} \cdot \left[\frac{9P_0^2}{\rho_m 16\eta\pi n_d R_0^3 c^2} + \frac{3P_0}{4\eta} \right].$$
(25)

This equation is similar to the one obtained by Collier *et al.* (2006). For an incompressible fluid, with $c \to \infty$, this expression could be simplified even more giving:

$$Q = \frac{3P_0}{4\eta\omega}.$$
(26)

The solutions for the compressible case (24) and the incompressible case (26), along with the low viscosity solution of Commander & Prosperetti (25) are presented in Figure 2a.

Magma rheology: Kelvin versus Maxwell visco-elasticity. The analytical NMF solutions allow a first peek into magma rheology. By examining the general equation (24), we may note that for low frequencies $Q \propto 1/\omega$ — the quality factor decreases with increasing frequency, and damping increases. This effect could be explained by Kelvin's model for a visco-elastic material (a spring and dashpot in parallel, see Fig. 3). In this model, the elastic response of the material is retarded and viscous response is more dominant in the initial moments of applied stresses. Therefore, when frequency increases the elastic component becomes less dominant, allowing viscous dissipation to control the oscillations and to enhance damping. At higher frequencies, damping decreases with increasing frequency and $Q \propto \omega$. This part of the solution indicates a behaviour according to Maxwell's model for a visco-elastic material (a spring and a dashpot in series, see Fig. 3). In this model, a visco-elastic material will initially react elastically before continuing its response by viscous flow. As the period of the applied stress is reduced (frequency increases), elastic response dominates the behaviour of the material and damping is reduced.

The full solution (24) is typical for a standard linear solid (Malvern 1969; Aki & Richards 2002) and may be presented by the three-element model (two springs and a dashpot, see Fig. 3). This solution reflects the double role of the shear viscosity in relation to the elastic properties of a fluid. High viscosity means high resistance to flow, and thus high loss of energy when the material is flowing. On the other hand, when a material becomes very viscous it approaches an elastic behaviour, allowing elastic waves to propagate without loss of energy.

At low frequencies, viscous flow is the main actor in melt deformation (Kelvin's model); therefore $Q(\omega) \propto 1/\eta$ and the role of viscosity is pronounced in the dissipation of elastic energy. At high frequencies, viscous flow hardly contributes to melt deformation and viscosity leads to domination of the elastic nature of the material (Maxwell's model) and reduction of energy loss. With this basic physical insight into magma rheology, we may proceed and add diffusion to our system. We first present analytical end-member solutions and then the results of the numerical simulations.

The Q Factor for mass flux conditions

Collier *et al.* (2006) obtained a numerical solution for Q in an incompressible, viscous, bubbly magma with diffusive mass flux between melt and bubbles. They adjusted the bubble growth numerical code of Lyakhovsky *et al.* (1996) to obtain the phase lag, ψ (6), and resulting Q factor. When mass flux between bubbles and melt is allowed, the three regimes controlling the change in bubble volume (viscous, diffusive and equilibrium) provide the following end-member analytical solutions:

Viscous regime. Viscous resistance controls growth when diffusion is efficient and succeeds in keeping the initial gas pressure in the bubble close to P_0. Under such conditions, the momentum equation (16) is the governing equation, and since $P_a(t) = P_0 + \delta P(t)$ it becomes

$$\frac{\dot{R}}{R} = -\frac{1}{4\eta}\delta P(t).$$
(27)

We differentiate (27) in respect to time and use (7) to obtain an expression for $\ddot{\delta R}$. We substitute this expression into (9) and obtain:

$$\frac{1}{c^2}\ddot{\delta P}(t) - \nabla^2 \delta P(t) = -\frac{\rho_m \pi n_d R_0^3}{\eta}\dot{\delta P}(t).$$
(28)

Applying (3) we obtain an expression for the complex phase velocity as:

$$\frac{1}{V_{bm}^2} = \frac{1}{c^2} + i\frac{\rho_m \pi n_d R_0^3}{\eta\omega}$$
(29)

and the related $Q(\omega)$ factor becomes

$$Q(\omega) = \frac{\omega\eta}{\rho_m \pi n_d R_0^3 c^2}.$$
(30)

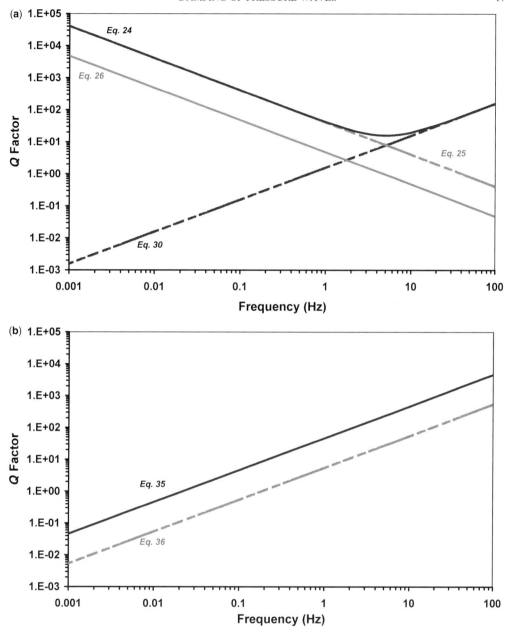

Fig. 2. Analytical solutions. (**a**) The NMF and viscous analytical solutions of $Q(f)$. The dashed black line stands for the viscous solution (equation 30); the grey dashed line for the low viscosity NMF solution (equation 25); the grey continuous line for the incompressible NMF solution (equation 26) and the black continuous line for the compressible NMF solution (equation 24). (**b**) The diffusive analytical solutions for $Q(f)$. The dashed grey line represents the solution for an incompressible melt (equation 36) and the continuous black line for a compressible melt (equation 35). *Magma properties:* $P_0 = 40$ MPa; $R_0 = 1$ μm; $D = 10^{-12}$ m^2 s^{-1}; $\eta = 10^6$ Pa · s; $n_d = 10^{14}$ m^{-3}.

This expression for Q is identical to the first term of (24) and is presented in Figure 2a. Here $Q \propto \eta$, as the role of viscosity is preventing bubble growth (or shrinkage), so it plays only one of its roles: pronouncing the elastic behaviour of the material (Maxwell).

Diffusive regime. The diffusive regime is reached when diffusion cannot compensate for growth and the gas pressure in the bubble approaches the ambient pressure. Using (12), for the water mass balance and (14), for the concentration gradient in

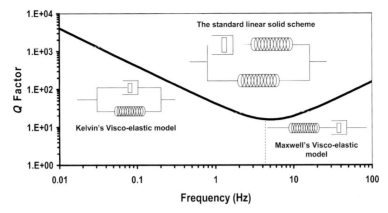

Fig. 3. Magma rheology: Kelvin versus Maxwell visco-elasticity. The visco-elastic model reflected by the compressible NMF solution and described by the spring and dashpot schemes. The two opposite trends ($Q \propto 1/\omega$ and $Q \propto \omega$) are related, respectively, to Kelvin's and Maxwell's visco-elastic model. The two trends add up to give the standard linear solid scheme. Magma properties are as in Figure 2.

the shell, we get the following expression:

$$\frac{d}{dt}\left((R_0^3 + 3R_0^2 \delta R)\rho_g\right) = 3D\rho_m(R_0 + \delta R)(C_0 - C_R).$$

$$(31)$$

For $C_0 - C_R$ we apply the solubility equation (11) approximating for $\delta P \ll P_0$, and obtain

$$C_0 - C_R \approx -C_0 \frac{\delta P}{2P_0}. \qquad (32)$$

Solving for small oscillations controlled by diffusive regime, we may approximate $P_g \approx P_a \approx P_0$, and $\rho_g \approx \rho_0$. Substitution of (15) and (32) in (31) leads to

$$\dot{\delta R} = -\frac{D\rho_m C_0}{R_0 \rho_0}\frac{\delta P}{2P_0} - \frac{R_0}{3P_0}\dot{\delta P}. \qquad (33)$$

Differentiation of (33) and substitution into (9) yields the following wave equation:

$$\frac{1}{c^2}\ddot{\delta P} - \nabla^2 \delta P$$

$$= \rho_m 4\pi n_d R_0^2 \left[-\frac{D\rho_m C_0}{R_0 \rho_0}\frac{\dot{\delta P}(t)}{2P_0} - R_0 \frac{\ddot{\delta P}(t)}{3P_0} \right] \qquad (34)$$

leading to

$$Q(\omega) = \omega \frac{2R_0^2 \rho_0}{3\rho_m DC_0} + \omega \frac{P_0 \rho_0}{2\pi c^2 \rho_m^2 DR_0 n_d C_0} \qquad (35)$$

$$= \omega \frac{2R_0^2 \rho_0}{3\rho_m DC_0}\left(1 + \frac{P_0}{c^2 \rho_m \phi}\right).$$

This solution reflects the major aspect regarding the addition of diffusion into the system. The process of diffusion increases the loss of energy in the system due to the additional energy consumed by the deformation of the viscous melt around the oscillating bubbles (Collier *et al.* 2006). Therefore, $Q(\omega) \propto 1/D$ and as frequency increases diffusion is less dominant leading to a decrease in damping and higher Q. Additional aspects of diffusion will be discussed below.

For an incompressible melt, the second term of (35) vanishes, and

$$Q(\omega) = \omega \frac{2R_0^2 \rho_0}{3\rho_m DC_0}. \qquad (36)$$

These $Q(\omega)$ solutions (equations 35, 36) are presented in Figure 2b.

Equilibrium regime. The equilibrium growth regime occurs whenever bubbles and melt manage to maintain chemical equilibrium, even if the ambient pressure varies. In the context of this study, equilibrium is approached for very low-frequency pressure waves propagating through a bubbly magma. At equilibrium, bubble radius is defined by the mass balance equation (10), using an expression analog to (32) for $C_i - C_f$:

$$3R_0^2 \delta R(t) = -\frac{S_i^3 C_i \rho_m}{2\rho_0 P_0}\delta P(t). \qquad (37)$$

Double differentiation of (37) and substitution into (9) yields the following wave equation:

$$\left[\frac{1}{c^2} + \frac{2\pi \rho_m^2 S_i^3 n_d C_i}{3\rho_0 P_0}\right]\ddot{\delta P} + k^2 \delta P = 0. \qquad (38)$$

Equation 38 describes a harmonic oscillator without any damping term; this is expected, when assuming the system is in chemical equilibrium. Therefore, the phase lag between applied stress and resulting strain approaches zero and $Q \to \infty$.

Numerical solutions

The numerical simulation of attenuation covers the whole range of Q between the analytical end-member solutions presented above. In order to improve our perception of the numerical solutions, we will proceed in two steps. First, we present a simulation of an incompressible, viscous melt and compare it with the analytical solutions in order to explain the numerical results of Collier et al. (2006). Secondly, we present the more general visco-elastic solution (where we account for melt compressibility) and explain it using the analytical solutions and the incompressible, viscous numerical solution.

An incompressible, viscous melt

We applied the numerical code (Lyakhovsky et al. 1996; Collier et al. 2006) and obtained the Q factor using the phase lag, ψ, between applied

stress and resulting strain (6). This version of the code is applicable only for an incompressible viscous melt, hence the results where compared with the analytical solutions for an incompressible melt (Fig. 4). As expected from the analytical solutions, at very low frequencies, when bubbles and melt manage to maintain chemical equilibrium, Q approaches infinity (to the left of the frequency range presented in Figure 4). As frequency increases and the efficiency of diffusion in attaining chemical equilibrium is reduced, more energy is consumed for each cycle and damping increases. This trend reaches a local minimum at frequencies close to $f_E = 1/\tau_E = D/S_i^2$, the equilibrium growth frequency-scale. Beyond this frequency, diffusion does not manage to maintain equilibrium over S_i, (S_i represents the approximate length-scale that diffusion must affect in order to approach chemical equilibrium). This behaviour reflects an additional role of diffusion in maintaining chemical equilibrium. Higher D leads to higher f_E and to higher Q values at frequencies below f_E.

Above f_E, the diffusion process becomes only an additional consumer of energy, and like in the diffusive analytical solution, as frequency increases so damping decreases. This trend continues up to the frequency where Q approaches the NMF analytical

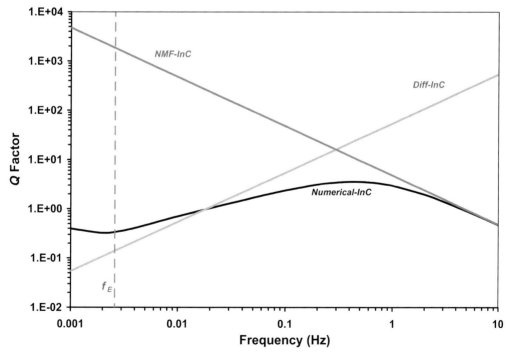

Fig. 4. The incompressible, viscous numerical solution for $Q(fr)$. The numerical solution corresponds to the incompressible analytical solutions and to f_E (see text for further explanation). *Magma properties*: $P_0 = 40$ MPa; $R_0 = 10$ μm; $D = 10^{-11}$ m^2 s^{-1}; $\eta = 10^6$ Pa · s; $n_d = 10^{12}$ m^{-3}. *Abbr: InC* – Incompressible; *Diff* – Diffusive.

solution. In the NMF solution, damping increases with increasing frequency, therefore, the numerical solution exhibits a local maximum before sticking to the NMF solution as frequency continues to increase.

The numerical solution for a visco-elastic melt

Accounting for melt compressibility requires a visco-elastic bubble growth formulation (Appendix B) and modification of the numerical code. Examination of the results emphasizes two end-member behaviours which are distinguished by their different values for the non-dimensional *Peclet* number. The *Peclet* number, *Pe*, is a measure of the relative importance of advection to diffusion. In our case, the only process controlling the NMF timescale is the viscous melt deformation, which in some sense may be viewed as an advective timescale. Therefore, the ratio between the diffusive timescale, R_0^2/D, and the NMF timescale, $\approx \eta/P_0$, could also be referred to as the *Peclet* number:

$$Pe = \frac{\tau_D}{\tau_{NMF}} = \frac{P_0 R_0^2}{D\eta}. \qquad (39)$$

High *Peclet* numbers indicate that bubble growth is dominated by melt deformation expressed by the NMF approximation, while low *Peclet* numbers indicate the dominant role of volatile diffusion.

*The high-*Peclet* end-member.* As can be seen in Figure 5a, the incompressible assumption is valid for very low frequencies, up to f_E, the equilibrium frequency-scale. At higher frequencies, compressibility becomes more important, introducing elastic deformation to the total strain of the bubble. This effect is responsible for the larger *Q* values exhibited by the visco-elastic numerical solution. As frequency increases, the role of diffusion in damping is decreased and *Q* increases like in the diffusive analytical solution. This trend continues until diffusion becomes negligible and *Q* approaches the compressible NMF analytical solution. In the case of plane waves, differing from radial oscillations, the numerical solution is applicable up to the frequency where the viscous-related dissipation of the melt (Eq. A21 in Appendix A) is larger than the bubble-related dissipation. For even higher *Peclet* numbers, the approximation of NMF becomes valid at even lower frequencies than shown in Figure 5a.

*The low-*Peclet* end-member.* This is the case for very efficient diffusion leading to dominance of visco-elastic resistance in the nature of damping. Therefore, melt compressibility comes into

consideration at lower frequencies than f_E, and the incompressible assumption is valid only for very low frequencies. So, for very low frequencies *Q* decreases with increasing frequency. However, as *Q* approaches the viscous analytical solution, it reaches a minimum (at a frequency smaller than f_E) and then increases with increasing frequency, approaching the viscous analytical solution. As shown in Figure 5b, for higher frequencies the viscous and the compressible NMF solutions converge and both are higher than the incompressible model. As in the high-*Peclet* end-member (Fig. 5a), for plane waves, the numerical solution is applicable up to the frequency where the viscous-related dissipation of the melt is larger than the bubble-related dissipation.

Discussion: the factors controlling attenuation

The roles of melt viscosity, elasticity, water diffusion, bubble size, bubble number density and the equilibrium pressure on the attenuation of pressure waves were examined by changing the parameters: η, μ, D, R_0, n_d and P_0, respectively (Table 2). We examined the *Q factor* as a function of frequency since the frequency provides the time-window for the various deformations to take place, interact and attenuate seismic waves. As magma is a compressible fluid, our coming discussion will naturally focus on the visco-elastic numerical solution.

The dependency of f_E (the equilibrium frequency-scale) on the number density

The equilibrium timescale, τ_E, describes the time required for diffusivity, D, to act along the diffusion length-scale, approximated by the initial shell radius, S_i. Since S_i depends on the number density, $n_d = 3/4\pi S_i^3$, it follows that low n_d leads to longer τ_E and lower f_E (the equilibrium frequency-scale which is the inverse of τ_E) (Fig. 6a). f_E is a good measure for the transition from equilibrium regime to diffusive regime for small vesicularities and negligible melt compressibility. For larger bubbles, the diffusion length-scale becomes smaller than S_i, and the transition from equilibrium regime to diffusive regime is at higher frequencies than f_E. We may assume incompressibility at high *Pe* values ($Pe > 1$) only if the frequencies are low enough for water transfer to take place and control deformation. For lower values of the *Peclet* number, corresponding to an efficient water transfer to and from the bubble, viscosity controls deformation, and the *Q factor* begins its approach to the viscous analytical solution at frequencies lower than f_E.

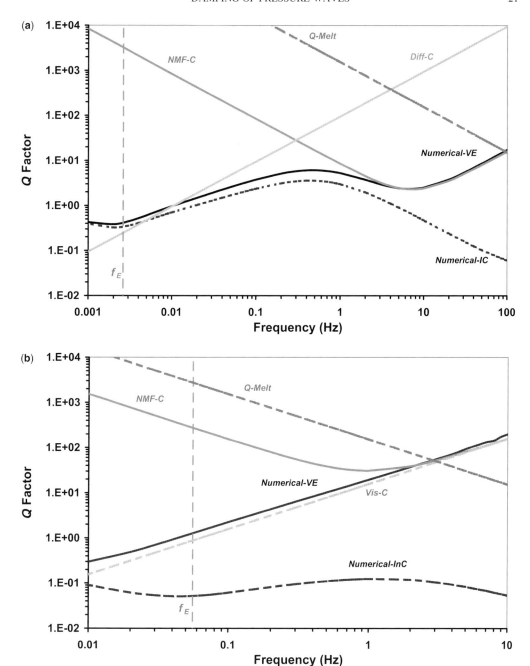

Fig. 5. The visco-elastic numerical solution for $Q(fr)$. The relations of the numerical solution to the compressible analytical solutions, the incompressible numerical solution, the viscosity-related Q-melt (see equation A21 in Appendix A) and f_E (see text for further explanation). (**a**) The solution for high-*Peclet* conditions. In this case, the viscous analytical solution is irrelevant. Magma properties are as in Figure 4. (**b**) The solution for low-*Peclet* conditions. In this case, the diffusive analytical solution is irrelevant. *Magma properties*: $P_0 = 80$ MPa; $R_0 = 1$ μm; $D = 10^{-11}$ m^2 s^{-1}; $\eta = 10^7$ Pa · s; $n_d = 10^{14}$ m^{-3}. *Abbr:* VE – Visco-elastic; C – Compressible; InC – Incompressible; Diff – Diffusive; Vis – Viscous.

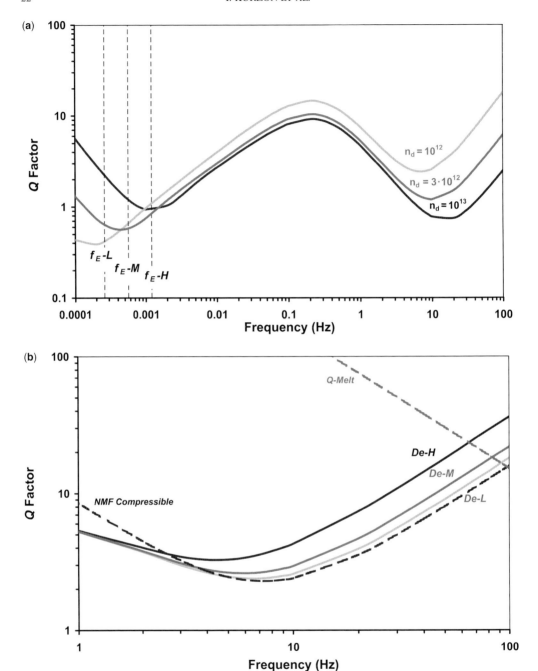

Fig. 6. (**a**) The influence of the number density, n_d, on the upper-frequency limit for reaching chemical equilibrium. For lower number densities, the shell becomes larger leading to longer times for obtaining equilibrium; hence, f_E becomes smaller. *Magma properties: $P_0 = 40$ MPa; $R_0 = 1$ μm; $D = 10^{-12}$ m^2 s^{-1}; $\eta = 10^6$ Pa · s.* (**b**) The influence of the Deborah number, *De*, on the *Q* factor for high frequencies. When the Deborah number approaches zero, elastic deformation has no influence on damping and the compressible NMF analytical solution is valid. Elastic deformation becomes important when the shear modulus of the melt is decreased, leading to an increase in *De* (see text for the full explanation). For plane-waves, the solutions are applicable up to the frequency where they intersect with the viscosity-related *Q-melt*. *Magma properties: $P_0 = 40$ MPa; $R_0 = 10$ μm; $D = 10^{-11}$ m^2 s^{-1}; $\eta = 10^6$ Pa · s. The variation in shear modulus: De-L: $\mu = 1$ GPa; De-M: $\mu = 0.3$ GPa; De-H: $\mu = 0.1$ GPa. Abbr: L* – low; *M* – moderate; *H* – high.

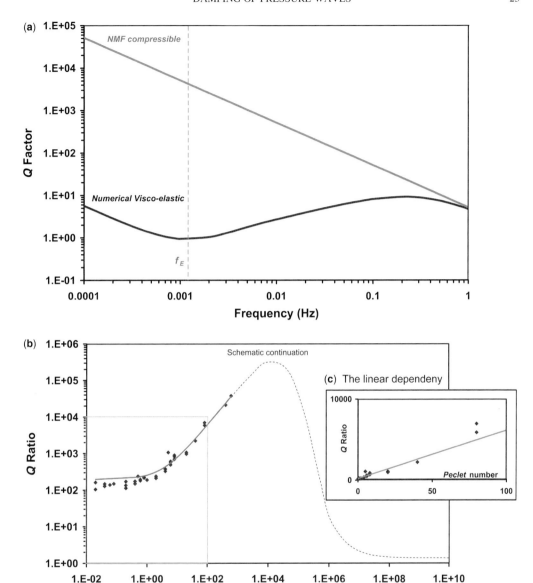

Fig. 7. The relation between the *Q factor* and the *Peclet* number for very low frequencies. (**a**) At very low frequencies, when the numerical solution is parallel to the NMF solution, we may produce a *Q ratio* defined as Q_{NMF}: $Q_{Numerical}$. (**b**) Plotting this ratio against the *Peclet* number (keeping n_d and μ constant), we obtain an interesting relation. The *Q ratio* is linear with the *Peclet* number for $Pe < 1000$ (the axes are in log scale). Beyond $Pe \approx 1000$, the *Q ratio* increases non-linearly with *Pe*, reaching a maximum and descends towards a value of 1 *as the numerical solution approaches the NMF solution.* (**c**) An enlargement of the linear section up to $Pe = 10$, showing the intersection with the *Q ratio* axis for very low values of *Pe*. This intersection provides a constant, representing the basic increase in damping as mass transfer is introduced into the system. *Here,* $n_d = 10^{15}$ m^{-3} and $\mu = 1$ GPa.

The role of elastic deformation

The shear modulus of silicic magmas at high temperatures is not well constrained. Experimental studies (Dingwell 1998) for very viscous silicic magmas show that Maxwell relaxation time does not exceed 0.01 seconds. We expect that for $\omega \approx 100$ Hz the *Deborah* number ($De = \eta/\mu\omega$) approaches one ($De \rightarrow 1$), and elasticity will take a larger role in the total deformation of the melt.

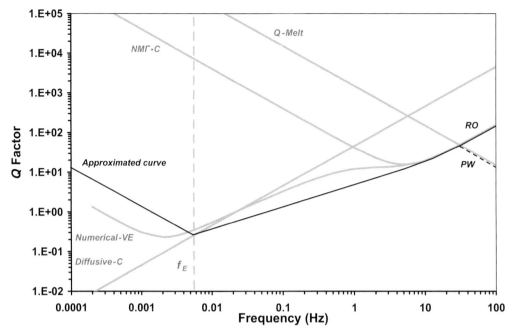

Fig. 8. The approximation of the numerical solution for Q when $1 < Pe < 1000$. Up to f_E, $Q \propto 1/\omega$ and runs parallel to the NMF solution, but is displaced to lower Q values according to the *Peclet* number. Beyond f_E, the Q curve ascends with increasing frequency, gradually approaching the compressible NMF solution (equation 24). *Abbr: C* – Compressible; *VE* – Visco-elastic; *PW* – Plane Waves; *RO* – Radial Oscillations.

This effect is shown in Figure 6b, where it is shown that as *De* increases, Q increases, and damping decreases.

In this case, the increase of *De* is done by decreasing the shear modulus (keeping all the other parameters constant). This, in fact, conceals a very interesting effect: reducing μ indicates weakening of the elastic components of the melt, which allows a larger elastic deformation to be taken by the melt; this may occur in hydrous, vesicular magmas ($\mu \approx 10^6$ Pa according to Mungall *et al.* 1996, Romano *et al.* 1996 and Dingwell 1998) leading to even higher Q values than the ones presented in Figure 6b.

Attenuation at very low frequencies

The *Peclet* number is also important for the understanding of damping at very low frequencies. At these low frequencies, the numerical solution is proportional to $1/\omega$, exactly like in the NMF analytical solution, differing only by a factor (Fig. 7a). This means that the addition of diffusion into the system can be accounted for by dividing the NMF solution by a factor. We found that this factor corresponds to the *Peclet* number. For a constant number density and for $Pe < 1000$, we found that the Q

ratio, $Q_{NMF}/Q_{Numerical}$, increases linearly with increasing *Peclet* numbers (Fig. 7b). Thus, for very low *Peclet* numbers, corresponding to a very efficient diffusion, the addition of diffusion is responsible for a fixed increase in damping (the intersection with the Q ratio axis in the enlargement — Fig. 7c).

Higher *Peclet* numbers mean that the diffusion process becomes less efficient and chemical equilibrium is harder to obtain. This leads to an increase in the actual damping, which means that $Q_{Numerical}$ decreases, leading to an increase in the Q ratio. On the other hand, very high *Peclet* numbers ($Pe \gg 1000$) mean that the NMF approximation is valid and therefore, $Q_{ratio} \rightarrow 1$. So as the *Peclet* number increases, the increase in the Q ratio reaches a maximum before decreasing into the NMF approximation (Figmm. 7b).

Analytical approximation for the visco-elastic numerical solution

For *De* < 1, we present a simple scheme for approximating analytically the Q *factor* according to the *Peclet* number of the system. For very high *Peclet* numbers (see Fig. 7b), we may approximate the Q

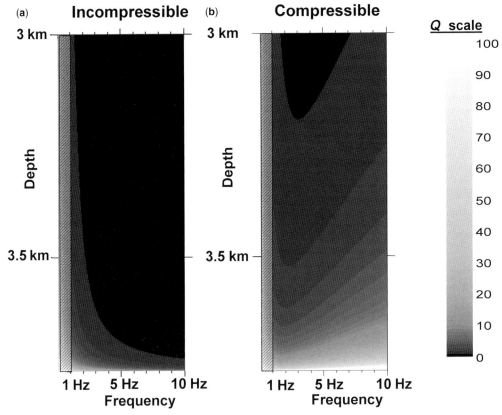

Fig. 9. The $Q(f, depth)$ conduit profiles for NMF conditions. We present a comparison between an incompressible profile (following Collier *et al.* 2006) and a visco-elastic profile, accounting also for melt compressibility. The dashed area for $f < 1$ Hz indicates that both profiles are a close approximation to their corresponding numerical solution only from *c.* 1 Hz and higher. The addition of compressibility leads to very different profiles than the ones presented in Collier *et al.* (2006). Here, we present the lower section of the conduit where compressibility has a major role in the damping properties of the magma. *Magma properties: $P_0 = 125$ MPa; $D = 2 \cdot 10^{-11}$ m^2 s^{-1}; $\eta = 10^7$ Pa · s; $n_d = 10^{12}$ m^{-3}; Initial water concentration $= 4.6\%$.*

factor using the compressible NMF analytical solution (equation 24).

$10^3 > Pe > 1$ — Up to f_E, $Q \propto 1/\omega$ and runs parallel to the NMF solution, but is displaced to lower Q values according to the *Peclet* number. Beyond f_E, the Q curve ascends with increasing frequency, gradually approaching the compressible NMF solution (equation 24; Fig. 8).

$Pe < \approx 1$ — At very low frequencies, $Q \propto 1/\omega$ according to a $Q_{NMF}/Q_{Numerical}$ ratio that can be approximated by statistical curve fitting. At higher frequencies, the numerical solutions for Q fall close to the viscous analytical solution (equation 30), they gradually approach it and Q increases with increasing frequency.

For plane waves, as opposed to radial oscillations (Fig. 8), the Q *factor* will decrease beyond the frequency where the viscous-related dissipation of the melt is larger than the bubble-related dissipation (see Eq. A21 in Appendix A).

The Q(f) *depth-profile*

Applying our model to a magma-filled conduit, we produced a $Q(f)$ depth-profile and compared it to the one presented by Collier *et al.* (2006). We find a major difference between the $Q(f)$ profile they obtained (see Figure 9 in Collier *et al.* 2006) and the $Q(f)$ profile obtained here (Fig. 9). While their profile shows that Q decreases with increasing frequency, simply following the incompressible NMF solution, our profile shows an opposite behaviour: the addition of melt compressibility leads to an increase in Q for increasing frequency. Our results indicate larger values of the visco-elastic Q, which become important in the lower section of the conduit. For frequencies of c. 1 Hz and above, we find that similar to Collier *et al.* (2006), showing that the incompressible NMF solution closely approximates the incompressible numerical solution, we show that in our profile the compressible NMF solution closely approximates the visco-elastic numerical solution.

This may be significant for low-frequency earthquakes (LFs). These volcanic earthquakes originate at the vicinity of the magma conduit, and are regarded as an important indicator prior to explosive eruptions (Neuberg *et al.* 2000). Their typical waveform is a narrowband signal presenting an initial stage of amplification followed by a long tail of attenuation. This signature reflects excitation in a resonating system (Chouet 1996). Two main approaches were advanced for explaining the resonating system: the fluid-filled crack resonator (Aki *et al.* 1977; Chouet 1986, 1988) and the magma-filled conduit resonator (Neuberg *et al.* 2000). In the case of a magma-filled conduit, pressure waves propagating through the bubbly magma are expected to be attenuated, resulting in an ineffective conduit resonator. According to Collier *et al.* (2006), there is a very limited depth-range where the Q values are high enough to allow efficient wave propagation in the bubbly magma. The present findings extend this depth-range, improving significantly the efficiency of the magma-filled conduit as a possible resonator of low-frequency earthquakes. Following decompression, magma may become supersaturated, leading to an accelerated bubble growth and releasing additional energy into the system. This energy may be partly converted into seismic energy, reducing damping or even leading to wave-amplification.

Conclusions

Based on the theoretical approach of Commander & Prosperetti (1989), we have developed a theory for damping of pressure waves in saturated bubbly magma and obtained analytical and numerical solutions. We have shown that the analytical end-member solutions are related to the incompressible and visco-elastic numerical solutions, and allow better understanding of physical processes controlling wave attenuation under various magma properties.

Furthermore, we found out that the dynamic behaviour of pressure waves propagating in bubbly magma can be obtained by the compressible NMF solution combined with the following three characteristic magma properties: (i) the *Peclet* number, defining the ratio between the diffusive and NMF timescales; (ii) the *Deborah* number, defining the visco-elastic nature of the melt; and (iii) the number density of bubbles in the melt.

We demonstrated analytically and numerically the dominant role of compressibility in reducing the attenuation of pressure waves; this role becomes even more significant for high frequencies (≥ 1 Hz). Applying our model to a magma-filled conduit, we demonstrated that the compressible NMF solution is apparently a good approximation for a wide range of natural conditions, and that it improves the resonating qualities of a magma-filled conduit.

We thank Jurgen Neuberg for useful discussions, the reviewers: Mie Ichihara and Dork Sahagian for constructive reviews, and the editors: Steve Lane and Jennie Gilbert. I. Kurzon thanks Amotz Agnon and Yizhaq Makovsky for giving advice and suggestions. Funding was provided by the EC MULTIMO project and by the USA–Israel Binational Science Foundation (grant 2004046).

Appendix A. The formulation of Commander & Prosperetti (1989)

The formulation of Commander & Prosperetti (1989) is based on the theory of Van Wijngaarden (1968). Here, we obtain their equation (1) starting with the equation of motion (Newton's 2nd law):

$$\rho \frac{\partial^2 u_i}{\partial t^2} = \frac{\partial \sigma_{ij}}{\partial x_j} \tag{A1}$$

where ρ is the density of the suspension, u_i is the total displacement and σ_{ij} is the stress tensor. By taking the divergence of (A1) and changing the order of derivatives, we get:

$$\rho \frac{\partial^2}{\partial t^2} \left(\frac{\partial u_i}{\partial x_i} \right) = \frac{\partial^2 \sigma_{ij}}{\partial x_i \partial x_j}. \tag{A2}$$

The total volume of the magma is a sum of melt and gas volumes. Hence, the total volumetric deformation ($\partial u_i / \partial x_i$) is a sum of two components:

$$\frac{\partial u_i}{\partial x_i} = \frac{\partial u_i^{(m)}}{\partial x_i} + \phi. \tag{A3}$$

The first term ($\partial u_i^{(m)} / \partial x_i$) corresponds to the melt compaction, and the second one to the change in vesicularity (relative to some background value). Compaction of the melt is proportional to the pressure, P:

$$\frac{\partial u_i^{(m)}}{\partial x_i} = -\frac{P}{K} \tag{A4}$$

where K is the bulk modulus of the melt. Finally, the volumetric deformation results in:

$$\frac{\partial u_i}{\partial x_i} = \frac{-P}{K} + \phi. \tag{A5}$$

The stress tensor could also be separated into volumetric (mean stress or pressure) and deviatoric components:

$$\sigma_{ij} = \frac{1}{3}\sigma_{nn}\delta_{ij} + \tau_{ij} = -P\delta_{ij} + \tau_{ij}. \tag{A6}$$

For small vesicularity, the melt pressure used in (A5) is equal to the ambient pressure, on the outer boundary of the shell, used in (A6). The second derivative of the stress tensor is:

$$\frac{\partial^2 \sigma_{ij}}{\partial x_i \partial x_j} = -\nabla^2 P + \frac{\partial^2 \tau_{ij}}{\partial x_i \partial x_j} = -\nabla^2 P + H \tag{A7}$$

where H is a scalar variable used for notation. Substituting (A5) and (A7) back into (A2) yields:

$$\frac{\rho}{K}\frac{\partial^2 P}{\partial t^2} - \nabla^2 P + H = \rho\frac{\partial^2 \phi}{\partial t^2}. \tag{A8}$$

Neglecting the term H related to the deviatoric stress and using the notation $c^2 = K/\rho$ for the speed of sound, we end up with the following wave equation:

$$\frac{1}{c^2}\frac{\partial^2 P}{\partial t^2} - \nabla^2 P = \rho\frac{\partial^2 \phi}{\partial t^2}. \tag{A9}$$

For small vesicularities ($\phi \ll 1$), we can approximate $\rho \approx \rho_m$ (where ρ_m is the melt density) and c is the speed of sound in the melt. These approximations lead to equation (1), the wave equation obtained by Van Wingaarden (1968), which is applied in our model. For $\partial^2 \phi/\partial t^2 = 0$, (A9) reduces to a linear wave equation for a non-dissipative medium.

Now we derive the conditions that allow neglecting the deviatoric stress (dropping H from equation 8). Following Maxwell's visco-elastic model, the total deviatoric strain rate, e_{ij}, is a sum of elastic and viscous components:

$$e_{ij} = \frac{1}{2\mu}\frac{\partial \tau_{ij}}{\partial t} + \frac{\tau_{ij}}{2\eta} \tag{A10}$$

where μ and η are the effective shear modulus and viscosity of the porous melt. Assuming small void fraction and deformable bubbles, we may use Mackenzie's expression $\eta = \eta_o\,(1 - 5\alpha/3)$, $\mu = \mu_o\,(1 - 5\alpha/3)$ (Ichihara & Kameda 2004). Taking the second derivative yields:

$$\frac{\partial^2 e_{ij}}{\partial x_i \partial x_j} = \frac{1}{2\mu}\frac{\partial H}{\partial t} + \frac{H}{2\eta}. \tag{A11}$$

Using the definition for the strain rate

$$e_{ij} = \frac{1}{2}\frac{\partial}{\partial t}\left(\frac{\partial u_i}{\partial x_j} + \frac{\partial u_j}{\partial x_i} - \frac{2}{3}\cdot\frac{\partial u_n}{\partial x_n}\delta_{ij}\right) \tag{A12}$$

and taking its second derivative leads to

$$\frac{\partial^2 e_{ij}}{\partial x_i \partial x_j} = \frac{2}{3}\nabla^2\left(\frac{\partial}{\partial t}\left(\frac{\partial u_n}{\partial x_n}\right)\right). \tag{A13}$$

Substituting (A5) into (A13) and then into (A11) yields

$$\frac{1}{2\mu}\frac{\partial H}{\partial t} + \frac{H}{2\eta} = \frac{2}{3}\nabla^2\left(\frac{\partial}{\partial t}\left(-\frac{P}{K} + \phi\right)\right). \tag{A14}$$

The H-value obtained from (A14) should be incorporated in (A8) to account for the effect of the deviatoric stress components propagation of P-waves. Comparing the first and the second term on the left-hand side of (A14), we discuss two end-member cases: (a) $\omega/2\mu \gg 1/2\eta$ corresponding to negligible role of the viscous strain components and (b) $\omega/2\mu \ll 1/2\eta$, corresponding to negligible role of the elastic strain components.

In the first case, H is equal to

$$H = \frac{4}{3}\mu\nabla^2\left(-\frac{P}{K} + \phi\right) \tag{A15}$$

leading to two extra terms in the wave equation (A8):

$$\frac{\rho}{K}\frac{\partial^2 P}{\partial t^2} - \nabla^2 P - \frac{4}{3}\frac{\mu}{K}\nabla^2 P + \frac{4}{3}\mu\nabla^2\phi = \rho\frac{\partial^2 \phi}{\partial t^2}. \tag{A16}$$

The additional term of $\nabla^2 P$ leads to a well-known expression for p-wave velocity in an elastic media ($\rho\,V_p^2 = K + 4/3\mu$). The ratio between the second term and the first term of $\nabla^2 P$ gives: $4\mu/3K$. The second, vesicularity-related term is of the order of $\mu\phi L^2$ (where L is the wavelength), which should be compared with the right-hand-side term, of the order of $\rho\phi\omega^2$ (where ω is the angular frequency). Using $L^2\omega^2 = V_p^2 \approx K/\rho$, the ratio between these two terms is also $4\mu/3K$, which is generally lower than one for magmas, and may get very low for hydrous, silicate melts ($\mu \approx 10^6$ Pa according to Mungall et al. 1996; Romano et al. 1996; Dingwell 1998, while $K \approx 10^{10}$ Pa). Hence, the elastic components in the deviatoric stresses can be ignored, leading to equation A9.

In the second case ($\omega/2\mu \ll 1/2\eta$), the elastic components are neglected, resulting in

$$H = \frac{4}{3}\eta\nabla^2\left(\frac{\partial}{\partial t}\left(-\frac{P}{K} + \phi\right)\right) \tag{A17}$$

which lead to two additional terms in the wave equation (A8):

$$\frac{\rho}{K}\frac{\partial^2 P}{\partial t^2} - \nabla^2 P - \frac{4}{3}\frac{\eta}{K}\frac{\partial}{\partial t}\left(\nabla^2 P\right) + \frac{4}{3}\eta\frac{\partial}{\partial t}\left(\nabla^2 \phi\right) = \rho\frac{\partial^2 \phi}{\partial t^2}. \tag{A18}$$

For $\phi = 0$, we may solve (A18), and obtain the role of melt viscosity in wave attenuation. Applying $\delta P(x, t) = A\exp(ikx - i\omega t)$, we obtain

$$-\frac{\omega^2}{c^2} + k^2 = i\frac{4}{3}\frac{\eta}{K}\omega k^2 \qquad (A19)$$

which for $\omega = kV$ may be re-organized as:

$$V^2 = c^2\left[1 - \frac{4\eta\omega}{3K}i\right]. \qquad (A20)$$

Substituting (A20) into the definition of Q (equation 6), we obtain

$$Q = \frac{3K}{4\eta\omega}. \qquad (A21)$$

This equation describes the melt-related Q, which for propagating plane-waves provides the upper frequency-limit of our model (see Figures 5, 6b and 8). The additional vesicularity-related term is of the order of $\eta\phi\omega/L^2$, while the right hand-side term, as in the previous elastic case, is of the order of $\rho\phi\omega^2$; their ratio is $4\eta\omega/3K$.

This means that in order to neglect the viscosity-related terms and to use equation A9, we require that $4\eta\omega/3K \ll 1$ and that the melt-related Q is higher than the bubble-related Q. For $K \approx 10^{10}$ and $\eta \approx 10^6$, the frequency of the pressure waves cannot exceed 100 Hz, and for $\eta \approx 10^7$ it cannot exceed 10 Hz.

Appendix B. The visco-elastic bubble growth model

Strain–stress relationship

We use Maxwell visco-elastic model that assumes superposition of deformations associated with different mechanisms. This assumption is expressed as

$$\varepsilon_{ij}^{tot} = \varepsilon_{ij}^{el} + \varepsilon_{ij}^{vis} \qquad (B1)$$

where ε_{ij}^{tot} is the total strain, ε_{ij}^{vis} is the viscous strain and ε_{ij}^{el} is the elastic strain. The total Cauchi strain tensor, ε_{ij}^{tot}, is related to the displacement, u_i by

$$\varepsilon_{ij}^{tot} = \frac{1}{2}\left(\frac{\partial u_i}{\partial x_j} + \frac{\partial u_j}{\partial x_i}\right). \qquad (B2)$$

The elastic strain tensor is related to the stress tensor, σ_{ij}^{el}, through Hooke's law:

$$\sigma_{ij} = \lambda\varepsilon_{kk}^{el}\delta_{ij} + 2\mu\varepsilon_{ij}^{el} \qquad (B3)$$

where λ and μ are Lame coefficients (and μ is also known as the shear modulus or the rigidity), and δ_{ij} is the unit tensor (Einstein summation convention is used).

The stress tensor could be represented as the sum of volumetric stress (diagonal components) and deviatoric stress. The pressure is equal to

$$P = -\frac{\sigma_{\alpha\alpha}}{3} = -\left(\lambda + \frac{2}{3}\mu\right)\varepsilon_{\alpha\alpha}^{el} \qquad (B4)$$

where $\sigma_{\alpha\alpha}$ is the sum of volumetric stress components and $\varepsilon_{\alpha\alpha}^{el}$ is the sum of the elastic strain components. By using the definitions $\varepsilon_{\alpha\alpha}^{el} = div(u_i)$ for the sum of the volumetric strain components and $K = \lambda + \frac{2}{3}\mu$ for the bulk modulus, we may re-express (B4) as:

$$P = -K \cdot div(u_i) \qquad (B5)$$

Similarly, the deviatoric stresses $\tau_{ij} = \sigma_{ij} - \frac{1}{3}\sigma_{\alpha\alpha}\delta_{ij}$ and the deviatoric strains $\tilde{\varepsilon}_{ij}^{el} = \varepsilon_{ij}^{el} - \frac{1}{3}\varepsilon_{\alpha\alpha}^{el}\delta_{ij}$ are proportional:

$$\tau_{ij} = 2\mu\tilde{\varepsilon}_{ij}^{el}. \qquad (B6)$$

The rate of viscous strain accumulation in Newotonian media with viscosity, η, is:

$$\frac{d\tilde{\varepsilon}_{ij}^{vis}}{dt} = \frac{1}{2\eta}\tau_{ij}. \qquad (B7)$$

Substituting (B6) and (B7) back into (B1) yields $\tilde{e}_{ij}$, the total deviatoric strain rate tensor:

$$\tilde{e}_{ij} \equiv \frac{d\tilde{\varepsilon}_{ij}^{tot}}{dt} = \frac{\tau_{ij}}{2\eta} + \frac{\dot{\tau}_{ij}}{2\mu}. \qquad (B8)$$

The ratio between the elastic and viscous terms in (B8) provides the non-dimensional *Deborah* number discussed earlier in this paper.

For constant shear coefficients, μ and η, the general solution of (B8) is

$$\tau_{ij}(t) = \exp\left(-\frac{t}{\tau_m}\right)\left[\int 2\mu\tilde{e}_{ij}\exp\left(\frac{t}{\tau_m}\right)dt + \tau_0\right] \qquad (B9)$$

where $\tau_m = \eta/\mu$ is the Maxwell relaxation time and τ_0 is the stress at $t = 0$.

3-D spherical symmetry

Assuming 3-D spherical symmetry, we search a solution for radial displacement u_r (r, t) as a function of the distance, r, from the centre of the bubble and time, t, with no tangential components of the displacement:

$$u_r(r) = a(t)r + \frac{b(t)}{r^2}. \qquad (B10)$$

Therefore, the components of the strain tensor in spherical coordinates are:

$$\varepsilon_{rr}(r) = \frac{\partial u_r}{\partial r} = a(t) - \frac{2b(t)}{r^3}$$

$$\varepsilon_{\theta\theta}(r) = \varepsilon_{\phi\phi}(r) = \frac{u_r}{r} = a(t) + \frac{b(t)}{r^3} \qquad (B11a, b)$$

and the radial velocity is

$$v_r(r) = \dot{a}(t)r + \frac{\dot{b}(t)}{r^2}. \tag{B12}$$

For $\dot{b}(t) = v_R R^2$ and $a(t) = const$, (B12) reduces to the radial distribution of melt velocity in an incompressible viscous fluid:

$$v_r(r) = v_R \frac{R^2}{r^2}. \tag{B13}$$

We may separate the strain tensor to deviatoric and volumetric components resulting in

$$\tilde{\varepsilon}_{rr} = -\frac{2b(t)}{r^3}$$

$$\tilde{\varepsilon}_{\theta\theta} = \tilde{\varepsilon}_{\phi\phi} = \frac{b(t)}{r^3} \tag{B14a-c}$$

$$div(u_i) \equiv \varepsilon_{\alpha\alpha} = 3a(t)$$

leading to the following shear strain rates

$$e_{rr} \equiv \dot{\varepsilon}_{rr}(r) = -\frac{2\dot{b}(t)}{r^3}$$

$$\tag{B15a,b}$$

$$e_{\theta\theta} = e_{\phi\phi} \equiv \dot{\varepsilon}_{\theta\theta}(r) = \dot{\varepsilon}_{\phi\phi}(r) = \frac{\dot{b}(t)}{r^3}.$$

Substitution of (B14c) into (B5) and of (B15a, b) into (B9) results with the following stresses:

$$P = -3Ka(t)$$

$$\tau_{rr} = \exp\left(-\frac{t}{\tau_m}\right)\left[\int_0^t \left(\left(-2\mu\frac{2\dot{b}(t)}{r^3}\right)\exp\left(\frac{t}{\tau_m}\right)\right)dt + \tau_{rr}^0\right]$$

$$\tau_{\theta\theta} = \tau_{\phi\phi} = \exp\left(-\frac{t}{\tau_m}\right)\left[\int_0^t \left(2\mu\frac{\dot{b}(t)}{r^3}\exp\left(\frac{t}{\tau_m}\right)\right)dt + \tau_{\theta\theta}^0\right].$$

$$\tag{B16a-c}$$

The time-dependent functions $a(t)$ and $b(t)$ could be found from the boundary conditions on the outer boundary of the shell ($r = S$):

$$(-P + \tau_{rr})|_{r=S} = -P_a \tag{B17}$$

and on the bubble-melt interface ($r = R$):

$$(-P + \tau_{rr})|_{r=R} = -P_g + \frac{2\gamma}{R} \tag{B18}$$

where P_a is the ambient pressure acting on the outer shell, and P_g is the gas pressure in the bubble. The term $2\gamma/R$ stands for the Laplace surface tension.

Substitution of the expressions for the pressure (B16a) and for the radial stress (B16b) into the two boundary conditions results in

$$3Ka - \exp\left(-\frac{t}{\tau_m}\right)\left[\int_0^t 4\mu\frac{\dot{b}(t)}{r^3}\exp\left(\frac{t}{\tau_m}\right)dt + \tau_{rr}^0(r)\right]\Bigg|_{r=S} = -P_a$$

$$\tag{B19}$$

$$3Ka - \exp\left(-\frac{t}{\tau_m}\right)\left[\int_0^t 4\mu\frac{\dot{b}(t)}{r^3}\exp\left(\frac{t}{\tau_m}\right)dt + \tau_{rr}^0(r)\right]\Bigg|_{r=R}$$

$$= -P_g + \frac{2\gamma}{R}. \tag{B20}$$

Assuming that the system is in equilibrium at $t = 0$, thus $\tau_{rr}^0 \equiv 0$, and ignoring surface tension, we may re-express the boundary conditions as

$$\begin{cases} 3Ka - \frac{4\mu}{S^3}\exp\left(-\frac{t}{\tau_m}\right)\int_0^t \dot{b}(t)\exp\left(\frac{t}{\tau_m}\right)dt = -P_a \\ 3Ka - \frac{4\mu}{R^3}\exp\left(-\frac{t}{\tau_m}\right)\int_0^t \dot{b}(t)\exp\left(\frac{t}{\tau_m}\right)dt = -P_g \end{cases}$$

$$\tag{B21a,b}$$

By multiplying (B21a) by S^3 and (B21b) by R^3, subtracting (B21a) from (B21b) and rearranging, we obtain the following expression for $a(t)$:

$$a(t) = \frac{R^3 P_g - S^3 P_a}{3K(S^3 - R^3)}$$

$$= -\frac{P_a}{3K} + \frac{1}{3K}\cdot\frac{R^3}{S^3 - R^3}(P_g - P_a). \tag{B22}$$

By subtracting (B21a) from (B21b), differentiation with respect to time and dividing by the common factor, $\exp(t/\tau_m)$ we get

$$b(t) = \dot{f}(t) + \frac{1}{\tau_m}f(t) \tag{B23}$$

where

$$f(t) = \frac{P_g - P_a}{4\mu} * \frac{R^3 S^3}{S^3 - R^3}.$$

Instead of calculating the radial velocity using (16) from the viscous bubble growth model, we can now calculate it according to (B12), where $\dot{a}(t)$ and $\dot{b}(t)$ are taken from (B22) and (B23); the result is given by the following equation:

$$v_r(r) = \frac{d}{dt}\left(\frac{R^3 P_g - S^3 P_a}{3K(S^3 - R^3)}\right)\cdot r + \frac{1}{r^2}$$

$$\times \left\{\frac{R^3 S^3}{4\mu(S^3 - R^3)}\left[\frac{d}{dt}(P_g - P_a) + \frac{1}{\tau_m}(P_g - P_a)\right]\right\}.$$

$$\tag{B24}$$

Equation B24 couples the viscous and elastic deformations of the melt surrounding the deforming bubble. This is the equation used in the visco-elastic numerical solution for $Q(\omega)$. For an infinite shell model, this monster (B24) gets a more elegant form:

$$v_r(r) = r \cdot \left[\frac{1}{4\eta}\left(P_g(t) - P_o(t)\right) + \frac{1}{4\mu}\frac{d\left(P_g(t) - P_a(t)\right)}{dt} \right.$$
$$\left. - \frac{1}{3K}\frac{dP_a(t)}{dt} \right] \tag{B25}$$

and on the bubble-melt interface it yields equation 18.

Appendix C. The finite shell model

We generalized our analytical solutions to the finite shell model, where the bubble 'sees' the far boundary of the shell. However, we still keep to small oscillations and $\delta R \ll R_0$.

For the NMF solution, we should multiply τ_{NMF} by a factor of $1 + R_0^3/S_i^3$ leading to

$$Q = \frac{\eta\omega}{\rho_m \pi n_d R_0^3 c^2} \cdot \frac{S_i^3}{R_0^3 + S_i^3}$$
$$+ \frac{1}{\omega} \cdot \left[\frac{9P_0^2}{\rho_m 16\eta \pi n_d R_0^3 c^2} + \frac{3P_0}{4\eta} \right] \cdot \left[1 + \frac{R_0^3}{S_i^3} \right]. \tag{C1}$$

For the viscous solution, we get:

$$Q = \frac{\eta\omega}{\rho_m \pi n_d R_0^3 c^2} \cdot \frac{S_i^3}{R_0^3 + S_i^3}. \tag{C2}$$

For the diffusive solution, we use the concentration gradient definition given by Lyakhovsky *et al.* (1996) for a finite shell model:

$$\left(\frac{\partial C}{\partial r}\right)_R = \frac{S_i^3(C_0 - C_R) - \frac{\rho_g}{\rho_m}R^3}{S_i^3 R - 1.5(S^2 - R^2)R^2} \tag{C3}$$

which for $R \ll S$ reduces to (14) and converges with the infinite shell model, discussed above. Using (C3) in the development of the diffusive solution, we obtain an expression for $Q(\omega)$ that is similar to (35) but is divided by a factor of

$$F_{diff} = \frac{S_i^3}{\left\{ S_i^3 - 1.5\left[\left(S_i^3 + R_0^3\right)^{2/3} - R_0^2 \right]R_0 \right\}}. \tag{C4}$$

However, the corrections for the finite shell case are relevant only for the upper limit of the vesicularity-range considered in our model.

References

AKI, K., FEHLER, M. & DAS, S. 1977. Source mechanism of volcanic tremor — fluid-driven crack models and their application to 1963 Kilauea eruption. *Journal of Volcanology and Geothermal Research*, **2**, 259–287.

AKI, K. & RICHARDS, P. G. 2002. *Quantitative Seismology — 2nd edition.* University Science Books.

BARCLAY, J., RILEY, D. S. & SPARKS, R. S. J. 1995. Analytical models for bubble growth during decompression of high viscosity magmas. *Bulletin of Volcanology*, **57**, 422–431.

CAFLISCH, R. E., MIKSIS, M. J., PAPANICOLAOU, G. C. & TING, L. 1985a. Effective equations for wave-propagation in bubbly liquids. *Journal of Fluid Mechanics*, **153**, 259–273.

CAFLISCH, R. E., MIKSIS, M. J., PAPANICOLAOU, G. C. & TING, L. 1985b. Wave-propagation in bubbly liquids at finite volume fraction. *Journal of Fluid Mechanics*, **160**, 1–14.

CHOUET, B. 1986. Dynamics of a fluid-driven crack in 3 dimensions by the finite-difference method. *Journal of Geophysical Research (Solid Earth and Planets)*, **91**, 13967–13992.

CHOUET, B. 1988. Resonance of a fluid-driven crack — radiation properties and implications for the source of long-period events and harmonic tremor. *Journal of Geophysical Research (Solid Earth and Planets)*, **93**, 4375–4400.

CHOUET, B. A. 1996. Long-period volcano seismicity: its source and use in eruption forecasting. *Nature*, **380**, 309–316.

COLLIER, L., NEUBERG, J. W., LENSKY, N., LYAKHOVSKY, V. & NAVON, O. 2006. Attenuation in gas-charged magma. *Journal of Volcanology and Geothermal Research*, **153**, 21–36.

COMMANDER, K. W. & PROSPERETTI, A. 1989. Linear pressure waves in bubbly liquids — comparison between theory and experiments. *Journal of the Acoustical Society of America*, **85**, 732–746.

DINGWELL, D. B. 1998. Recent experimental progress in the physical description of silicic magma relevant to explosive volcanism. *In*: GILBERT, G. S. & SPARKS, R. S. G. (eds) *The Physics of Explosive Volcanic Eruptions.* Geological Society, London, Special Publication, **145**, 9–26.

HURWITZ, S. & NAVON, O. 1994. Bubble nucleation in rhyolitic melts — experiments at high-pressure, temperature, and water-content. *Earth and Planetary Science Letters*, **122**, 267–280.

ICHIHARA, M. & KAMEDA, M. 2004. Propagation of acoustic waves in a visco-elastic two-phase system: influences of the liquid viscosity and the internal diffusion. *Journal of Volcanology and Geothermal Research*, **137**, 73–91.

ICHIHARA, M., OHKUNITANI, H., IDA, Y. & KAMEDA, M. 2004. Dynamics of bubble oscillation and wave propagation in viscoelastic liquids. *Journal of Volcanology and Geothermal Research*, **129**, 37–60.

KELLER, J. B. & KOLODNER, I. I. 1956. Damping of underwater explosion bubble oscillations. *Journal of Applied Physics*, **27**, 1152–1161.

KELLER, J. B. & MIKSIS, M. 1980. Bubble oscillations of large-amplitude. *Journal of the Acoustical Society of America*, **68**, 628–633.

LENSKY, N. G., LYAKHOVSKY, V. & NAVON, O. 2001. Radial variations of melt viscosity around growing bubbles and gas overpressure in vesiculating magmas. *Earth and Planetary Science Letters*, **186**, 1–6.

LENSKY, N. G., LYAKHOVSKY, V. & NAVON, O. 2002. Expansion dynamics of volatile-supersaturated liquids and bulk viscosity of bubbly magmas. *Journal of Fluid Mechanics*, **460**, 39–56.

LENSKY, N. G., NAVON, O. & LYAKHOVSKY, V. 2004. Bubble growth during decompression of magma: experimental and theoretical investigation. *Journal of Volcanology and Geothermal Research*, **129**, 7–22.

LEZZI, A. & PROSPERETTI, A. 1987. Bubble dynamics in a compressible liquid. 2. 2nd-order theory. *Journal of Fluid Mechanics*, **185**, 289–321.

LYAKHOVSKY, V., HURWITZ, S. & NAVON, O. 1996. Bubble growth in rhyolitic melts: experimental and numerical investigation. *Bulletin of Volcanology*, **58**, 19–32.

MALVERN, L. E. 1969. *Introduction to the Mechanics of a Continuous Medium*. Prentice-Hall.

MOURTADA-BONNEFOI, C. C. & MADER, H. M. 2001. On the development of highly-viscous skins of liquid around bubbles during magmatic degassing. *Geophysical Research Letters*, **28**, 1647–1650.

MUNGALL, J. E., BAGDASSAROV, N. S., ROMANO, C. & DINGWELL, D. B. 1996. Numerical modelling of stress generation and microfracturing of vesicle walls in glassy rocks. *Journal of Volcanology and Geothermal Research*, **73**, 33–46.

NAVON, O. & LYAKHOVSKY, V. 1998. Vesiculation processes in silicic magmas. *In*: GILBERT, G. S. & SPARKS, R. S. G. (eds) *The Physics of Explosive Volcanic Eruptions*. Geological Society, London, Special Publication, **145**, 27–50.

NEUBERG, J., LUCKETT, R., BAPTIE, B. & OLSEN, K. 2000. Models of tremor and low-frequency earthquake swarms on Montserrat. *Journal of Volcanology and Geothermal Research*, **101**, 83–104.

O'CONNELL, R. & BUDIANSKY, B. 1978. Measures of dissipation in viscoelastic media. *Geophysical Research Letters*, **5**, 5–8.

PROSPERETTI, A., CRUM, L. A. & COMMANDER, K. W. 1988. Nonlinear bubble dynamics. *Journal of the Acoustical Society of America*, **83**, 502–514.

PROSPERETTI, A. & LEZZI, A. 1986. Bubble dynamics in a compressible liquid. 1. 1st-order theory. *Journal of Fluid Mechanics*, **168**, 457–478.

PROUSSEVITCH, A. A. & SAHAGIAN, D. L. 1996. Dynamics of coupled diffusive and decompressive bubble growth in magmatic systems. *Journal of Geophysical Research (Solid Earth)*, **101**, 17447–17455.

PROUSSEVITCH, A. A., SAHAGIAN, D. L. & ANDERSON, A. T. 1993. Dynamics of diffusive bubble-growth in magmas — isothermal case. *Journal of Geophysical Research (Solid Earth)*, **98**, 22283–22307.

ROMANO, C., MUNGALL, J. E., SHARP, T. & DINGWELL, D. B. 1996. Tensile strengths of hydrous vesicular glasses: An experimental study. *American Mineralogist*, **81**, 1148–1154.

VAN WIJNGAARDEN, L. 1968. On the equations of motion for mixtures of liquid and gas bubbles. *Journal of Fluid Mechanics*, **33**, 465–474.

Numerical simulation of the dynamics of fluid oscillations in a gravitationally unstable, compositionally stratified fissure

ANTONELLA LONGO[1], DAVID BARBATO[1], PAOLO PAPALE[1], GILBERTO SACCOROTTI[1,2] & MICHELE BARSANTI[1,3]

[1]*Istituto Nazionale di Geofisica e Vulcanologia, Sezione di Pisa, Via della Faggiola 32, 56126 Pisa, Italy (e-mail: longo@pi.ingv.it)*

[2]*Osservatorio Vesuviano, Via Diocleziano 328, 80124 Napoli, Italy*

[3]*Dipartimento di Matematica Applicata, Università di Pisa, Via F. Buonarroti 1/c, 56127 Pisa, Italy*

Abstract: We have simulated the dynamics of convection, mixing and ascent of two basaltic magmas differing in their volatile and crystal content, giving rise to a gravitationally unstable configuration along a dyke or fissure. Numerical simulations are performed by a recently developed code which describes the transient 2D dynamics of multicomponent fluids from the incompressible to the compressible regime, and the initial and boundary conditions are inspired to the paroxysmal eruption which occurred at Stromboli in 2003 (D'Auria *et al.* 2006). Multicomponent ($H_2O + CO_2$) saturation is accounted for by modelling the non-ideal equilibrium between the gas phase and the melt. The numerical results show the formation of a rising bulge of light magma, and the sink of discrete batches of dense magma towards deep fissure regions. Such dynamics are associated with a complex evolution of the pressure field, which shows variations occurring over a wide spectrum of frequencies. A first order analysis of the propagation of such pressure disturbances through the country rocks shows that the pre-eruptive fissure dynamics are able to produce mm-size, mainly radial deformation of the volcano, and a detectable seismic signal with spectral peaks at periods of about 50 s.

Magmatic systems are often characterized by complex compositional heterogeneities (Jaupart & Tait 1995; Folch *et al.* 1998; Kuritani 2004), which translate in heterogeneous distribution of physical properties. The origin of such heterogeneities is to be found in a variety of processes including pressure–temperature-dependent crystallization and gas exsolution, liquid-crystal separation and open system degassing, heat and mass transfer with heterogeneous country rocks, recharge of the magmatic system from below, and so on. In some cases, the above processes lead to configurations which are gravitationally stable, as for example in long-living magma chambers where liquid-crystal segregation and associated volatile concentration often result in more chemically evolved, gas-rich magma layers overlying more primitive, crystal-rich magma (Kuritani 2004; Bergantz 2000). However, it is also possible that denser magma layers develop on top of lighter magma, giving origin to a gravitationally unstable configuration. This is the case with many open conduit basaltic volcanoes, in which open system degassing on top of the magmatic column, and consequent crystallization, lead to the formation of crystal-rich, gas-poor magma layers overlying crystal-poor, gas-rich magma. Stromboli volcano is a well-known example of this type. Here, the discharge of the deeper gas-rich magma occurs during major eruptions or paroxysms, while the shallow crystal-rich and gas-poor magma feeds the 'normal' low-intensity persistent Strombolian activity (Bertagnini *et al.* 2003; Metrich *et al.* 2003; Landi *et al.* 2004; Chouet *et al.* 2003).

When the magmatic processes are such that a gravitationally unstable configuration develops, the magmatic body is subject to forces which tend to re-establish a stable configuration. This reorganization occurs through the development of convection, in which the light magma moves up while the dense magma moves down under the action of gravity. The processes observed on top of lava lakes, with sinking of cold, degassed magma and arrival of gas-rich magma producing gas burst and fountaining (e.g. Jellinek & Kerr 2001; Aster *et al.* 2003; Harris *et al.* 2005), display the action of magma convection in continuously trying to re-establish gravitational equilibrium against degassing- and cooling-induced disequilibrium.

Here we examine the basic physics of convective motion in a vertically stratified, gravitationally unstable volcanic fissure or dyke, by means of

From: LANE, S. J. & GILBERT, J. S. (eds) *Fluid Motions in Volcanic Conduits: A Source of Seismic and Acoustic Signals.* Geological Society, London, Special Publications, **307**, 33–44.
DOI: 10.1144/SP307.3   0305-8719/08/$15.00 © The Geological Society of London 2008.

numerical simulations of the 2D, transient dynamics of compressible multicomponent magma. Numerical simulations are carried out by the use of the GALES code, which has been developed and successfully applied to simulate the dynamics of magmatic systems (Longo *et al.* 2006). The numerical results are processed in order to investigate particularly the mixing processes and the pressure changes occurring within the fissure and close to the conduit walls. Pressure oscillations are found to develop over a range of frequencies encompassing those typical of both static and dynamic rock deformation observed at basaltic volcanoes. A first order analysis of the characteristics of the expected deformation and seismic changes shows that buoyancy-induced magma convection in a volcanic fissure is capable of producing signals that can be observed at the surface.

Description of the physico-mathematical model

The physical model describes the isothermal motion in a vertical fissure of a fluid made of two miscible magmatic end-member components having the same temperature. The governing two-dimensional Navier–Stokes equations are the following:

mass of components:

$$\frac{\partial \rho y_k}{\partial t} + \nabla \cdot (\rho v y_k) = -\nabla \cdot (\rho D_k \nabla y_k), \ k = 1, 2 \quad (1)$$

momentum:

$$\frac{\partial \rho v}{\partial t} + \nabla \cdot (\rho v \otimes v + pI) = \nabla \cdot$$

$$\left(\mu (\nabla v + \nabla v^T) - \frac{2}{3} \mu (\nabla \cdot v) I \right) + \rho g \quad (2)$$

In the above equations, t is time, ρ is fluid density, v is velocity vector, y is weight fraction, D is the diffusion coefficient, p is pressure, I is the identity matrix, μ is the Newtonian mixture viscosity, g is gravity acceleration, and the subscript k denotes the components.

Each of the two end-member components is characterized by its composition in terms of ten major oxides plus total H_2O and CO_2 (where 'total' refers to the volatile mass in the gas and liquid phase with respect to the total mass of the gas and liquid phases) and volume-density distribution of crystals. The non-ideal multicomponent $H_2O + CO_2$ saturation model of Papale *et al.* (2006) is used to calculate gas-liquid partition of volatiles, according to the local composition conditions. Liquid and liquid + crystal density is also

calculated on the basis of local composition conditions, by using the Lange (1994) EOS for the liquid phase and simple mixing rules for the effect of crystals.

The multiphase magma is assumed to behave as a Newtonian pseudo-fluid. Newtonian liquid viscosity for each of the two end-members at the assumed constant temperature is taken from viscosity measurements on separated glasses from the crystal-rich and crystal-poor magma types discharged at Stromboli (Giordano *et al.* 2006), the compositions of which are used as the volatile-free end-member components in the present simulations (see Table 1). Due to lack of experimental data, the effect of dissolved H_2O on liquid viscosity is modelled by analogy with that for Etnean basalts (Giordano & Dingwell 2000). The effects of crystals and non-deformable gas bubbles on Newtonian viscosity are modelled as in Papale (2001).

Component mass diffusion fluxes due to concentration gradients are computed with the Curtiss and Hirschfelder law for multicomponent fluids (Hirschfelder *et al.* 1969), while mass diffusion due to pressure gradients is assumed to be negligible. Surface tension effects in mass diffusion are also neglected.

The resulting set of model equations is solved through the GALES code, which consists in a finite element C++ scheme for the dynamics of 2D multicomponent compressible-to-incompressible flow (Longo *et al.* 2006). The numerical algorithm is based on space–time discretization, and on the Galerkin formulation with a least-squares term for numerical stability less in the streamline direction, and a discontinuity-capturing operator which stabilizes the numerical solution along cross-wind direction (Hauke & Hughes 1998). This system of equations is

Table 1. *Composition of the liquid phase for the two CRGP and CPGR magmas employed in the simulation*

	CRGP	CPGR
SiO_2	52.9	50.71
TiO_2	1.7	0.91
Al_2O_3	16.3	17.80
FeO	8.3	6.79
Fe_2O_3	1.4	1.12
MgO	3.5	5.61
CaO	7.2	10.61
Na_2O	3.5	2.93
K_2O	4.5	2.36
P_2O_5	0.7	0.58

From Bertagnini *et al.* (2003) and Landi *et al.* (2004). Temperature for both magmas is taken to be 1400 K.

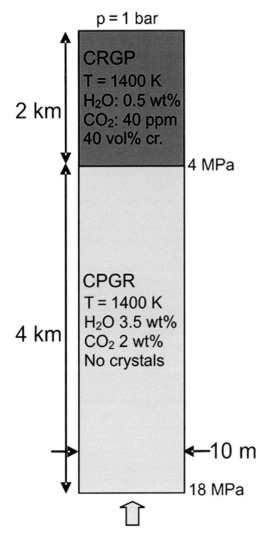

Fig. 1. Configuration of the simulated systems (cases A–D), initial and boundary conditions.

Table 2. *Conditions at time 0 resulting from the assumed system configuration*

Pressure at interface (MPa)	4.2
ρ_+ (kg/m^3)	1000
ρ_- (kg/m^3)	165
$\Delta\rho$ (kg/m^3)	835
μ_+ (Pa s)	200
μ_- (Pa s)	575
Gas volume (%) at 1 atm (top of system)	99.3

The subscripts + and – refer to conditions determined immediately above and below the interface, respectively.

advanced in time with a predictor multi-corrector algorithm, linearized by a Newton–Raphson method, and solved within each corrector pass with a block-diagonal preconditioned GMRES (Shakib *et al.* 1989).

Simulated system

In order to investigate the physics of magma convection in a volcanic fissure filled by gravitationally unstable, vertically stratified multicomponent magma, we have designed the system depicted in Figure 1 and detailed in Tables 1 and 2. This system is a simplified representation of a possible natural one, and it has been selected with inspiration from the case of Stromboli (Bertagnini *et al.* 1999, 2003; Métrich *et al.* 2005). A 6 km long, 10 m wide fissure is considered. The shallower 2 km of the fissure is filled with dense magma having 40 vol% crystals, and total (exsolved in the gas phase plus dissolved in the liquid magma) H_2O and CO_2 contents of 0.5 wt% and 40 ppm, respectively. This kind of magma is therefore crystal-rich and gas-poor (CRGP). The deeper 4 km of the fissure contains instead magma with no crystals and total H_2O and CO_2 contents of 3.5 and 2 wt%, respectively. This second kind of magma is therefore crystal-poor and gas-rich (CPGR). The liquid composition of the top magma reflects crystallization from the original bottom one, while the different volatile assemblages are taken to reflect previous open system degassing in the shallow magma type. Pressure at the fissure top is fixed at 1 atm. The resulting density contrast at the CRGP/CPGR interface is about 800 kg/m^3, positive upward. This represents a gravitational instability, the consequences of which are explored by numerical simulation.

The initial condition depicted above is not intended to represent an initially stable configuration in the volcanic fissure; rather, it is designed as a first approximation of the state immediately following contact between the two considered magma types, leading to gravitational destabilization and to convection and mixing dynamics. An upward flow of CPGR magma at the base of the simulated system corresponding to 2×10^4 kg/s is estimated with reference to the low-intensity phase of the 5 April 2003 paroxysm at Stromboli (Rosi *et al.* 2006), and by considering a not-simulated third dimension 100 m long, with analogy to the length of the active crater system at the surface. Such an upward flow, which emerges from the observations, is maintained during the whole simulation and represents the simplest way to provide the trigger mechanism for the destabilization of the compositional interface at 2 km depth, which could be destabilized by small perturbations. Note that the 2D assumption implies translational symmetry of the flow along the third dimension and neglect of side effects close to its edges. Consistently, the width of the considered fissure is one order of magnitude smaller than its extent along the direction perpendicular to the sketch in Figure 1. No slip conditions are adopted at fissure walls.

The domain of the fissure is divided into 2706 rectangular finite elements with 1 m width and height less variable from 1–25 m. Dynamic properties at the interface between the shallower and deeper layers were accurately investigated using the closer mesh (1×1m), in order to focus on the processes generated by gravitational instability.

The homogeneous (or one-fluid) assumption implies perfect coupling between the magmatic phases. This means that the separated ascent of large gas bubbles is not considered in the simulation. Note that the presence of volatiles in the shallow magma, though in small amount, implies a very large gas volume fraction when approaching 1 atm pressure at the upper system boundary (Table 2). As a consequence, the bulk magma density and the resulting pressure gradient in the system are quite small. The simulated system is therefore unlikely to exist in nature with the characteristics assumed, since the difference at depth between the magmastatic and lithostatic pressure turns out to be too high to be realistic. A more consistent picture in future simulations may consider therefore a totally degassed magma extending to a larger depth, so to keep a higher pressure gradient and a lower pressure difference between the magmatic system and the country rocks.

Numerical results

Figure 2 shows the time evolution of the simulated system, in terms of composition and density distributions. At time zero (Fig. 2a), the two CRGP and CPGR magmas are assumed to lie undisturbed one over the other, producing an inversion in the density gradient in correspondence to the compositional interface. After 30 s, an instability develops at the interface with the CRGP, and more viscous upper magma which starts to sink at the centre of fissure while the CPGR magma rises at its margins (Fig. 2b). After 90 s (Fig. 2c), the CRGP magma has reached a depth of about 300 m from the interface. After 200 s (Fig. 2d), this depth has become about 500 m. The CPGR magma initially rising along the two sides now tends to form a single bulge in the central portion of the fissure, which has risen about the same length as the descending magma. After 400 s (Fig. 2e), the situation is represented by discrete batches of CRGP magma

sinking through the CPGR magma down to a depth of about 1300 m from the position of the original interface, and by a single bulge of CPGR magma rising through the CRGP magma up to about 900 m above the interface. Finally, after 650 s corresponding to the maximum time reached by the simulation, the CRGP magma has reached a depth of 2500 m from the interface, while the CPGR magma has risen to 1100 m above the interface. During the simulated process, the vertical velocity of magma at fissure top progressively increases, therefore a portion of magma leaves the system, with maximum achieved velocities of the order of a few cm/s as a consequence of the high viscosity of the crystal- and bubble-rich mixture close to system top. The arrival of new magma into the system is therefore mostly accommodated by the pressure (and density) changes described below.

It is worth noting the visual scale effect in Figure 2, whereby the discrete sinking batches of CRGP magma seem to be thin and wide. Actually, their vertical dimension is about one order of magnitude larger than the horizontal one (as an example, the lowermost batch at 420 s in Figure 2 is 5 m wide and about 50 m high).

Figure 3 shows the difference between the initial and local pressure along the fissure walls as a function of time. The convective motion generated by the gravitational instability and by the entrance of new magma from the fissure bottom produces the development of three main zones. Above the original interface, a positive pressure change originates, with maximum values up to about 1 MPa in an area as thick as about 500 m and centred about 700 m above the original interface. Instead, below the interface a negative pressure change occurs, with maximum values of about 0.9 MPa at about 5 m below the original interface position. Finally, a region of positive pressure change characterizes the lower fissure region, reflecting compression due to new magma entrance from below.

The pressure variations depicted in Figure 3 span the whole time interval encompassed by our simulation. In order to highlight the features of the system occurring over shorter temporal scales, we high-pass filter the time histories of the pressure at the conduit wall using a 2-pole, 0-phase shift Butterworth filter (MATLAB$^{\text{©}}$) with a corner at a period of 100 s.

Figure 4 shows the filtered pressure signal along the fissure walls as a function of time. Soon after the destabilization of the compositional interface, we observe the downward propagation of a compression wave, and the upward propagation of a rarefaction wave, both travelling at an average speed of 160 m/s approaching the speed of sound in the multiphase magma. While the upward-moving rarefaction wave soon dissipates, the

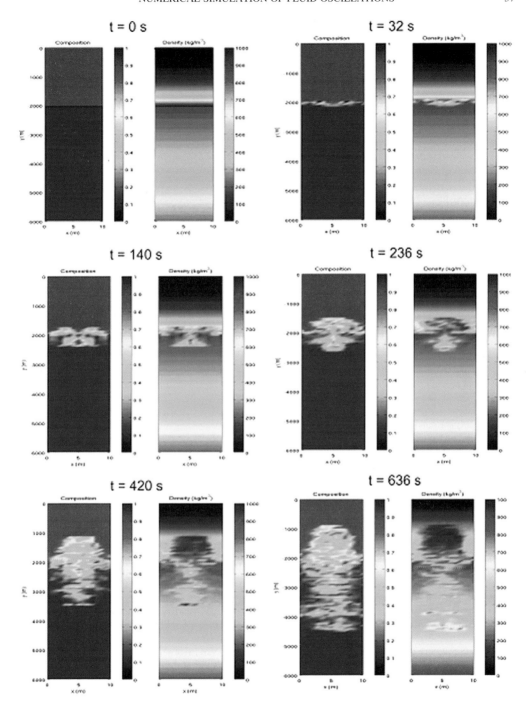

Fig. 2. Simulated evolution of composition and density at six different times.

downward-moving compression wave reaches the bottom of the computational domain after about 30 s from the beginning of the simulation. At this time, a spurious reflection occurs, originating a train of compressive waves reflecting between the domain bottom and the sinking CRGP magma.

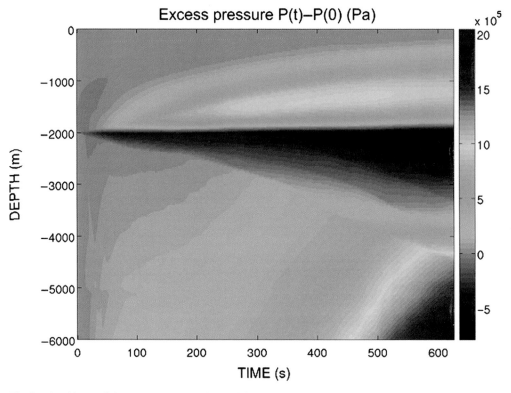

Fig. 3. Time history of the excess pressure at the conduit's wall. The excess pressure is derived by subtracting from the calculated pressure profile the pressure value at the same level corresponding to the initial condition, determined from the magmastatic pressure distribution.

A complex series of compression and rarefaction waves is instead generated in correspondence to the sinking CRGP and rising CPGR magma batches, with maximum amplitude of about 60 (compression) and 30 (rarefaction) kPa. The slope of the lowermost and uppermost waves reflects the velocity of the sinking (average 3.8 m/s) and rising (average 2.2 m/s) magma batches, respectively. It is interesting to note that the slope of the compression front associated with the sinking magma batch changes with time. This indicates variations in the sinking velocity, probably associated with time-varying changes in the local density and pressure distributions.

Ground deformation due to magma convection

In this section, we use the numerical results reported in Figure 2 to investigate the propagation of the pressure disturbances through the country rocks and to relate them to observable signals of geophysical interest. In particular, these calculations aim at evaluating the potential for the convection dynamics described above to produce measurable ground deformation by seismic monitoring networks around the volcano.

From examination of the simulation's results, we note that the deviatoric terms of the stress tensor are generally 3–5 orders of magnitude lower than the isotropic component. Therefore, we parameterize the pressure field at the fluid–rock interface as a succession of horizontal forces acting on a set of point sources distributed along the fissure wall with a constant spacing of 25 m. For each time step, the magnitude of these forces is retrieved from spatial interpolation of the pressure field at the fissure wall, assuming that the fissure extends 100 m in the third not-simulated direction. Using the classic Stokes solution (Aki & Richards 2002), we then calculate the horizontal (radial) and vertical components of ground displacement at a set of 11 observers located at the same elevation as the top of the conduit (or fissure vent), and spanning an epicentral distance range of 2000 m with a constant spacing of 200 m. For this calculation, we assumed the propagation to

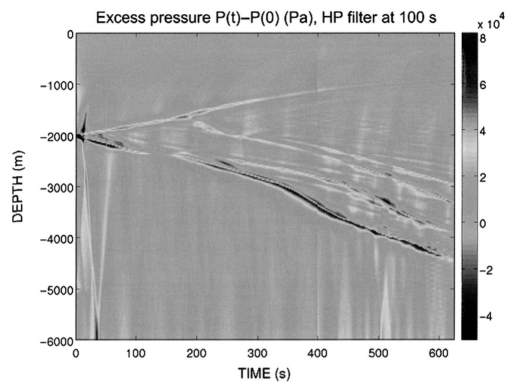

Fig. 4. The same as in Figure 3, but after high-pass filtering using a 2-pole Butterworth filter with cut-off frequency of 0.01 Hz.

occur in an elastic, homogeneous, isotropic infinite medium having compressional wave velocity v_P of 3000 m/s, shear wave velocity $v_S = v_P/\sqrt{3}$, and density of 2500 kg/m^3.

In the above procedure, the main assumptions regard the homogeneity of the rock model and the neglect of the effects associated with the free Earth's surface. The first assumption is expected to have little influence on the reliability of our results. This is because the largest computed pressure variations occur over large timescales. This implies that the associated ground displacement has wavelengths which are much longer than the distance covered by the propagating waves, and therefore the propagation medium may be safely considered homogeneous. Conversely, neglecting the effects associated with the free surface induces an underestimation of the actual ground motion. Therefore, our results should represent a lower limit for the effective ground displacement produced as a response to the simulated magma convection dynamics.

Figure 5 shows the horizontal and vertical ground displacement as a function of time and horizontal offset. The maximum displacements as a function of horizontal distances are instead depicted in Figure 6. The largest displacement is observed for the horizontal component in the near-source region, and amounts to about 1 mm. Following the geometry of the source–receiver settings and the dominance of horizontal forces at fissure walls, the ground displacement observed on the vertical component is about five times smaller than that associated with the horizontal component, and its maximum value lies 600–800 m apart from the fissure vent.

Figure 7 illustrates an example of ground displacement at a given distance from the vent, specifically at a horizontal offset of 800 m. The calculated time-dependent ground displacement has been high-pass filtered at 100 s, thus simulating the signal that would be recorded by a standard broadband seismometer. The figure shows that oscillations with periods of order tens of seconds are associated with the simulated convective dynamics. The largest displacements, observed on the horizontal component, amount to about 2 μm. In principle, such small values are easily detectable

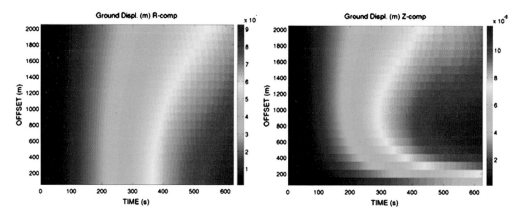

Fig. 5. Horizontal (left) and vertical (right) ground displacement as a function of time and horizontal offset from fissure vent.

by modern seismological devices. However, for the frequency band taken into account, the background noise (mainly of marine origin at Stromboli volcano) is generally of the same order of magnitude, or even larger (e.g. Havskov & Alguacil 2004). Therefore, the detection and discrimination of such weak signals would be extremely difficult in real cases.

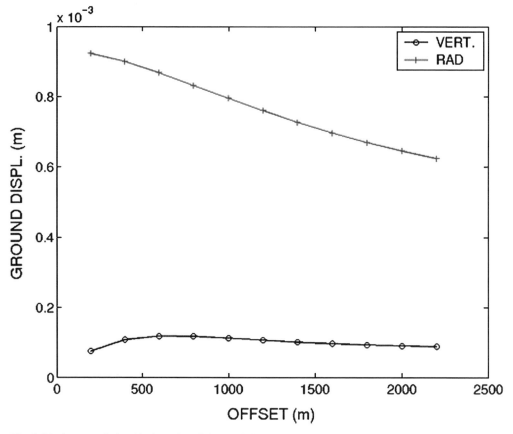

Fig. 6. Maximum vertical and horizontal–radial ground displacement as a function of distance from the fissure vent.

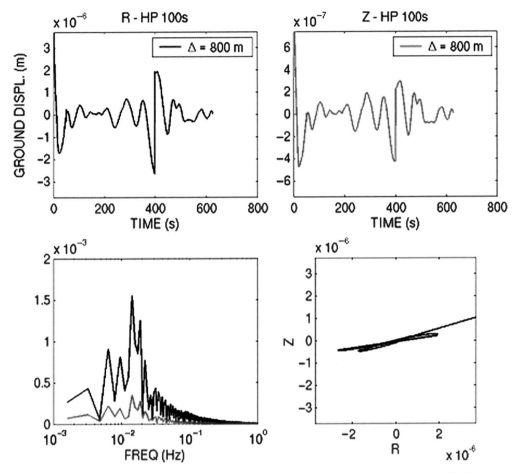

Fig. 7. Radial (top left) and vertical (top right) time histories of ground displacement after high-pass filtering at 100 s. The two seismograms simulate the ground motion which would be recorded by a standard broadband seismometer located at 800 m from the fissure vent. In this particular frequency range, the ground displacement is about three orders of magnitude lower than that associated with the quasi-static deformation. At the bottom, the amplitude spectra (left) and particle motion trajectory (right) are over the radial-vertical plane. The vent is located at the left of the plot.

Discussion and conclusions

The results of the numerical simulation presented in this work depict the dynamics of fluid flow oscillations associated with convection in a compositionally stratified, gravitationally unstable volcanic fissure. Fluid flow oscillations are found to be associated with the rise of deep CPGR magma and sink of shallow CRGP magma. The simulation describes a dynamic picture whereby discrete batches of the upper CRGP magma migrate towards the bottom of the fissure, while a large buoyant bulge of CPGR magma develops above the original CRGP–CPGR magma interface level. The resulting convective motion produces local pressure fluctuations visible as short-period oscillations in the general time–pressure distribution.

It is worth noting again that the initial system conditions resulting from the two-layer magma distribution adopted in the numerical analysis and from the homogeneous fluid approach of the present modelling is not likely to be stable in nature. In fact, integration of the density profile obtained by assuming even nearly totally degassed magma in the upper fissure region leads to quite low pressures within the fissure compared to lithostatic pressure distribution. It is therefore expected that gas bubbles in the upper magma are far from mechanical equilibrium with the liquid, thus not contributing to the magmastatic pressure gradient.

This may be realized in the natural system though the development of permeable paths allowing gas flow through the liquid. In order to account properly for such a system, a more advanced multiphase separated flow model is required. Therefore, our analysis should be regarded as a basic approach aimed at establishing a link between magma movements within a volcanic system, associated fluid flow oscillations, and signals detectable at the Earth's surface.

Fluid flow oscillations occurring in the convecting magma are revealed by the present numerical analysis. Such oscillations are expected to produce mechanical solicitations on the containing walls. As a consequence, the wall rocks deform elastically, oscillating with periods which depend on the applied time-dependent stress and on the characteristics of the rock system. In the real case, the fluid and the rocks would vibrate coherently, influencing each other and producing a more complex pattern of pressure waves and fluctuations (Chouet 2003; Jousset *et al.* 2003, 2004). The results in this work do not account for such complex fluid–structure interactions, which are expected to result in surface waves on the margin of the conduit emitted mainly at reflection points (Jousset *et al.* 2004). Additionally, we do not account for free Earth surface and rock property changes which may result in complex patterns of reflections and refractions. Instead, the pressure oscillations revealed by the present numerical analysis are strictly linked to the convective dynamics inside the fluid system. The micron-size ground displacement oscillations in Figure 7 should therefore not be regarded as the expected seismic signal, but rather as the contribution from a hypothetically decoupled fluid system to the seismic signal associated with the simulated magma convection dynamics. For the frequency band of seismological interest ($f > 10^{-2}$ Hz), our findings reveal harmonic oscillations with periods of 30–70 s. Unfortunately, the amplitude of such oscillations in the particular case studied here seems too small to be clearly detected on most real seismograms, although a conclusion in this sense needs to wait for a simulation with fully coupled magma-rock or fluid-solid dynamics, and a realistic assessment of propagation media. Such more sophisticated simulations will allow better evaluation of the potential for the study of the seismic source in volcanic environments, and particularly of the observed harmonic excitation functions triggering the resonance of fluid-filled, buried cavities (e.g. Nakano *et al.* 2003).

The calculated overall pressure trend is also affected in principle by coupling with the country rocks. However, in this case the large changes of MPa order in the calculated fluid pressure suggest a minor capability of wall rock deformation to modify significantly the calculated trends. The pressure pattern depicted in Figure 3 shows a complex evolution with three main areas: (i) above the original CRGP–CPGR magma interface, where expansion of the rising gas-rich magma results in pressurization; (ii) below that level, where sinking and pressurization of the CRGP magma results in overall pressure decrease; and (iii) at fissure bottom, where the entrance of new magma produces compression and pressurization. Note that the zero pressure change level slowly migrates towards the top, from close to the interface to about 200 m above it after 600 s from the beginning of convection. The quasi-static ground deformation that as a first approximation is expected from the above pressure distribution has a dominant radial component with amplitude of order 1 mm after 600 s of simulation (Figs 5 and 6). Such a deformation would be barely detectable by standard techniques, while it may be revealed by more advanced instruments such as borehole strain meters (e.g. Scarpa *et al.* 2004). We note here that a clearly visible positive radial deformation with period of minutes has been observed immediately preceding the 5 April 2003 paroxysmal eruption at Stromboli (D'Auria *et al.* 2006).

As a final remark, we must point out that the above calculations were conducted using only forces acting at a single wall of the conduit. Another possible representation would be that of considering forces acting at both sides of the conduit (i.e. co-linear force dipoles whose separation distances along the horizontal axis would equal the cross-section of the conduit; 10 m in the present case). Under this circumstance, the simulated ground motion would depict a much-complicated pattern and maximum amplitudes about two orders of magnitude lower than those calculated for the single-force case.

It is worth noting that the present results pertain to the specific case investigated here, and should not be considered general or be extended to significantly different conditions. Particularly, the assumption of homogeneous gas–liquid mixture brings about a low pressure gradient which is likely to be unrealistic. The present results should therefore not be regarded as the solution of the convection and mixing dynamics at Stromboli, but rather just as a first example of fluid flow oscillations generated by such kinds of processes and captured by sophisticated simulations of multicomponent magma flow dynamics. In the present case, the mass of magma involved in the convection dynamics and the horizontal dimension of the simulated fissure are both small, having an influence on the magnitude of the associated fluid flow oscillations. The use of a vertical fissure with smooth walls in the present case is also likely to have a significant effect on the calculated dynamics. It is thus still possible that different system configurations can produce more pronounced flow changes and more distinct signals in a seismic and ground deformation network. Particularly, ongoing numerical simulation where the CRGP magma is assumed to carry zero volatiles reproduces a reliable density/pressure distribution and results in much larger acceleration of the rising bulge of CPGR magma, in large exit velocities, and in ground displacement well above the detection limit. This is part of additional work still in progress, which is being carried on for future investigation aimed at more realistic simulations of conduit dynamics during paroxysmal eruptions at Stromboli.

Much more work is still needed to confidently simulate the dynamics of a volcanic system and predict the associated signals of geophysical relevance. The present contribution is mainly aimed at gaining insights into the basic dynamics of magma convection, as due to the existence of a gravitationally unstable CRGP-CPGR magma interface within a volcanic fissure fed from below. Future developments should include a more thorough description of the wall rock characteristics and visco-elasto-dynamics, extend the simulations to include thermal effects and non-Newtonian rheology, and explore different magmatic conditions and system geometries.

This work has been funded by the Italian Dipartimento della Protezione Civile and INGV-Istituto Nazionale di Geofisica e Vulcanologia in the frame of Project 2005-06/V2, and by European Commission, 6th Framework Project – 'VOLUME', Contract No. 08471. The authors are grateful to Antonella Bertagnini for her support in the definition of a consistent set of initial conditions inspired by Stromboli Volcano. Oleg Melnik and an anonymous reviewer are thanked for having contributed with their comments and suggestions to significant improvement of an original manuscript version.

References

AKI, K. & RICHARDS, P. G. 2002. *Quantitative Seismology*, 2nd edn. University Science Books, Sausalito, CA.

ASTER, R., MAH, S., KYLE, P. *ET AL.* 2003. Very long period oscillations of Mount Erebus Volcano. *Journal of Geophysical Research*, **108**, doi:10.1029/2002JB002101.

BERGANTZ, G. W. 2000. On the dynamics of magma mixing by reintrusion: implications for pluton assembly processes. *Journal of Structural Geology*, **22**, 1297–1309.

BERTAGNINI, A., METRICH, N., LANDI, P. & ROSI, M. 2003. Stromboli volcano (Aeolian Archipelago, Italy): an open window on the deep-feeding system of a steady state basaltic volcano. *Journal of Geophysical Research*, **108**, doi:10.1029/2002JB002146.

CHOUET, B. A. 2003. Volcano seismology. *Pure and Applied Geophysics*, **160**, 739–788.

CHOUET, B. A., DAWSON, P., OHMINATO, T. *ET AL.* 2003. Source mechanisms of explosions at Stromboli Volcano, Italy, determined from moment-tensor inversions of very-long-period data. *Journal of Geophysical Research*, **108**, doi:1029/2002JB001919.

D'AURIA, L., GIUDICEPIETRO, F., MARTINI, M. & PELUSO, R. 2006. Seismological insight into the kinematics of the 5 April 2003 vulcanian explosion at Stromboli volcano (southern Italy). *Geophysical Research Letters*, **33**, L08308, doi:10.1029/2006GL026018.

FOLCH, A., MARTI, J., CODINA, R. & VASQUEZ, M. 1998. A numerical model for temporal variations during explosive central vent eruptions. *Journal of Geophysical Research*, **103**, 20883–20899.

GIORDANO, D. & DINGWELL, D. B. 2000. Viscosity of hydrous Etna basalt: implications to the modeling of Plinian basaltic eruptions. *Bulletin of Volcanology*, **65**, 8–14.

GIORDANO, D., MANGIACAPRA, A., POTUZAK, M., RUSSELL, J. K., ROMANO, C., DINGWELL, D. B. & DI MURO, A. 2006. An expanded non-Arrhenian model for silicate melt viscosity: a treatment for metaluminous, peraluminous and peralkaline liquids. *Chemical Geology*, **229**, 42–56.

HARRIS, A. J. L., CARNIEL, R. & JONES, J. 2005. Identification of variable convective regimes at Erta Ale Lava Lake. *Journal of Volcanology and Geothermal Research*, **142**, 207–223.

HAUKE, G. & HUGHES, T. J. 1998. A comparative study of different sets of variables for solving compressible and incompressible flows. *Computer Methods in Applied Mechanics and Engineering*, **153**, 1–44.

HAVSKOV, J. & ALGUACIL, G. 2004. *Instrumentation in Earthquake Seismology*. Springer.

HIRSCHFELDER, J. O., CURTISS, C. F. & BIRD, R. B. 1969. *Molecular Theory of Gases and Liquids*. Wiley, New York.

JAUPART, C. & TAIT, S. 1995. Dynamics of differentiation in magma reservoirs. *Journal of Geophysical Research*, **100**, 17615–17636.

JELLINEK, A. M. & KERR, R. C. 2001. Magma dynamics, crystallization, and chemical differentiation of the 1959 Kilauea Iki lava lake, Hawaii, revisited. *Journal of Volcanology and Geothermal Research*, **110**, 235–263.

JOUSSET, P., NEUBERG, J. & STURTON, S. 2003. Modelling the time-dependent frequency content of low-frequency volcanic earthquakes. *Journal of Volcanology and Geothermal Research*, **128**, 201–223.

JOUSSET, P., NEUBERG, J. & JOLLY, A. 2004. Modelling low-frequency volcanic earthquakes in a visco-elastic medium with topography. *Geophysical Journal International*, **159**, 776–802.

KURITANI, T. 2004. Magmatic differentiation examined with a numerical model considering thermodynamics and momentum, energy and species transport. *Lithos*, **76**, 117–130.

LANDI, P, METRICH, N., BERTAGNINI, A. & ROSI, M. 2004. Dynamics of magma mixing and degassing recorded in plagioclase at Stromboli (Aeolian Archipelago, Italy). *Contributions to Mineralogy*

and   Petrology, **147**, 213–227. doi:10.1007/
   s00410-004-0555-5.
LANGE, R. A. 1994. The effect of H_2O, CO_2 and F on the
   density and viscosity of silicate melts. *In*: CARROLL,
   M. R. & HOLLOWAY, J. R. (eds) *Volatiles in
   Magmas*. Mineralogical Society of America,
   Reviews in Mineralogy, **30**, 331–369.
LONGO, A., VASSALLI, M., PAPALE, P. & BARSANTI,
   M. 2006. Numerical simulation of convection
   and mixing in magma chambers replenished with
   CO_2-rich magma. *Geophysical Research Letters*, **33**,
   doi: 10.1029/2006GL027760.
MATLAB SIGNAL PROCESSING TOOLBOX USER
   MANUAL. The MathWorks. http://www.mathworks.
   com
METRICH, N., BERTAGNINI, A., LANDI, P. & ROSI, M.
   2003. Crystallization driven by decompression and
   water loss at Stromboli volcano (Aeolian Islands,
   Italy). *Journal of Petrology*, **42**, 1471–1490.
NAKANO, M., KUMAGAI, H. & CHOUET, B. A. 2003.
   Source mechanism of long-period events at
   Kusatsu–Shirane Volcano, Japan, inferred from wave-
   form inversion of effective excitation functions.
   Journal of Volcanology and Geothermal Research,
   122, 149–164.

PAPALE, P. 2001. Dynamics of magma flow in volcanic
   conduits with variable fragmentation efficiency and
   nonequilibrium pumice degassing. *Journal of Geophysical Research*, **106**, 11043–11066.
PAPALE, P., MORETTI, R. & BARBATO, D. 2006. The
   compositional dependence of the saturation surface
   of $H_2O + CO_2$ fluid in silicate melts. *Chemical
   Geology*, **229**, 78–95 doi:10.1016/j.chemgeo.
   2006.01.013.
ROSI, M., BERTAGNINI, A., HARRIS, A. J. L., PIOLI, L.,
   PISTOLESI, M. & RIPEPE, M. 2006. A case history of
   paroxysmal explosion at Stromboli: timing and
   dynamics of the April 5, 2003 event. *Earth and Planetary Science Letters*, **243**, 594–606.
SCARPA, R., RUGGIERO, L., AMORUSO, A.,
   CRESCENTINI, L., MARTINI, M., LINDE, A. &
   SACKS, S. 2004. Borehole strain measurements
   on Mt. Vesuvius, Italy. *Geophysical Research
   Abstracts*, **6**, 02646, SRef-ID: 1607-7962/gra/
   EGU04-A-02646.
SHAKIB, F., HUGHES, T. J. & JOHAN, Z. 1989. A multi-
   element group preconditioned GMRES algorithm for
   nonsymmetric systems arising in finite element
   analysis. *Computer Methods in Applied Mechanics
   and Engineering*, **75**, 415–456.

The feasibility of generating low-frequency volcano seismicity by flow through a deformable channel

A. C. RUST[1], N. J. BALMFORTH[2] & S. MANDRE[3]

[1]*University of Bristol, Wills Memorial Building, Bristol BS8 1RJ, UK*
(e-mail: Alison.Rust@bristol.ac.uk)

[2]*Department of Mathematics & Department of Earth & Ocean Sciences, University of British Columbia, Vancouver, Canada*

[3]*School of Engineering and Applied Sciences, Harvard University, Cambridge, USA*

Abstract: Oscillations generated by flow of magmatic or hydrothermal fluids through tabular channels in elastic rocks are a possible source of low-frequency seismicity. We assess the conditions required to generate oscillations of approximately 1 Hz *via* hydrodynamic flow instabilities (roll waves), flow-destabilized standing waves set up on the elastic channel walls (wall modes), and unstable normal modes ringing in an adjacent fluid reservoir (clarinet modes). Stability criteria are based on physical and dimensional arguments, and discussion of destabilized elastic modes is supplemented with laboratory experiments of gas flow through a channel in a block of gelatine, and between a rigid plate and a rubber membrane. For each of the mechanisms considered, oscillations are generated if flow speeds exceed a critical value. Roll waves are waves of channel thickness variation that propagate in the direction of flow and are equivalent to traveling crack waves. The convective instability criterion is that the flow is faster than those travelling waves. Similarly, wall modes and clarinet modes require that the flow speed exceeds a critical value related to a wave speed (e.g. elastic or acoustic wave) multiplied by a geometrical factor. Flow destabilized modes offer a plausible explanation for low-frequency volcano seismicity, but there are limitations on what kind of standing waves comprises them.

Continuous seismic tremor, as well as transient 'long-period' (LP) seismic events, are important indicators of unrest at volcanoes, and as such are used to evaluate eruption hazards and to invert for the geometry and nature of fluids in volcanoes (Chouet 1996, 2003; McNutt 2005). It is widely accepted that this low-frequency volcano seismicity emanates from fluid channels encased in rock, such as fluid-filled hydrofractures, dykes and cylindrical conduits. However, the precise mechanisms responsible for generating those signals are still under debate, particularly for harmonic tremor characterized by well-defined spectral peaks at integer multiples of a fundamental frequency.

One prominent explanation for these harmonic signals is that they stem from the resonant excitation of standing waves in the fluid channel. Individually, both the solid and fluid support their own kind of waves (elastic and acoustic, respectively). But these wave types propagate at relatively high speeds, and if the standing waves have either elastic or acoustic origin, then the resonating body may need to be implausibly large to match typical tremor frequencies. However, a coupled fluid–solid system might also support a variety of interfacial waves that can propagate much more slowly than the elastic or acoustic wave speeds (Ferrazzini & Aki 1987). These 'crack waves', as they have been called, are related to Stoneley waves, and persist even when the elastic and acoustic wavespeeds are made infinitely large in comparison to the crack wavespeed. In that limit, the crack waves are simply incompressible sloshing modes of the fluid in the channel, as permitted by variations in its thickness and the restoring forces exerted by the elastic walls (see Appendix A). In other words, the crack modes are analogous to the seiches of shallow fluid layers (with gravity playing the same role as the elastic forces), as encountered in harbours, lakes and a variety of other problems in hydraulic engineering. Connections between possible mechanisms for volcanic tremor and hydraulic transients were recognized already by Ferrick *et al.* (1982).

The relatively low speeds of standing crack waves make them attractive ingredients in the resonance mechanism. Indeed Chouet (1986, 1988) showed that synthetic seismograms generated by such modes of a fluid-filled tabular crack are remarkably similar to some seismograms from volcanoes. Nevertheless, the trigger of the resonance is often neither identified nor specified. There are many plausible origins for fluid pressure transients that could drive resonance impulsively and

From: LANE, S. J. & GILBERT, J. S. (eds) *Fluid Motions in Volcanic Conduits: A Source of Seismic and Acoustic Signals.* Geological Society, London, Special Publications, **307**, 45–56.
DOI: 10.1144/SP307.4   0305-8719/08/$15.00 © The Geological Society of London 2008.

thereby explain LP events (e.g. Konstantinou & Schlindwein 2002) but the source of energy driving tremor must be sustained for minutes or much longer.

An alternative perspective presented by Julian (1994) involves a fluid moving through a channel interacting with its deformable rock walls. Flow-induced oscillations arise in many areas of engineering and science and explain phenomena as diverse as the flapping of flags, the sounds of some musical instruments, and noise generation by flow in blood vessels and lung passageways (e.g. Backus 1963; Grotberg & Jensen 2004; Pedley 1980). Julian (1994) proposed that a similar instability could be a source mechanism for tremor and LP events at active volcanoes. The flow-induced oscillations could provide the triggering pressure fluctuations to drive resonance with a standing crack wave, but they may also generate harmonic (and non-harmonic) seismic signals by themselves. The mechanism is promising for tremor because the excitation lasts as long as flow is sufficiently fast, and can explain observations of nonlinear phenomena such as period doubling and amplitude-dependent frequency (Julian 1994, 2000; Konstantinou 2002).

To explore the mechanism, Julian (1994) constructed a 'lumped-parameter model' in which the channel opened and closed uniformly along its length and the rocks were represented by masses on springs with dashpots. Balmforth *et al.* (2005) elaborated further on Julian's ideas and developed a two-dimensional model coupling semi-infinite elastic blocks with a viscous incompressible channel flow to assess whether hydrodynamic instabilities could arise under volcanological conditions. Here we review and build upon the work of Julian (1994, 2000) and Balmforth *et al.* (2005). More specifically, we consider how flow-induced oscillations might arise in the coupled fluid–solid system illustrated in Figure 1.

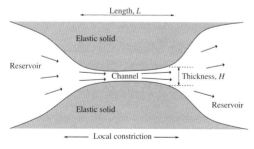

Fig. 1. Schematic illustration of a channel between two reservoirs. Flow is driven through the channel by a pressure difference maintained between the reservoirs. The channel is tabular, and its walls deform elastically in response to pressure changes in the fluid.

The geometry consists of two reservoirs that are connected by a relatively narrow channel and maintained at different pressures so that fluid is driven through the channel. For most of the source mechanisms we assess, the details of the reservoirs are not important, and all the action takes place in the flow through the channel, which is relatively vigourous owing to its narrowness.

We consider three specific types of flow-induced oscillations: hydrodynamic flow instabilities, flow-destabilized standing waves set up on the channel walls, and unstable normal modes ringing in one of the fluid reservoirs (Fig. 2). The hydrodynamic instabilities are analogous to what fluid dynamicists call roll waves (e.g. Balmforth & Mandre 2004), and exist when elastic and acoustic wavespeeds are infinite. Normal modes of elastic origin that are localized in the channel walls can be destabilized by the flow in a manner similar to the mechanism behind the operation of the vocal cords (Ishizaka & Flanagan 1972). Normal modes in one of the reservoirs can also be made unstable *via* coupling to the channel flow; in this case, the mechanism has many common points with the operation of musical instruments like the clarinet (cf. Lesage *et al.* 2006). For each type of oscillation, we summarize the physics involved and determine a criterion for instability on physical and dimensional grounds. Destabilized elastic modes are further studied through laboratory experiments of gas flow through a narrow channel in a block of gelatine, and between a rigid plate and a rubber membrane. With these results, as well as consideration of the factors that set oscillation frequency, we evaluate the feasibility of flow through a deformable channel generating low-frequency volcano seismicity.

Hydrodynamic flow instabilities

Small perturbations in velocity and pressure naturally occur in all flows. If the system is stable then the perturbations shrink away and do not disrupt the flow, but under some conditions the perturbations may grow with time, forming hydrodynamic instabilities. This article focuses on seismicity related to deformable channel walls; however, hydrodynamic flow instabilities in channels are not limited to fluid interactions with a moving wall. For example, at high Reynolds number, shear instability can occur in the wake of an irregularity in the rigid channel. The resulting eddy shedding has been suggested as a source of flow transients that could trigger low-frequency volcanic resonance (Hellweg 2000). However, because this mechanism requires high Reynolds number flow, it has limited volcanic application.

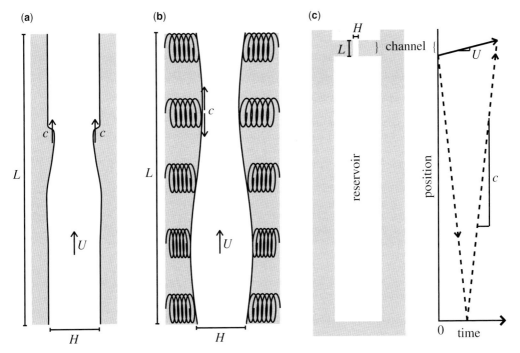

Fig. 2. Diagrams of the three mechanisms of flow-induced oscillations discussed in this paper. In each case, oscillations of the deformable channel may develop if the flow speed, U, exceeds a critical value related to a wavespeed, c. The channel has length L and thickness H. (**a**) Roll waves, a type of hydrodynamic instability, are travelling waves of channel thickness that propagate in the direction of flow at a speed similar to the flow speed. (**b**) Elastic normal modes in the channel walls form as elastic waves (e.g. Rayleigh waves), which are destabilized by flow, form standing waves by travelling up and down the conduit as shown, or back and forth along the width (perpendicular to page). This is similar to models for sound generation by vocal cords. A variation of this model for the volcanic application replaces elastic waves with slower 'crack waves'. (**c**) Reservoir modes occur when the timescale for oscillations in the channel (akin to clarinet reed) are set by waves in an attached reservoir (akin to clarinet body). Oscillations are generated if the time for fluid to travel through the channel (path illustrated with solid line and arrow in plot on right) is similar to the time for a pressure perturbation in the fluid to travel through the reservoir and back to the channel at the acoustic wavespeed (path illustrated with dashed line).

For flow through a deformable channel, there are instabilities at lower Reynolds numbers that are mathematically similar to the roll waves that develop on thin sheets of water flowing down slopes (e.g. Balmforth & Mandre 2004; Fig. 2a). These waves are shock-like flow disturbances with phase speeds similar to the background fluid speed. Analogous instabilities occur in blood flow through deformable veins (Pedley 1980; Brook et al. 1999) and in slug formation in bubbly two-phase flow (Woods et al. 2000). The essential ingredients for roll waves include a force driving flow, drag, and a restoring force that flattens disturbances in the free surface (such as gravity, surface tension or, in the volcanic context discussed here, the rock elasticity).

Balmforth et al. (2005) examined theoretically the generation of roll waves for fluid flow through a thin channel with elastic walls. They treated the fluid as incompressible and viscous, and the walls as semi-infinite linear elastic solids. Generally, magmatic fluid flow is much slower than shear or compressional wave propagation in rocks, and it is reasonable to consider the limit of infinite elastic wave speeds. Assuming periodic inlet–outlet flow conditions, Balmforth et al. found that the roll wave instability requires a finite critical flow speed ($U_{crit\ roll}$) given by

$$U_{crit\ roll} \approx \beta \sqrt{\frac{\rho_s}{\rho_f}} \varepsilon. \tag{1}$$

where β is the shear wave speed in the rock, ρ_s/ρ_f the rock-to-fluid density ratio, and ε is the channel aspect ratio (thickness/length $\equiv H/L$) with $\varepsilon \ll 1$.

This criterion can also be arrived at by simple dimensional arguments. Roll waves are characterized by timescale L/U, where L is the length of the crack and U is the characteristic flow speed. The pressure change related to the elastic solid due to a displacement, ξ, of the wall is of order $\mu\xi/L$, where μ is the shear modulus of the solid (i.e. Hooke's law). The elastic force induces fluid acceleration and advection of order Uv/L, where v is the perturbation in flow speed. Moreover, conservation of mass demands that Hv be of the order of $U\xi$. Thus, balance is achieved when $\mu\xi/(\rho_f L^2) \sim Uv/L \sim U^2\xi/(HL)$. Rearranging and substituting $\beta = \sqrt{\mu/\rho_s}$ gives the condition of equation (1).

Note that the presence of the elastic wavespeed in the expression for $U_{crit\,roll}$ is misleading, as there are no elastic waves in this problem; β appears because it also characterizes the restoring force from the elastic walls. In fact, the instability condition does contain a wavespeed, but it is not elastic: for a thin channel of fluid with an elastic wall, waves of thickness variation exist with the wavespeed $\sqrt{\mu kH/\rho_f} \equiv \beta\sqrt{kH\rho_s/\rho_f}$, where $k \sim L^{-1}$ is the wavenumber. This wavespeed is the limit of the dispersion relation of Ferrazzini & Aki (1987) when the elastic and acoustic wavespeeds are relatively large (see Appendix A). In other words, $U_{crit\,roll}$ is the speed of travelling crack waves.

The critical condition (1) implies that roll waves are destabilized by fast flow of dense fluid through long, thin channels. The limitations of roll wave instabilities as a source of volcanic tremor are illustrated in Figure 3, which shows $U_{crit\,roll}$ versus ϵ at several values of ρ_f for typical rock properties of $\beta = 1$ km/s and $\rho_s = 2500$ kg/m^3. Aspect ratios of magma dykes are usually 10^{-3} to 10^{-2}, and

even for a magma with density as high as 3000 kg/m^3, sustained flow exceeding 10 m/s is required for generating tremor by roll waves. Such high flow rates are problematic, given constraints on the size of the dyke from the frequencies, f, of volcanic tremor. Because the wavespeed is of the order of U, f is order (U/L). Generally $f \sim 1$ Hz, which for flow at 20 m/s and $\varepsilon = 10^{-3}$ implies a dyke with $L \sim 20$ m and $H \sim 2$ cm (point A on Fig. 3). Even flow of a low viscosity magma with $\eta = 10$ Pa s through a dyke of these dimensions requires pressure gradients of order 10^7 Pa/m to overcome viscous drag. The situation is improved for very long period tremor (e.g. 0.1 Hz), which is occasionally observed at volcanoes (e.g. Kawakatsu *et al.* 1994) because it allows an order-of-magnitude larger dyke and thus a substantial decrease in viscous drag. However, even for low-viscosity basalt the possibility of generating tremor by roll waves is marginal, and it is impossible for more viscous (i.e. crystal-bearing or more silicic composition) magmas to flow sufficiently fast in thin dykes.

The least viscous fluids at volcanoes are gases, but roll-wave development in gas-filled channels is impeded by the low densities (equation 1). Despite the small aspect ratios of hydrofractures (typically $10^{-5} < \varepsilon < 10^{-4}$), for gas with $\rho_f = 1$ kg/m^3, roll-wave destabilization requires fluid speeds in excess of 100 m/s (point B on Fig. 3) through cracks hundreds of metres long (for $f \sim 1$ Hz). There are, however, fluids at volcanoes of both intermediate density and viscosity (compared to magma and gases at atmospheric pressure) such as liquid or supercritical H$_2$O- or CO$_2$-rich fluids, or gases with substantial fractions of suspended rock or magma fragments. The best candidates for roll waves in volcanoes are hot, high-pressure H$_2$O- and CO$_2$-rich fluids, which have low kinematic viscosities (η/ρ_f). For example, H$_2$O at 500 °C and 50 MPa has $\rho_f \approx 300$ kg/m^3 and $\eta \approx 4 \times 10^{-5}$ Pa s (Wagner & Overhoff 2006). With such fluids, roll waves of $f \sim 1$ Hz could be generated with reasonable pressure gradients and fracture geometries but still require that high flow speeds of order 10 m/s be sustained for the duration of tremor (e.g. point C on Fig. 3).

To this point, we have assumed typical rock properties of $\beta = 1$ km/s and $\rho_s = 2500$ kg/m^3. The flow speeds required to generate roll waves are decreased for porous rocks (reduced ρ_s), partially molten or fluid-saturated rocks (reduced β). Also low oscillation frequencies (<1 Hz as assumed in Fig. 3) increase the range of feasible fluid viscosities because longer timescales permit larger channels and reduced viscous drag. However, we still conclude that roll waves require extreme natural conditions and do not provide an explanation for most volcanic tremor.

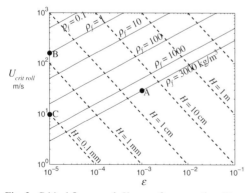

Fig. 3. Critical flow speed, $U_{crit\,roll}$, from equation (1) for roll waves versus channel aspect ratio for several fluid densities (solid lines) for constant rock properties of $\beta = 1$ km/s and $\rho_s = 2500$ kg/m^3. Dashed lines indicate crack thickness for $f \sim U/L \sim 1$ Hz. Points A, B and C refer to conditions discussed in the text.

To make matters worse, a major shortcoming of the Balmforth *et al.* (2005) analysis, on which the above discussion is based, is that the channel has spatially periodic inlet–outlet boundary conditions. This allows roll waves to grow continually as they cycle repeatedly through the domain. In a finite channel, perturbations might be flushed out the end before roll waves have time to develop. One way to characterize the flushing action of the basic flow is in terms of the notion of 'convective' and 'absolute' instability. Convective instabilities grow exponentially as one moves with the disturbance (until nonlinearity becomes important), or equivalently if the domain is periodic. At any given fixed position in a non-periodic channel, however, the disturbance only grows as the instability propagates towards the fixed observer, but is then completely advected past and thereafter decays; i.e. the disturbance is flushed out of the system and can be sustained only if continually fed by external perturbations upstream. By contrast, an absolute instability is one that grows exponentially even at a fixed position; it is impossible for flow to sweep out this instability, which amplifies at every point in the channel until quenched by nonlinearity.

Further analyses of the linearized stability problem using a method attributed to Briggs (1964), numerical computations with idealized models, and laboratory experiments in shallow water (Liu *et al.* 1993; Mandre 2006) all suggest that roll waves are convective instabilities. We illustrate the essential aspects of the problem in Figure 4, which shows numerical solutions of non-linear roll waves in a model of flow through an elastic-walled channel. Details of the equations of motion and boundary conditions involved in the calculations are relegated to Appendix B. The first result, shown in Figure 4a, is for periodic boundary conditions, so that waves that reach the end of the channel (e.g. at $t \sim 100$) reappear at the start of the channel and continue to grow with time. This is in contrast to the computations for a finite channel (Fig. 4b,c). Figure 4b shows the result for an initial-value problem beginning with random perturbations (at all x) about the equilibrium flow. Those perturbations seed the growth of roll waves, which subsequently reach the outlet and disappear. Perturbations that begin near the outlet barely grow before reaching the end; perturbations initially near the inlet do grow into roll waves but when they are flushed out at $t \sim 175$, the roll-wave transient

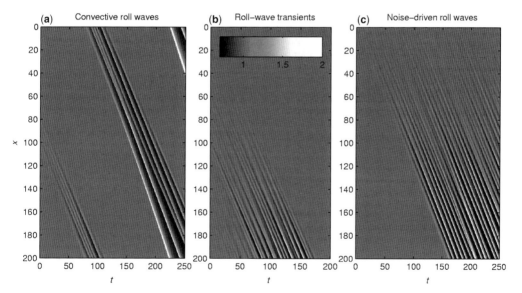

Fig. 4. Numerical solutions for the simple two-dimensional flow model described in Appendix B, showing roll wave development. The three panels show how the local channel thickness h varies with downstream position (x) and time (t). The shading indicates h as shown in the legend in **(b)**, with $h = 1$ corresponding to the unperturbed, equilibrium thickness of the channel. In **(a)**, the boundary condition is periodic in x, simulating an infinitely long channel, and the initial condition consists of random perturbations about the equilibrium flow, concentrated near mid-channel, with a peak-to-peak amplitude of less than 10^{-2}. Panel (b) shows a finite domain with the boundary conditions described in Appendix B. There is a random initial condition similar to that in panel (a), but here the perturbations are evenly distributed throughout the channel length. Panel **(c)** shows a computation in a finite domain with the same boundary conditions as (b), except that roll waves are continually seeded by random noise added to the equilibrium flow at the inlet $(x = 0)$ for the entire simulation duration.

is over. The third computation (c) shows an example in which the initial state is the equilibrium flow but it is continually and randomly perturbed at the inlet. The perturbations at the boundary now feed the convective roll waves to generate unsteady flow downstream. How big the roll waves grow depends on the initial level of excitation, the length of the channel and the roll wave growth rate.

LP events could resemble roll wave transients (Fig. 4b), but to generate tremor via roll wave instabilities, there must be a continuous source of agitation at the inlet (Fig. 4c; i.e. a trigger for the trigger). Such agitation would be present in most physical systems but we further need the channel to be long enough that the roll waves seeded by the noise amplify sufficiently that they become detectable. At the very least, this would require long channels and even higher flow speeds than $U_{crit\,roll}$. Taken together, the instability condition in equation (1) and the lack of an absolute instability lead us to conclude that roll waves could rarely be a source of volcanic tremor.

Elastic normal modes in the channel walls

Much like Julian's (1994) lumped-parameter formulation, simple models of the vocal cords combine a finite, spatially uniform channel with elastically sprung walls (Ishizaka & Flanagan 1972). The fluid flow destabilizes the oscillations of the walls to generate sound, and the frequencies that become excited are closely connected to the natural oscillation frequency of the walls and springs. Our idealization of the problem shown in Figure 1 can have analogous instabilities if elastic normal modes can somehow be excited in the channel walls (Fig. 2b).

Normal modes are easily set up in finite elastic blocks because standing waves are established in their bounded geometry. Their eigenfrequencies are dictated by the elastic wavespeeds and the block dimensions. A semi-infinite elastic block, on the other hand, does not support normal modes because the compressional and shear waves propagate off to infinity and are never reflected back to generate a standing wave. However, the interface does support localized Rayleigh surface waves that can set up standing waves on the channel walls if there is sufficient reflection either from the ends or sides of the channel. These elastic 'channel modes' have frequencies determined by the Rayleigh wavespeed and the length, L, or width, W, of the channel.

For either configuration, the elastic normal modes can be destabilized by fluid flow, as in the vocal cords, if that flow is sufficiently strong. The precise criterion for instability can be established *via* linearized stability theory, as for roll waves, and depends sensitively on the boundary conditions at the flow inlet and outlet (Mandre 2006). Simple estimates for convective instability, which ignore those conditions, indicate that the elastic normal modes become unstable when

$$U > U_{crit\,wall\,mode} \sim fL \qquad (2)$$

where f is the modal frequency. The dimensional argument behind this result is that the flow destabilizes the mode when the flow time down the channel L/U becomes of the same order as the period of the mode, f^{-1}. Although a comparably simple absolute instability criterion is more difficult to determine, the analysis indicates that destabilized elastic modes have an absolute nature.

We observe destabilized elastic modes in laboratory experiments of gas flow (1) between a rigid plate and a latex membrane, and (2) through a thin channel cut through a block of gelatine (Figs 5 & 6). In both cases, the elastic solid begins to ring persistently and harmonically once the flow rate exceeds some threshold, in line with the preceding arguments (see Fig. 5). The fact that oscillations are connected to elastic normal modes is verified by the fact that the frequency depends linearly on the dimensions of the elastic body (cf. Fig. 5d) and also changes with its properties (e.g. gelatine concentration or membrane tension), but is insensitive to flow speed and fluid density. Note that the threshold in flow speed in the experiment of Figure 5b is roughly 6 m/sec, whereas the mode frequency is about 300 Hz and the channel length is 8 cm, which is in agreement with the order of magnitude estimate of $U_{crit\,wall\,mode}$. In a set of experiments with constant L and H (Fig. 5c), $U_{crit\,wall\,mode}$ is nearly proportional to f as expected.

The unsteady flow through the air channels of both the latex-membrane and gelatine-block experiments generates audible acoustic signals that can be used to characterize the normal-mode dynamics. Spectra for three gelatine experiments are shown in Figure 6. Just beyond the onset of instability, the oscillations are periodic, although the spectrum contains a rich array of harmonics (Fig. 6b), as in some measurements of volcanic tremor (note that the recording device acts as a high-pass filter, shifting the power maxima to higher frequency). As the flow rate increases, the oscillations become more nonlinear, and phenomena such as frequency gliding and period-doubling occur, sometimes even during one experiment (see Fig. 6c, which shows period doubling). At the highest flow rates, the periodicity of the signals breaks down as the sides of the channel oscillate so violently that they 'slap'

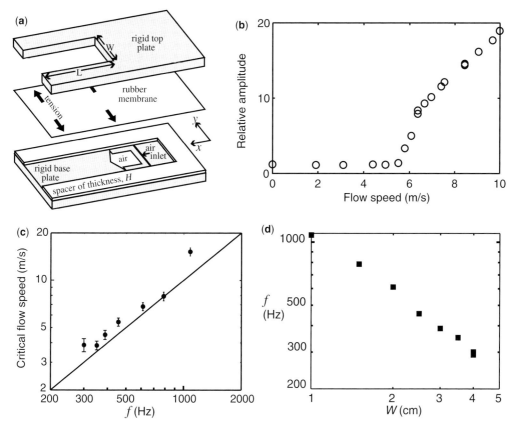

Fig. 5. (**a**) The apparatus for experiments of gas flow between an elastic membrane and a rigid plate. The components illustrated were laid on top of each other and clamped together, exposing a membrane of size W by L stretched parallel to the y-axis. Compressed air, nitrogen or helium was input from below, flowed in the x-direction, and exited to the left at the free boundary. (**b**) The amplitude of the sound recorded, relative to ambient levels, versus air flow speed for an elastic membrane experiment with $L = 8$ cm, $W = 4$ cm and $H = 0.19$ mm. The critical flow speed for the onset of elastic oscillations is between 5.5 and 5.8 m/s. (**c**) Log-log plot of the critical air flow speed to generate vibrations versus the frequency of the vibrations, for a set of experiments with $L = 8$ cm, $H = 0.106$ mm and W varied to change the frequency (see d). Error bars mark the range of possible critical flow speeds: from the highest speed recorded without vibrations, to the lowest speed recorded with vibrations. The diagonal line is $U_{crit} = f/100$. (**d**) The frequency of oscillations against W for the same experiments as in (c).

together intermittently. These wall collisions destroy the coherence of the elastic mode and generate large amounts of high-frequency noise (Fig. 6d).

Note that, because the modal frequency is order β/Δ, where Δ is the dimension of the block along which the standing waves are set up,

$$U_{crit\ wall\ mode} \sim \beta\frac{L}{\Delta} \qquad (3)$$

(for rock, the compressional, shear and Rayleigh waves all generally travel at roughly comparable speeds). Thus, if the critical flow speed is to be much lower than the elastic wavespeed, the block dimension, Δ, should be much greater than the length of the channel. For semi-infinite blocks, this can only be achieved if the channel is much wider than it is long ($W \gg L$), and the elastic mode is composed of lateral, standing Rayleigh waves. (This requirement resonates with one of Julian's (1994) assumptions.) The gelatine block experiment contrasts sharply with this geometrical constraint because shear wave speeds in this material are of the order of m/s, speeds that are easily surpassed by the airflow.

Clarinet modes

For both roll waves and elastic channel modes, the frequency of flow-induced oscillations is set by the

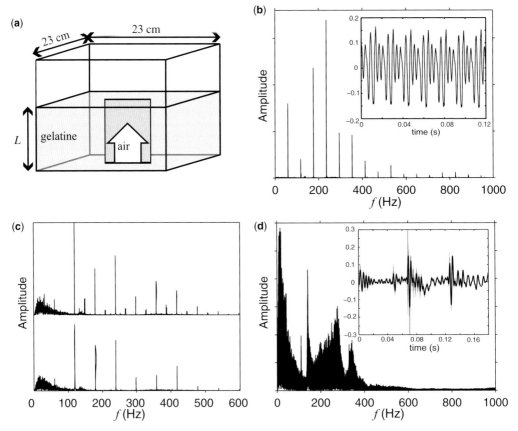

Fig. 6. (a) The apparatus for gelatine-block experiments. Air flows up a vertical slit cut in the block. The length of the slit, L, is the height of the gelatin, which varied from 6 to 12 cm. (b) Frequency spectrum and time series (inset) of the audio signal generated by a pressure drop of 1.5 kPa driving air through a 10 wt% gelatine block with $L = 6$ cm. The time series is filtered to remove freqencies greater than 1200 Hz. (c) Frequency spectra illustrating period doubling. Each of the spectra shows the frequency content for 10 seconds of the continuous experiment. The upper spectrum, which is shifted up for clarity, shows small peaks half-way between the larger peaks seen in both spectra. The period doubling occurred without obvious changes in the experimental conditions (7.5 wt% gelatine, $L = 12$ cm, air pressure drop of 3.6 kPa). (d) Frequency spectrum and time series for a signal generated by 5 wt% gelatine block with $L = 12$ cm and an air pressure drop of 2.8 kPa. The unfiltered time series is grey and the low-pass filtered signal (<1200 Hz) is in black. The strongest slapping event is at about 0.7 s.

dimensions of the channel and the wavespeed within it. Another possibility is that the oscillation timescale is set by the magmatic plumbing system. That is, the channel acts like a clarinet reed that excites and interacts with standing waves in an adjacent reservoir (Fig. 2c). In that musical instrument, the sound produced is not simply a result of a resonance between the frequencies generated by the oscillating reed and a mode in the neighbouring air column; by itself the reed vibrates at much higher frequencies. Instead, coupling between the flow in the reed and the feedback from the resonating air column at its outflow produces the sound as an intrinsic instability (e.g. Backus 1963).

By combining a shallow-channel flow theory like that of Appendix B with a compressible fluid column to represent the reservoir, one is again able to formulate a simple mathematical model to explore the feasibility of this mechanism in the volcanic setting. A convective stability analysis with the model then implies that the threshold for instability takes the form,

$$U_{crit\ reservoir\ mode} \sim \frac{c_{acoustic}L}{D} \qquad (4)$$

where $c_{acoustic}$ is the sound speed in the fluid and D is the wavelength of acoustic signal excited. This wavelength is determined by purely geometrical considerations and in many circumstances scales with the largest dimension of the reservoir. For example, similar to an organ pipe, the wavelength of the fundamental mode of a long cylindrical channel scales with its length. However, for certain shapes of the resonating reservoir these modes can have a wavelength much larger than any dimension of the reservoir. An example of such a case is a Helmholtz resonator, which is conceptually similar to a bottle. It has an enclosing volume which opens to the atmosphere or to other reservoirs through a narrow neck (Strutt 1871). For such objects, the wavelength scales as

$$D \sim \sqrt{\frac{V l_n}{A_n}} \qquad (5)$$

where V is the volume of the reservoir, while l_n and A_n are the length and area of cross-section of the neck. In fact, the critical velocity appearing in equation (4) can be written in terms of the frequency of the acoustic mode $f_{acoustic} = c_{acoustic}/D$; however, we have retained the description in terms of the wavelength, D, for its simple connection with the geometry of the resonating reservoir.

Once more, there is a justification for this condition on physical and dimensional grounds: A pressure perturbation at the end of the channel will be transmitted by acoustic waves through the reservoir and return to the channel after a time of order $\sim D/c_{acoustic}$. For $L \ll D$, the pressure perturbation at the channel outlet will affect flow at the inlet essentially instantaneously, so to couple effectively the channel and reservoir and build the required feedback, one must match the timescale for flow through the channel, $\sim L/U$, with the timescale for acoustic waves to traverse the reservoir, $\sim D/c_{acoustic}$ (Fig. 2c). Note the channel could be upstream or downstream of the reservoir. In practice, viscous forces and incomplete reflections will cause damping of the reservoir modes, and must be overcome by the fluid driving at the reed in order to set up and sustain tremor.

To excite oscillations, the flow speed U must exceed a fraction of $c_{acoustic}$. For fluids in volcanoes, $c_{acoustic}$ varies from less than 10^2 m/s in magma with 30–70% bubbles to more than 10^3 m/s in bubble-free magma (H_2O- and CO_2-rich fluids having intermediate sound speeds; Morrissey & Chouet 2001). Although $c_{acoustic}$ is high compared to flow speeds, L/D can be extremely small, so that reasonable flow speeds could destabilize the reservoir acoustic mode. Nevertheless, the acoustic wavelength must still be well over 100 m in order for the modal frequency to match typical tremor. As illustrated by the Helmholtz resonator

example, this does not necessarily require a reservoir length matching the wavelength: the choice of a suitable shape of the resonator provides flexibility in matching the frequency of tremor, with only weak constraints on the physical dimensions.

Discussion

In this paper, we have reviewed three mechanisms by which oscillations can be generated by flow in an elastic channel: hydrodynamic instabilities (roll waves), destabilized elastic wall modes, and unstable normal modes in an adjacent reservoir. Roll waves are unlikely in the geological context because they require relatively high flow speeds and may require seeding by external perturbations in long channels so that they are not flushed out. These disturbances are unstable waves of thickness variation in the channel, propagating in the direction of flow. As such, they are equivalent to travelling crack waves, modified by the mean flow. The (convective) instability criterion for these disturbances is equivalent to the requirement that the flow is faster than those travelling waves.

Elastic modes in the channel walls are destabilized by the flow according to a similar stability criterion, $U > \beta (L/\Delta)$, with Δ being the dimension of the elastic body along which the elastic standing waves are set up. Such flow speeds may be achieved in the geological context if the channel is much wider than it is long. However, the mode frequency f is of order β/Δ, and for the $O(1 \text{ Hz})$ frequencies characteristic of tremor, the relevant dimension of the elastic body must then be of order one kilometre, and is implausibly large (this is the same argument that dismisses any resonant *elastic* body as the origin of tremor, and motivates the introduction of a fluid-filled channel). Equivalently, if the elastic body has a realistic length, then there is a problem with its natural timescale. This is unfortunate, given the relative ease with which the elastic modes can be destabilized and their nonlinear properties, which resemble tremor observations.

The lumped-parameter model for volcanic tremor put forward by Julian (1994) does not capture propagating disturbances like roll waves, and is much like the models of the vocal cords (or blood vessels; Pedley 1980) in which elastic modes are destabilized. Indeed, his tremor frequencies are primarily set by the natural oscillation frequency of the wall and springs. Julian is also careful to draw a distinction between his model and resonant crack modes. Thus, one might, at first sight, think that the instabilities in his model may be destabilized elastic modes, in which case there is a timescale problem. However, Julian's model also incorporates unsteady fluid motions which affect the frequency of fluid-induced oscillations. Julian refers to the effect as 'added mass', giving the impression that his modes are fluid-modified elastic modes. In fact, those unsteady fluid motions correspond to thickness variations of the channel, and provide a natural oscillation even if the inertia of the walls is discarded (removing the normal mode in the elastic walls). In other words, Julian's model also contains a type of crack mode. And from his prescription of the restoring force, it is clear that these modes are standing crack waves across the width of the channel. Altogether, this suggests that the timescale problem of destabilized channel modes might be avoided if these modes are not of elastic origin, but are hydrodynamic crack, or 'sloshing' modes, like seiches.

Channel modes are not required whatsoever in the clarinet-type mechanism, which sets up the flow-destabilized oscillation in an adjacent reservoir. In the musical instrument, the reservoir mode is of acoustic origin, and so the sound speed and reservoir length set the timescale. Because the sound speed can be an order of magnitude smaller than the elastic wave speed in rock, acoustic reservoir modes have less of a timescale problem than elastic channel modes. One still remains, however, because some magmas can have quite high sound speeds. Moreover, acoustic waves could be strongly damped in a viscous fluid reservoir, raising the flow speed required to cause instability. All these problems might again be avoided if the reservoir mode is not actually acoustic, but a crack or sloshing mode.

The theory for destabilized sloshing modes in either the channel or a reservoir remains to be worked out. Our expectation is that the critical flow speeds required to drive oscillations are related to crack-wave speeds together with a geometrical factor, as in the three mechanisms discussed here. If this is borne out, and since the relatively slow crack-wave speed could resolve the timescale problem, such modes might provide the most plausible explanation of long-period volcanic seismicity.

Given that the mechanism and mode type might eventually be the same for both unstable channel and reservoir modes, the distinction between them boils down chiefly to one of geometry, and one wonders how one might choose between them seismologically. In this regard, a key geometrical detail is source location within the resonant body: the source for the channel modes occurs throughout its width, although one might imagine that flow speeds are greatest, and therefore instability strongest, closest to the midline. By contrast, the reservoir modes are always driven from one end. Along the lines suggested by Chouet (1988), one might then be able to use seismic signature to

tell the difference between reservoir and channel modes.

Finally, it is intriguing that there are observations of non-volcanic tremor in singing icebergs (Müller *et al.* 2005) and hydrocarbon reservoirs (Dangel *et al.* 2003). These other contexts can be more accessible or provide other constraints and thereby offer critical tests of the ideas and theory.

We thank S. De Angelis and B. Julian for their reviews. We also thank Keith Bradley and the WHOI GFD lab for support for the elastic membrane experiments. A.C.R. was supported by a Royal Society URF.

Appendix A. Slow crack waves

If we neglect viscous drag, linear motions of two-dimensional incompressible fluid in a thin, uniform channel can be described by the shallow-water-like model (cf. Appendix B):

$$\frac{\partial \xi}{\partial t} + H\frac{\partial u}{\partial x} = 0, \qquad \frac{\partial u}{\partial t} = -\frac{1}{\rho_f}\frac{\partial p}{\partial x}$$

where ρ_f is the fluid density, H is the equilibrium channel thickness, ξ is the variation in channel thickness ($\xi \equiv h - H$), and u and p are the associated flow speed and pressure perturbations. Assuming wave-like disturbances of the form $\exp ik(x - ct)$, where k is wavenumber and c is wavespeed, we find

$$c^2\xi = \frac{H}{\rho_f}p.$$

The pressure is related to the wall displacement according to the mechanics of the elastic wall. For semi-infinite blocks and elastic wavespeed much greater than c,

$$p = \frac{k\mu\xi}{2(1-\sigma)}$$

where k is the wavenumber, μ is the solid shear modulus and σ is Poisson's ratio. Waves therefore travel with the phase speed,

$$c = \sqrt{\frac{\mu k H}{2\rho_f(1-\sigma)}}.$$

This result is identical to the Ferrazzini & Aki's (1987) long-wavelength dispersion relation for symmetrical crack modes in the limit of high elastic and acoustic wavespeeds. Note that the 'crack stiffness' parameter, which uses the ratio of solid and fluid bulk moduli (e.g. Aki *et al.* 1977; Ferrazzini & Aki 1987; Chouet 1986), is not the natural parameter to describe these slow crack waves in this limit, because the fluid motions are incompressible.

Appendix B. Details of mathematical model

When the fluid channel is much thinner than it is wide, variations in the fluid motion across the channel thickness are much greater than variations in the flow direction and along the slot. Taking a 'thin-channel' approximation (e.g. Balmforth *et al.* 2005), our fluid model consists of slot-averaged equations in the local thickness, $h(x, t)$, and speed, $u(x, t)$, which express conservation of mass and momentum along the slot (the x-direction):

$$\frac{\partial h}{\partial t} + \frac{\partial(uh)}{\partial x} = 0,$$
$$\frac{\partial u}{\partial t} + u\frac{\partial u}{\partial x} = \frac{C}{h^2}(U - u) - \frac{1}{\rho_f}\frac{\partial p}{\partial x}. \quad (6)$$

Here, C is a viscous drag coefficient (constant), U is the mean flow speed (the base flow does not change with time), and $p(x, t)$ is the pressure perturbation induced in the fluid stemming from the motion of the wall. The model assumes $p = \Gamma(h - H)$, where Γ is a constant, and H is the equilibrium channel thickness. These choices correspond to the wall responding like a simple elastic foundation (a mattress) and flow resistance stemming from an approximation of laminar viscous drag. In Figure 4, we choose units such that $\Gamma/\rho_f = 1/5$, $C = 1/5$, $U = H = 1$ and $L = 200$.

For boundary conditions, we use either periodic conditions on u and h (Fig. 4a), or fixed flux (hu) and pressure (equivalently h) at the inlet, and $\partial(hu)/\partial x = 0$ at the outlet (Fig. 4b,c).

References

AKI, K., FEHLER, M. & DAS, S. 1977. Source mechanism of volcanic tremor: fluid-driven crack models and their application to the 1963 Kilauea eruption. *Journal of Volcanology and Geothermal Research*, **2**, 259–287.

BACKUS, J. 1963. Small-vibration theory of the clarinet. *Journal of the Acoustical Society of America*, **35**, 305–313.

BALMFORTH, N. J. & MANDRE, S. 2004. Dynamics of roll waves. *Journal of Fluid Mechanics*, **514**, 1–33.

BALMFORTH, N. J., CRASTER, R. V. & RUST, A. C. 2005. Instability in flow through elastic conduits and volcanic tremor. *Journal of Fluid Mechanics*, **527**, 353–377.

BRIGGS, R. J. 1964. *Electron-Stream Interaction with Plasmas*. MIT Press, Cambridge, MA.

BROOK, B. S., PEDLEY, T. J. & FALLE, S. A. 1999. Numerical solutions for unsteady gravity-driven flows in collapsible tubes: evolution and roll-wave instability of a steady state. *Journal of Fluid Mechanics*, **396**, 223–256.

CHOUET, B. 1986. Dynamics of a fluid-driven crack in three dimensions by the finite-difference method. *Journal of Geophysical Research*, **91**, 13967–13992.

CHOUET, B. 1988. Resonance of a fluid-driven crack: radiation properties and implications for the source of long-period events and harmonic tremor. *Journal of Geophysical Research*, **93**, 4375–4400.

CHOUET, B. 1996. Long-period volcano seismicity: its source and use in eruption forecasting. *Nature*, **380**, 309–316.

CHOUET, B. 2003. Volcano seismology. *Pure and Applied Geophysics*, **160**, 739–788.

DANGEL, S., SCHAEPMAN, M. E., STOLL, E. P., CARNIEL, R., BARZANDJI, O., RODE, E.-D. & SINGER, J. M. 2003. Phenomenology of tremor-like signals observed over hydrocarbon reservoirs. *Journal of Volcanology and Geothermal Research*, **128**, 135–158.

FERRAZZINI, V. & AKI, K. 1987. Slow waves trapped in a fluid-filled infinite crack: implication for volcanic tremor. *Journal of Geophysical Research*, **92**, 9215–9223.

FERRICK, M. G., QAMAR, A. & ST. LAWRENCE, W. F. 1982. Source mechanism of volcanic tremor. *Journal of Geophysical Research*, **87**, 8675–8683.

GROTBERG, J. B. & JENSEN, O. E. 2004. Biofluid mechanics in flexible tubes. *Annual Review of Fluid Mechanics*, **36**, 121–147.

HELLWEG, M. 2000. Physical models for the source of Lascar's harmonic tremor. *Journal of Volcanology and Geothermal Research*, **101**, 183–198.

ISHIZAKA, K. & FLANAGAN, J. L. 1972. Synthesis of voiced sounds from a two-mass model of the vocal cords. *Bell Systems Technology Journal*, **51**, 1233–1268.

JULIAN, B. R. 1994. Volcanic tremor: nonlinear excitation by fluid flow. *Journal of Geophysical Research*, **99**, 11859–11877.

JULIAN, B. R. 2000. Period doubling and other nonlinear phenomena in volcanic earthquakes and tremor. *Journal of Volcanology and Geothermal Research*, **101**, 19–26.

KAWAKATSU, H., OHMINATO, T. & ITO, H. 1994. 10s-period volcanic tremors observed over a wide area in southwestern Japan. *Geophysical Research Letters*, **21**, 1963–1966.

KONSTANTINOU, K. I. 2002. Deterministic non-linear source processes of volcanic tremor signals accompanying the 1996 Vatnajkull eruption, central Iceland. *Geophysical Journal International*, **148**, 663–675.

KONSTANTINOU, K. I. & SCHLINDWEIN, V. 2002. Nature, wavefield properties and source mechanism of volcanic tremor: a review. *Journal of Volcanology and Geothermal Research*, **119**, 161–187.

LESAGE, P., MORA, M. M., ALVARADO, G. E., PACHECO, J. & METAXIAN, J.-P. 2006. Complex behaviour and source model of the tremor at Arenal volcano, Costa Rica. *Journal of Volcanology and Geothermal Research*, **157**, 49–59.

LIU, J., PAUL, J. D. & GOLLUB, J. P. 1993. Measurements of the primary instabilities of film flows. *Journal of Fluid Mechanics*, **250**, 69–101.

MANDRE, S. 2006. *Two studies in hydrodynamic stability. Interfacial instabilities and applications of bounding theory*. PhD thesis, University of British Columbia.

MCNUTT, S. R. 2005. Volcanic seismology. *Annual Review of Earth and Planetary Sciences*, **33**, 461–491.

MORRISSEY, M. M. & CHOUET, B. A. 2001. Trends in long-period seismicity related to magmatic fluid compositions. *Journal of Volcanology and Geothermal Research*, **108**, 265–281.

MÜLLER, C., SCHLINDWEIN, V., ECKSTALLER, A. & MILLER, H. 2005. Singing icebergs. *Science*, **310**, 1299.

PEDLEY, T. J. 1980. *Fluid Mechanics of Large Blood Vessels*. Cambridge University Press.

STRUTT, J. W. 1871. On the theory of resonance. *Philosophical Transactions of the Royal Society*, **161**, 77–118.

WAGNER, W. & OVERHOFF, U. 2006. *Extended IAPWS-IF97 Steam Tables*. Springer-Verlag, Berlin, Heidelberg.

WOODS, B. D., HURLBURT, E. T. & HANRATTY, T. J. 2000. Mechanism of slug formation in downwardly inclined pipes. *International Journal of Multiphase Flow*, **26**, 977–998.

Shallow-conduit dynamics at Stromboli Volcano, Italy, imaged from waveform inversions

BERNARD CHOUET[1], PHILLIP DAWSON[1] & MARCELLO MARTINI[2]

[1]US Geological Survey, 345 Middlefield Rd., Menlo Park, CA 94025, USA
(e-mail: chouet@usgs.gov)

[2]Osservatorio Vesuviano, 328 Via Diocleziano, I-80124 Napoli, Italy

Abstract: Modelling of Very-Long-Period (VLP) seismic data recorded during explosive activity at Stromboli in 1997 provides an image of the uppermost 1 km of its volcanic plumbing system. Two distinct dyke-like conduit structures are identified, each representative of explosive eruptions from two different vents located near the northern and southern perimeters of the summit crater. Observed volumetric changes in the dykes are viewed as the result of a piston-like action of the magma associated with the disruption of a gas slug transiting through discontinuities in the dyke apertures. Accompanying these volumetric source components are single vertical forces resulting from an exchange of linear momentum between the source and the Earth. In the dyke system underlying the northern vent, a primary disruption site is observed at an elevation near 440 m where a bifurcation in the conduit occurs. At a depth of 80 m below sea level, a sharp corner in the conduit marks another location where the elastic response of the solid to the action of the upper source induces pressure and momentum changes in the magma. In the conduit underlying the southern vent, the junction of two inclined dykes with a sub-vertical dyke at 520 m elevation is a primary site of gas slug disruption, and another conduit corner 280 m below sea level represents a coupling location between the elastic response of the solid and fluid motion.

Some of the most promising advances in our knowledge of processes occurring within volcanoes are emerging from seismology. Seismic events produced during eruptive activity provide information about the physical properties of the magma as well as the fluid-flow and degassing mechanisms driving the eruptive process. Through an increased use of broadband seismometers on volcanoes, we now understand that Very-Long-Period (VLP) signals (with typical periods in the range 2–100 s) commonly accompany magma and associated fluid movement. A critically important element in the analysis of VLP seismic energy is the quantification of the source location and source mechanism of such energy, from which diagnostic information can be derived about the geometries of volcanic conduits and their volumetric variations during pre- and syn-eruptive phenomena.

A general kinematic description of seismic sources in volcanoes is commonly based on a moment-tensor and single-force representation of the source (Aki & Richards 1980). The seismic-moment tensor consists of nine force couples, with each force couple corresponding to one set of opposing forces (dipoles or shear couples). This symmetric second-order tensor allows a description of any generally oriented discontinuity in the Earth in terms of equivalent body forces. For example, slip on a fault can be represented by an equivalent

force system involving a superposition of two force couples of equal magnitudes — a double couple. Similarly, a tensile crack has an equivalent force system made of three vector dipoles, with dipole magnitudes with ratios $1:1:(\lambda+2\mu)/\lambda$, in which λ and μ are the Lamé coefficients of the rock matrix (Chouet 1996), and where the dominant dipole is oriented normal to the crack plane. Injection of fluid into the crack will cause the crack to expand and act as a seismic source. In general, magma movement between adjacent segments of conduit can be represented through a combination of volumetric sources of this type. Because mass-advection processes can also generate forces on the Earth, a complete description of volcanic sources commonly requires the consideration of single forces in addition to the volumetric source components expressed in the moment tensor. For example, a volcanic eruption can induce a force system that includes a contraction of the conduit/reservoir system in response to the ejection of fluid, and a reaction force from the eruption jet (Kanamori et al. 1984). Some volcanic processes can be described by a single-force mechanism only. An example is the single-force mechanism attributed to the massive landslide observed at the start of the 1980 eruption of Mount St. Helens (Kanamori & Given 1982; Kawakatsu 1989). A single-force source model was also proposed by

From: LANE, S. J. & GILBERT, J. S. (eds) Fluid Motions in Volcanic Conduits: A Source of Seismic and Acoustic Signals. Geological Society, London, Special Publications, **307**, 57–84.
DOI: 10.1144/SP307.5   0305-8719/08/$15.00 © The Geological Society of London 2008.

Uhira *et al.* (1994) to explain the mechanism of dome collapses at Unzen Volcano, Japan. Another example of single-force mechanism in nature is the stick-slip, downhill sliding of a glacial ice mass (Ekström *et al.* 2003).

Waveform inversions solving for the source-time histories of the moment-tensor and single-force components of the source have become the primary tool to identify and understand the source mechanisms generating VLP signals recorded by broadband seismometers (Uhira & Takeo 1994; Kaneshima *et al.* 1996; Ohminato *et al.* 1998; Kawakatsu *et al.* 2000; Nishimura *et al.* 2000; Legrand *et al.* 2000; Kumagai *et al.* 2001, 2003; Chouet *et al.* 2003, 2005; Ohminato *et al.* 2006; Ohminato 2006). When the wavelengths of observed seismic waves are much longer than the spatial extent of the source, the source may be approximated by a point source and the force system represented by the moment tensor and single force components is localized at this point. The standard approach to estimate source-time histories of the moment and force components at the source is based on the Green's function, which describes the signal that would be observed at a receiver if the source-time function were a perfect impulse. Once the Green's functions are known, the source-time histories of the six independent components of the moment tensor and three components of force can be retrieved through least-squares inversion of the VLP waveform (e.g. Chouet *et al.* 2003).

The present study describes results of detailed analyses of VLP signals observed during explosive activity at Stromboli in September 1997. We begin with a brief description of the methodology commonly used in the quantification of seismic source mechanisms in volcanoes, and follow with a summary of results obtained by Chouet *et al.* (2003) from inversions of VLP data recorded by the broadband network in operation at Stromboli at the time. We then proceed with a reconstruction of the geometry of the conduit system beneath Stromboli based on systematic waveform inversions of VLP waveforms recorded, and conclude with a discussion of the implications of this conduit model for the source processes driving Strombolian eruptions.

Stromboli Volcano

Stromboli is the northernmost active volcano of the Aeolian Island arc in the Tyrrhenian Sea. The island of Stromboli rises approximately 3000 m from the sea floor, and its summit stands 924 m above sea level (Fig. 1). The twin peaks forming the summit of the volcano are southern remnants of older edifices that have been truncated by sector collapses

of the volcano to the northwest (Tibaldi 2001). The current activity originates in vents located within a 250 m-long by 150 m-wide crater on a terrace northwest of, and about 130 m below, the northern peak. The crater is buttressed by nearly vertical walls on the southeastern side and merges into a long talus slope on the northwestern side. This talus partially fills a large sector graben called the 'Sciara Del Fuoco' that extends from the summit to the sea. The formation of the 'Sciara Del Fuoco' is attributed to a giant sector collapse that occurred less than 5000 years ago (Pasquaré *et al.* 1993; Tibaldi 2001).

Stromboli is considered one of the most active volcanoes in the world, and its persistent but moderate explosive activity, termed 'Strombolian', is only interrupted by occasional episodes of more vigorous activity accompanied by lava flows, as last occurred in 2002–2003 (Calvari *et al.* 2005). Its normal eruptive behaviour, first described by Aristotle 2000 years ago, is characterized by mild, intermittent explosive activity, during which well-collimated jets of incandescent gases laden with molten fragments burst in short eruptions, each lasting 5–15 s and occurring at a typical rate of 3–10 events per hour (Chouet *et al.* 1974).

Persistent explosive activity and ease of access make this volcano ideally suited for detailed measurements of the seismic wave fields radiated by Strombolian activity. To improve our understanding of the seismic source mechanisms for these explosions, detailed broadband measurements were carried out at Stromboli in September of 1997 by Chouet *et al.* (2003). The seismic data were recorded by a network of 21 three-component broadband (0.02–60 s) seismometers (Fig. 1). This network remained in operation for one week.

Figure 2a shows examples of broadband signals and their associated VLP waveforms for two types of explosions observed during the experiment. Eruptive activity at that time was limited to two distinct vents located near the northern and southern perimeters of the crater (vents 1 and 2 in Fig. 1), and the two types of waveforms illustrated in Figure 2a are representative of eruption signals from these two vents. The signals of Type-1 events are representative of eruptions from the northern vent (vent 1), characterized by canon-like blasts typically lasting a few seconds and producing well-collimated jets of incandescent gases laden with molten fragments. In contrast, Type-2 eruptions from the southern vent (vent 2) were much less impulsive than those from vent 1; these typically lasted longer (up to 20 s) and produced wider fans of ejecta and significant amounts of ash. The close similarities among the VLP waveforms from explosions occurring at different times from the same vent (Fig. 2b) clearly reflect the repetitive action of a non-destructive process at

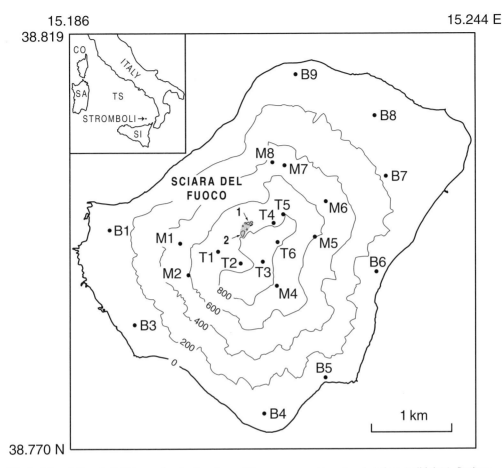

Fig. 1. Map of Stromboli Volcano showing locations of three-component broadband stations (solid dots). Stations prefixed by 'T' denote those of the 'T' ring of sensors, by 'M' those of the 'M' ring, and by 'B' those of the 'B' ring, located at the crater level, mid-level, and base of the volcano, respectively. The crater is marked by the shaded area, which encompasses distinct vents. The arrows point to two eruptive vents that were active at the time of the seismic experiment in September 1997. Contour lines represent 200 m contour intervals. The inset shows the location of Stromboli in the Tyrrhenian Sea (TS) in relation to Italy, Sicily (SI), Sardinia (SA) and Corsica (CO).

the source. As the operative source processes are essentially stationary with time within the bandwidth of the VLP data, an analysis of representative events is adequate to fully describe the overall source dynamics.

Waveform inversion method

The displacement field generated by a point source is described by the representation theorem, which is expressed through the convolution (Chouet 1996; Ohminato *et al.* 1998; Chouet *et al.* 2003, 2005; Auger *et al.* 2006; Ohminato 2006)

$$u_n(t) = \sum_{i=1}^{N_m} m_i(t) * G_{ni}(t), \quad n = 1, \ldots, N_t \quad (1)$$

where $u_n(t)$ is the n-component of seismic displacement at a receiver at time t, $m_i(t)$ is the time history of the i-th moment-tensor or single-force component at the source, $G_{ni}(t)$ are the Green's functions or their spatial derivatives corresponding to each of the respective moment-tensor and single-force components, N_m is the number of source mechanism components, N_t is the number of seismic traces, and the subscript n stands for both receiver location and component of motion. The source-mechanism components include the six independent components of the moment-tensor and three components of force. Once the Green's functions are known, the nine $m_i(t)$ in equation (1) are retrieved through least-squares inversion expressing the standard linear problem $\mathbf{d} = \mathbf{Gm}$ in equation (1) (Ohminato *et al.* 1998; Chouet *et al.* 2003).

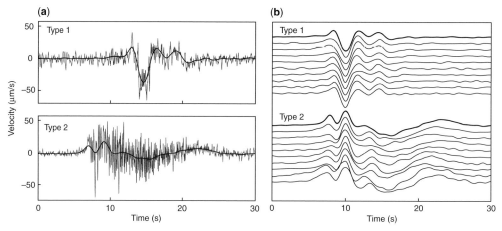

Fig. 2. (**a**) Typical records obtained for Type-1 and Type-2 explosions. The red trace illustrates the broadband signal recorded on the east component of ground velocity at station T6 (see Fig. 1), and the bold black trace shows the VLP waveform obtained by band-pass filtering the broadband data (red trace) with a 2-pole zero-phase-shift Butterworth filter. (**b**) Normalized east component of velocity seismograms recorded at station T6 for 10 Type-1 events and 10 Type-2 events selected from a seven-hour-long record. The traces are filtered between 2 and 20 s (Type 1) or 2 and 30 s (Type 2). Bold traces represent the events shown in the two left panels. Notice the similarities among waveforms of different events of the same type. The records in (a) and (b) are representative of true ground motion after correction for instrument response.

For an accurate determination of the source location and associated source mechanism, one needs to compare observed seismic data to synthetic data calculated for a realistic model of the volcanic edifice. A standard approach is to use a discretized representation of the edifice based on a digital elevation model. Using this discretized model, synthetics may then be calculated by the three-dimensional finite-difference method. The procedure involves a calculation of Green's functions for individual moment and single-force components applied at a preset position representing the anticipated location of the point source in the edifice. As the actual position of the source is unknown *a priori*, the calculations are repeated for different source positions and the best-fitting point source is determined by minimizing the residual error between calculated and observed seismograms. Calculations are usually carried out for point sources distributed over a uniform mesh encompassing the source region under study and the number of point sources considered may be quite large. A dramatic reduction in computation time is achieved by using the reciprocal relation between source and receiver (Aki & Richards 1980; Chouet *et al.* 2005; Auger *et al.* 2006; Ohminato 2006), and further speedup in calculations is achieved by performing the waveform inversion in the frequency domain (Auger *et al.* 2006).

The selection of an optimum solution is based on variance reduction and relevance of the free parameters used in the model. We use the following definition of squared error to evaluate results (Ohminato *et al.* 1998):

$$E = \frac{1}{N_r} \sum_{n=1}^{N_r} \left[\frac{\sum_{1}^{3} \sum_{p=1}^{N_s} (u_n^0(p\Delta t) - u_n^s(p\Delta t))^2}{\sum_{1}^{3} \sum_{p=1}^{N_s} (u_n^0(p\Delta t))^2} \right] \times 100 \quad (2)$$

where $u_n^0(p\Delta t)$ is the p-th sample of the n-th data trace, $u_n^s(p\Delta t)$ is the p-th sample of the n-th synthetic trace, N_s is the number of samples in each trace, and N_r is the number of three-component receivers. In this formulation, the squared error is normalized by receiver, so that receivers with weak-amplitude signals contribute equally to the squared error as receivers with large-amplitude signals.

To test the significance of the number of free parameters, each source model is evaluated by calculating Akaike's Information Criterion (AIC) (Akaike 1974) defined as

$$\text{AIC} = N_{obs} \ln E + 2N_{par} \quad (3)$$

where $N_{obs} = N_t N_s$ is the number of independent observations, E is the squared error defined in equation (2), and N_{par} is the number of free parameters used to fit the model. In our case, N_{par} is the number of source mechanisms considered times the number of spectral components used in the frequency-domain inversion. Additional free parameters in the source mechanism are considered to be physically relevant when both the residual error and AIC are minimized.

Original results obtained from VLP waveform inversions

Analyses of VLP waveforms radiated by Strombolian explosions (Chouet *et al.* 2003) demonstrated that a moment and force representation of the source consistently yields the minimum AIC and optimum variance reduction in the waveform matches obtained for the two types of events observed at Stromboli, supporting the contention that this is the most appropriate model describing the source mechanism of these events. In their analyses, Chouet *et al.* (2003) restricted their attention to data from the top two rings of receivers (T and M rings in Fig. 1) where the largest signal amplitudes were recorded. Their inversions considered VLP signals band-pass filtered in the 2–20 s band for Type-1 event, or 2–30 s band for Type-2 event. The two point sources that best fitted the waveform data associated with eruptions from the two vents were found to be located at elevations of 520 m (Type-1) and 480 m (Type-2), approximately 160 m northwest of the vents (Chouet *et al.* 2003). Figures 3a and 4a show the source-time functions of moment and force components obtained by Chouet *et al.* (2003) for Type-1 and Type-2 events. The principal axes of the moment tensor obtained by eigenvalue decomposition of the source-time functions of moment components are illustrated in Figures 3b and 4b. The three eigenvectors identified by thick red lines in these figures are obtained from measurements of the maximum peak-to-trough amplitudes in the individual tensor components. These represent the main deflation phase of the source seen during the interval 7–10 s in Type-1 event (Fig. 3a), or interval 30–40 s in Type-2 event (Fig. 4a). These vectors are shown again in Figures 3c and 4c as bold black arrows.

The force system in Figures 3c and 4c consists of three dipoles with amplitude ratios $1:0.8:2$ and $1.1:1:2$ in Type-1 and Type-2 events, respectively. These ratios closely match the amplitude ratios $1:1:(\lambda + 2\mu)/\lambda$ for a crack, in which $\lambda = 2\mu$ is assumed — a value appropriate for volcanic rock at temperature close to liquidus. A simple crack model, illustrated by the red-coloured planes in Figures 3c and 4c, therefore constitutes an appropriate first-order representation of the source mechanism producing the moment components shown in Figures 3a and 4a. The imaged crack for Type-1 event dips 63° to the northwest and strikes northeast–southwest along a direction parallel to the elongation of the volcanic edifice (see Fig. 1). The imaged crack for Type-2 event displays a slightly shallower northwest dip of 62° with a strike that differs by four degrees from that of the crack resolved for Type-1 event. Both crack azimuths basically parallel the trends of exposed dykes and a known zone of weakness in the volcanic edifice.

Accompanying the volumetric source components in Figures 3a and 4a is a dominantly vertical single force, which represents an exchange of linear momentum between the source and the rest of the Earth (Takei & Kumazawa 1994). The force is initially down, then up in both event types. In the Type-1 event, the downward force is synchronous with the initial inflation of the source volume while the following upward force coincides with the deflation of the source volume. Although less clear, a similar synchronicity is manifest in the Type-2 event. A downward force can be explained as the reaction force on the Earth associated with either an upward acceleration or downward deceleration of the centre of mass of the source volume. Similarly, an upward force may result from either a downward acceleration or upward deceleration of the centre of mass of the source volume. The observed forces are thus suggestive of a piston-like action of the liquid magma associated with the disruption of a gas slug passing through this particular location in the conduit system.

Reconstruction of single-point source mechanisms

Although the crack models elaborated for Type-1 and Type-2 events are adequate for a first-order description of the source of these events, a detailed sampling of the eigenvectors (Figs 3b and 4b) reveals complexities in the source mechanisms of explosions that do not fit this simple picture of the source. Some scatter is noticeable in the eigenvectors obtained at different times in the source-time histories of Type-1 event (Fig. 3b), and more pronounced scatter is seen for eigenvectors representing the source of Type-2 event (Fig. 4b). Also, the dipole ratios obtained from peak-to-trough measurements in the source-time functions are not exactly $1:1:2$, suggesting the presence of some distortion in the waveforms of source-time functions associated with different moment components. A plausible explanation for this observation is that

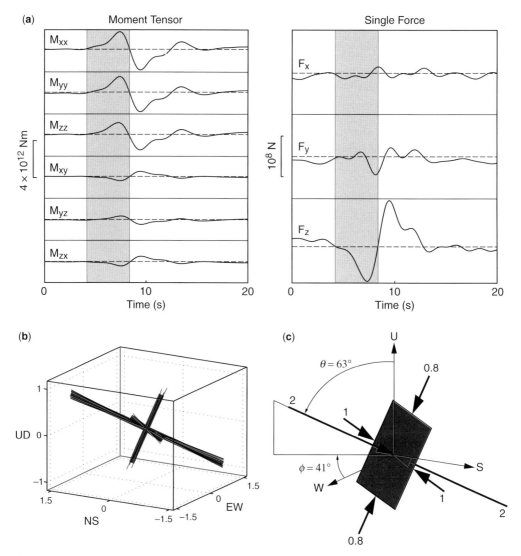

Fig. 3. (a) Source-time functions obtained for Type-1 event, in which six moment-tensor components and three force components are assumed for the source mechanism (after Chouet *et al.* 2003). Shading marks the interval during which the initial volumetric expansion of the source occurs. **(b)** Source mechanism imaged by Chouet *et al.* (2003) for Type-1 event. The reference coordinates for the eigenvectors are EW (east–west), NS (north–south), and UD (up–down). The eigenvectors are obtained from the moment-tensor solution shown in (a), in which eigenvectors are sampled every 0.3 s during the time interval 0–20 s. **(c)** The three eigenvectors obtained by measurements of maximum peak-to-trough amplitudes in the individual tensor components within the interval 7–10 s. In both (b) and (c), eigenvectors are normalized to a maximum length of 2 and in (b) no distinction is made between expansion and contraction.

the underlying mechanism represents a composite of two intersecting cracks, in which one crack contributes the dominant component of radiation and the other crack the subdominant component. This idea is tested below assuming a point source composed of two intersecting cracks.

Our calculations of Green's functions assume a homogeneous medium and include consideration of the topography and bathymetry of Stromboli. The discretized model of Stromboli used in our source reconstruction is similar to that used by Chouet *et al.* (2003). Our model uses the same

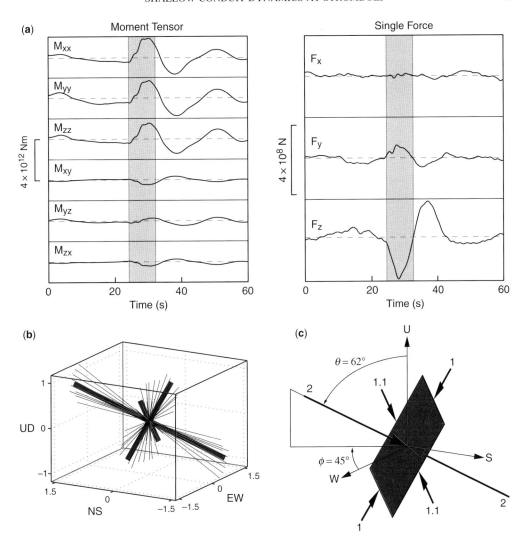

Fig. 4. (a) Source-time functions obtained for Type-2 event, in which six moment-tensor components and three force components are assumed for the source mechanism (after Chouet *et al.* 2003). Shading marks the interval during which the downward force is synchronous with the initial volumetric expansion of the source. **(b)** Source mechanism imaged by Chouet *et al.* (2003) for Type-2 event. The reference coordinates for the eigenvectors are EW (east–west), NS (north–south), and UD (up–down). The eigenvectors are obtained from the moment-tensor solution shown in (a), in which eigenvectors are sampled every 1 s during the time interval 20–45 s. **(c)** The three eigenvectors obtained by measurements of maximum peak-to-trough amplitudes in the individual tensor components within the interval 30–40 s. In both (b) and (c), eigenvectors are normalized to a maximum length of 2 and in (b) no distinction is made between expansion and contraction.

grid of 40 m in a uniform mesh centred on Stromboli, but considers a larger computational domain with lateral dimensions of 13.8 × 13.8 km and vertical extent of 8 km. We assume a compressional wave velocity $V_p = 3.5$ km/s, shear wave velocity $V_s = 2$ km/s, and density $\rho = 2650$ kg/m^3. Using this model, synthetics are calculated by the three-dimensional finite-difference method of Ohminato & Chouet (1997).

The strategy used in our reconstruction of the source mechanism is the same as that used by Chouet *et al.* (2005) in their reconstruction of the source of VLP signals observed at Popocatépetl Volcano, Mexico. Specifically, our approach considers a point source composed of two cracks whose orientations are allowed to vary independently of each other. A search for the best-fitting model is carried out by systematically varying the azimuth ϕ and polar angle θ defining the orientation of the normal vector to each crack plane. An initial search for the best set of angles is carried out for the same source position as that of the best-fitting point source obtained by Chouet *et al.* (2003). Once a best-fitting set of angles has been obtained at this location, the search is repeated for adjacent point sources within a small volume centred on the original point source location to guarantee that the parameter space has been fully explored and that a minimum in residual error between data and synthetics has been obtained. As in the original inversions of Chouet *et al.* (2003), only data from the T and M rings are included in these fits. Data from the B-ring receivers are excluded because of evidence for contributions from deeper source components in the signals from these receivers as compared to data from the T and M rings (Chouet *et al.* 2003).

The minimum in residual error for Type-1 event yields a point source located 80 m below and 40 m north of the original centroid location. For Type-2 event, the point source obtained from reconstruction remains at the same location as the original source centroid determined by Chouet *et al.* (2003). The difference in residual error between the reconstructed source and original source obtained by Chouet *et al.* (2003) amounts to 1.3% in Type-1 event, and 0.7% in Type-2 event. Figures 5 and 6 illustrate the waveform matches obtained from reconstruction of the composite mechanism made of two intersecting cracks, along with the original fits obtained by Chouet *et al.* (2003) from inversion assuming six moment-tensor and three single-force components. Overall, the differences between the fits obtained for the reconstructed source and fits obtained by Chouet *et al.* (2003) are negligible, hence the two models may be viewed as equivalent.

Figure 7 shows the results of our reconstruction of the source mechanism of Type-1 event. The moment-tensor and single-force components representing the two-crack mechanism of the composite source very closely match the moment and single-force components obtained by Chouet *et al.* (2003) from waveform inversion assuming six moment components and three single force components (compare with Figure 3a). The volume changes in the two cracks are obtained from the amplitude $(\lambda + 2\mu)\Delta V$ of the principal dipole component representing each crack in the composite source. These results assume $\mu = 7\,\text{GPa}$ and Poisson ratio $\nu = 1/3$ ($\lambda = 2\mu$) (Chouet *et al.* 2003).

The source-time histories of volume changes in the two cracks show that the source mechanism is dominated by a crack whose orientation is within *c.* 2° of the approximate crack mechanism inferred from the original waveform inversion of Chouet

et al. (2003) (see Fig. 3c). The dominant crack, coloured red, contributes 90% of the overall volume change in the composite source. The secondary crack, shown in grey, is roughly orthogonal to the dominant crack and contributes the remaining 10% of the overall volume change in the source. The source-time histories of volume change point to a sequence of inflation, deflation, inflation in both cracks, a picture fully consistent with the earlier result obtained by Chouet *et al.* (2003).

The source mechanism of the composite source in Type-2 event (Fig. 8) also displays features that are similar to those seen in the original source mechanism obtained by Chouet *et al.* (2003) (compare with Figure 4a), yet includes subtle variations in details that become apparent upon close examination of the source-time histories in Figures 8 and 4a. For example, the main oscillations in the source-time histories of moment components are shorter-lasting in the composite source compared to the original source. Oscillations are also less apparent in the north component of force F_y, and more apparent in the east component of force F_x in the reconstructed source compared to the earlier model. The composite source model points again to a dominant crack (coloured red), in this case contributing 84% of the volume change, and subdominant crack generating the remaining 16% of the source volumetric component. The two cracks display similar dips, but differ in strike by 44°. The orientation of the dominant crack is within *c.* 8° of the approximate crack mechanism inferred by Chouet *et al.* (2003) (see Fig. 4c).

As stated above, an explicit assumption in our source reconstruction is that the observed mechanism can be represented by a composite of two intersecting cracks. This assumption was primarily based upon the observation that the dipole ratios derived from the original inversion by Chouet *et al.* (2003) showed slight deviations from the ratios $1:1:2$ expected for a pure crack mechanism. To assess the validity of this assumption, we conducted systematic reconstructions for models consisting of a single pipe, a single crack, two intersecting pipes, two intersecting cracks, and an intersecting pipe and crack for both Type-1 and Type-2 events. Each model consists of a point source co-located with the reconstructed source and is composed of one of these combinations. The azimuthal and polar orientations of individual source components in the models with dual mechanisms were allowed to vary independently of each other. For each model, we determined the minimum residual error, and the mean dominant dipole orientation and dipole ratios calculated from a point-by-point eigenvalue decomposition of the reconstructed mechanism. These values were then compared to the minimum residual

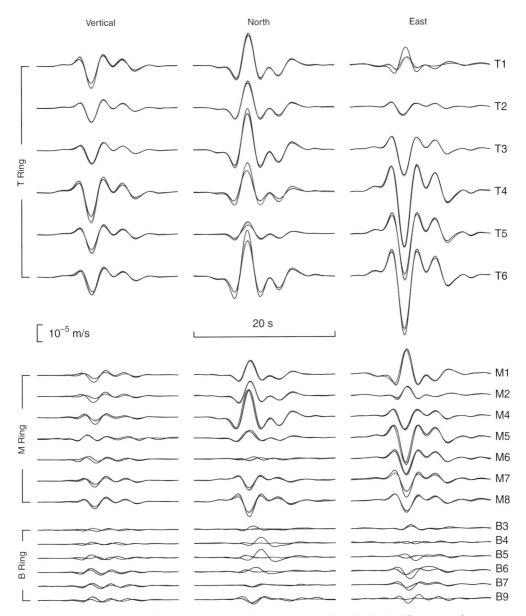

Fig. 5. Observed velocity waveforms of Type-1 event (black lines), waveform fits obtained from composite source consisting of two intersecting cracks combined with three single-force components (red lines), and waveform fits obtained from original inversion by Chouet *et al.* (2003) assuming six moment-tensor and three single-force components (green lines). The fits obtained from the two different approaches are virtually indistinguishable from each other. Though shown (bottom of figure), the data from the B ring receivers are not used in these inversions. The station codes are indicated at the right and the station components are indicated at the top of the figure, respectively.

error and mean dominant dipole orientation and dipole ratios derived from the free inversion assuming six moment-tensor and three single-force components.

At the reconstructed source location for the Type-1 event, the mean dipole ratios derived from the statistics of the free inversion are $1 : 1.1 : 2$ and the mean azimuth and polar angles are within $1°$ of those shown in Figure 3c. In this case, the single crack and single pipe models have residual errors that are noticeably higher ($>6\%$) than the free inversion. The single crack model reproduces

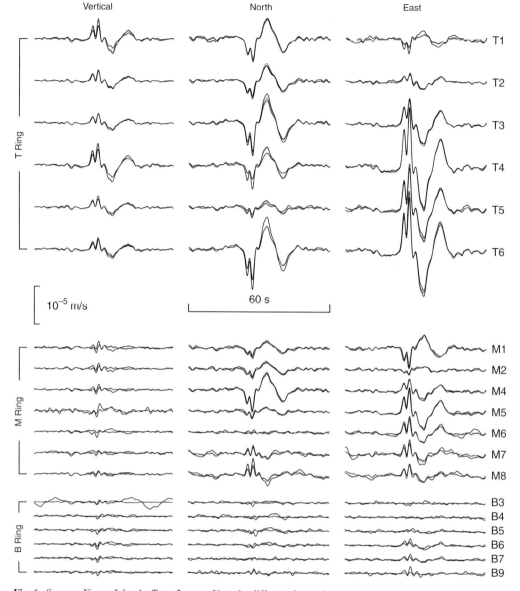

Fig. 6. Same as Figure 5 for the Type-2 event. Note the different timescale.

the mean azimuth to within a few degrees and the mean polar angle to within 11° and has dipole ratios 1 : 1 : 2. The single pipe model does not reproduce the mean azimuth or polar angles and has dipole ratios 1.3 : 2 : 2. The composite models consisting of pipe-crack, crack-crack and pipe-pipe mechanisms have residual errors less than 3% higher than the free inversion. The pipe-crack model has the lowest residual error of the three types, but the mean azimuth and polar angle for this model vary by more than 12° from the free

inversion model and the mean dipole ratios are 0.8 : 1.2 : 2. The pipe-pipe model does not reproduce the azimuth or polar angles and has mean dipole ratios of 0.5 : 1.5 : 2. The crack-crack model reproduces the azimuth and polar angles to within 2° and has mean dipole ratios 1 : 1.1 : 2.

The assessment of the statistical distributions of dipole orientations and ratios in addition to residual errors for various mechanisms and combinations of mechanisms provides a strong constraint upon the choice of an appropriate model.

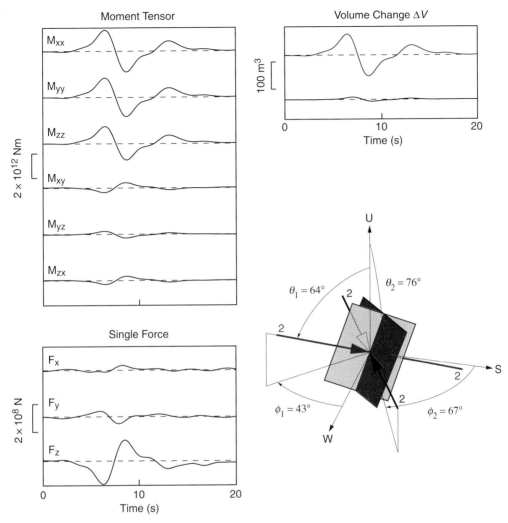

Fig. 7. Results of a reconstruction of the source mechanism of Type-1 event based on the assumption that the source is a composite of two cracks. The source-time histories of the two cracks, shown at the upper right, are colour-coded with the same colours as the corresponding cracks, shown at the bottom right; the dominant crack is coloured red.

Our analysis demonstrates that the reconstruction of two intersecting cracks best reproduces the results from the Type-1 and Type-2 free inversions.

Two-point source models

As seen in Figures 5 and 6, the waveform fits obtained on the T and M rings for the composite source mechanism are excellent overall. These results, however, do not yet provide a complete picture of the source of VLP signals at Stromboli because of the observed misfits on the receivers of the B ring. As noted by Chouet *et al.* (2003),

these misfits point to other source components that remain to be identified. Our next objective is to quantify these other components of the source.

Our procedure to identify further components of the source consists in the following steps. (1) Keeping the positions of the two cracks in the original point source fixed, a search is conducted for another composite mechanism at a second point source. In this search, the second point source is again assumed to be a composite of two intersecting cracks combined with three single force components, and data from the entire network are used. Our search for the best-fitting

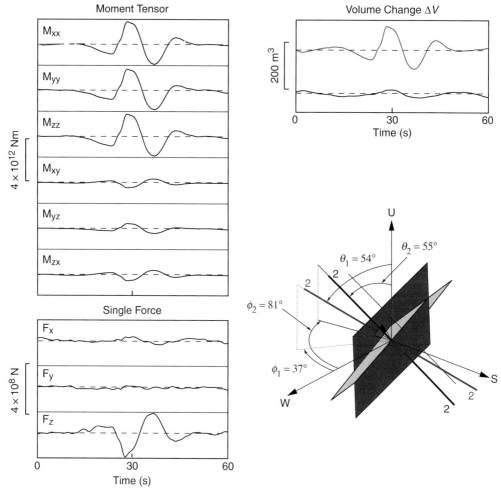

Fig. 8. Same coding as Figure 7 but here for Type-2 event. Notice the different timescale.

signal is carried out for trial point sources spaced 40 m apart in a uniform mesh sampling a total source volume of 6 km³. At each trial position of the secondary source, the orientations of the two cracks are allowed to vary independently of each other. The source-time histories of the two fixed cracks and three force components in the original point source are then recalculated along with the source-time histories of the two cracks and three force components in the secondary point source. (2) Once the position and mechanism of the second point source have been identified, we perform a fine adjustment of the coupled mechanisms of the two point sources to find the absolute minimum of residual error between fitted synthetics and data. In this step, the positions of the two point sources remain fixed. (3) A similar analysis of coupled mechanisms is also carried out at point

sources adjacent to each of the point sources imaged in step (2) to ensure that the solution yielding the absolute minimum in residual error has been obtained.

Figure 9 shows the locations of the point sources and distributions of residual errors calculated with equation (2) for the two types of events. In Type-1 event, the final position of the upper source remains unchanged from the solution obtained in our earlier reconstruction of mechanism in the single-point source model. In Type-2 event, the upper source location is shifted 40 m above its original position (Chouet *et al.* 2003). The error minimum for Type-1 event yields an upper source at elevation of 440 m above sea level and lower source 80 m below sea level. The upper source epicentre is located 120 m northwest of the northern vent area, and the lower source epicentre is

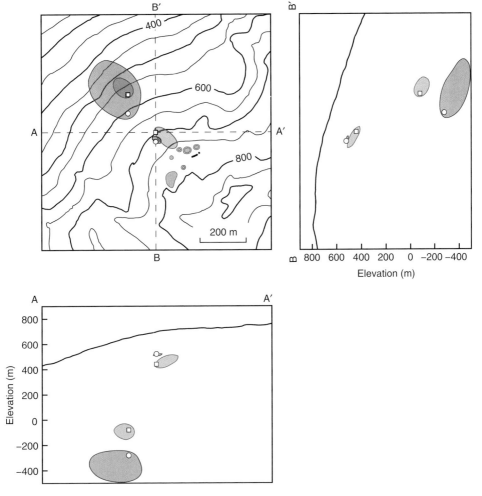

Fig. 9. Horizontal, east–west, and north–south vertical cross-sections through the northwest quadrant of Stromboli showing the locations of the point sources representing the two types of events analysed in this paper. The planes of the cross-sections are indicated by dashed lines (AA′ and BB′) in map view. Vents are identified by contours filled in with grey shading. Open squares and open circles mark the locations of the sources for Type-1 and Type-2 events, respectively. Coloured patches are projections of the source location uncertainties associated with error increments of 0.1% above the absolute minimum error (red for Type-1 and blue for Type-2 event).

located 200 m northwest of the upper source epicentre. The error minimum for Type-2 event yields an upper source at elevation of 520 m above sea level, and a lower source 280 m below sea level roughly 170 m northwest of the upper source epicentre. The upper source location is approximately 170 m north-northwest of the southern vent, and 80 m above and 40 m south of the upper source of Type-1 event.

The distributions of residual errors in Figure 9 represent misfit values E (eq. 2) obtained at each grid node assuming fixed mechanisms at both upper and lower sources. The mechanisms considered are those corresponding to the minimum misfit. Although the misfit distributions obtained in this manner are the results of only a partial search (through the eight dimensions of the parameter space defining the crack orientations), these nonetheless provide an adequate representation of the error volume in light of the stability of the mechanisms imaged by our inversion. The shape of the misfit distribution around the minimum residual is defined by the iso-surface representing a misfit level of 0.1% above the minimum misfit (coloured patches in Fig. 9). The shape and extent of this surface represent the uncertainty in source location under the stated assumption of fixed mechanisms. In Type-1 event, the uncertainty in the upper source position is *c.* 120 m in the northwest–southeast direction, *c.* 60 m in the northeast–southwest direction, and *c.* 130 m in the vertical direction. Much better resolution is achieved for the upper source in Type-2 event, where the position is known to within 40 m. Larger uncertainties are attached to the positions of the lower sources in both events. The uncertainty for the lower source in Type-1 event amounts to *c.* 100 m, *c.* 70 m, and *c.* 130 m in the northwest–southeast, northeast–southwest, and vertical directions, respectively. Resolution is worse in the lower source in Type-2 event, where the uncertainty is *c.* 270 m in the northwest–southeast direction, *c.* 200 m in the northeast–southwest direction, and *c.* 240 m in the vertical direction. The northwest–southeast-trending error patterns reflect the lack of receiver coverage in the Sciara Del Fuoco. Within the 0.1% range encompassed by these error regions, the waveform fits are indistinguishable from one another, hence all the solutions obtained within this range may be viewed as equivalent.

Figures 10 and 11 show the waveform matches obtained with two point sources for Type-1 and Type-2 events, respectively. The fits include data from the entire network and are representative of the best fit source centroids (Fig. 9). The fits are excellent overall as demonstrated by the small values of residual errors listed in Table 1. Compared to the residual errors and AIC values obtained for the original point-source model, including six moment-tensor and three single-force components, the two-point source model composed of four cracks and six force components is a clear improvement. In addition to the well-matching waveforms in the B ring, we also note that some of the fits obtained on the T and M rings are further enhanced compared to the original fits shown in Figures 5 and 6. In the Type-1 event for example, the vertical and north components at receiver T1, vertical component at T5, north component at M4, east component at M5, and vertical components at M6 and M7 are all noticeably improved compared to the fits displayed in Figure 5. Similar improvements are observed in the vertical components in the top two rings in Type-2 event. Note that the vertical component at receiver B3 was not used in our inversion of waveforms for this event as this trace shows evidence of contamination by noise during this event.

To assess the robustness of the models elaborated for Type-1 and Type-2 events, we examined the following alternative models at the same source locations: (1) an upper source including two intersecting cracks and three single-force components, and lower source composed of two intersecting cracks only; (2) an upper source composed of two intersecting cracks only, and lower source including two intersecting cracks and three single-force components; and (3) two sources, each including two intersecting cracks, and no single-force components. The residual errors calculated with equation (2), and corresponding AIC calculated with equation (3) are listed in Table 1 for each model considered. The errors and AIC are both significantly larger than values obtained with our model in all three cases considered. These tests confirm that the two-point source model including four cracks and six force components is consistently better than any of these other models.

The source mechanisms associated with the fits in Figure 10 are illustrated in Figure 12. A striking aspect of the mechanisms imaged at the two point sources is the presence of dominantly vertical single-force components with common-looking source-time histories, except for a polarity reversal in one source compared to the other. The upper source (a) displays an initially downward force followed by an upward force, while an upward force followed by a downward force is manifest in the lower source (b). These force components compensate each other so that total momentum in the overall source volume is preserved. Note also that the waveform shape of the vertical force in the upper source remains generally consistent with the shape of the force obtained by Chouet *et al.* (2003) (Fig. 3a), or shape of the force derived through reconstruction of the original source mechanism (Fig. 7).

Both upper and lower sources include a dominant crack intersected by a subdominant crack. Comparing the orientations of the two cracks in the upper source with those shown in Figure 7, one observes that the main dipole component defining the orientation of the secondary crack in this source is markedly rotated compared to the dipole orientation seen in Figure 7. The dip of the dominant crack is also 8° steeper. These apparent rotations of the two cracks are the result of the coupling of the mechanisms of the upper and lower sources in the model.

The volume changes in Figure 12 are obtained from the amplitude of the principal dipole component representing each crack in each composite mechanism. These volume changes are estimated assuming $\mu = 7$ GPa and Poisson ratio $\nu = 1/3$ ($\lambda = 2\mu$) (Chouet *et al.* 2003). The dominant crack in the upper source carries a maximum volume change of 147 m^3, that is roughly 75% of the overall volume change in this source, while the subdominant crack carries the remaining 25% with a maximum volume change of 50 m^3. As in the single-point source solution (Fig. 7), the signal from the subdominant crack is delayed by a few seconds with respect to

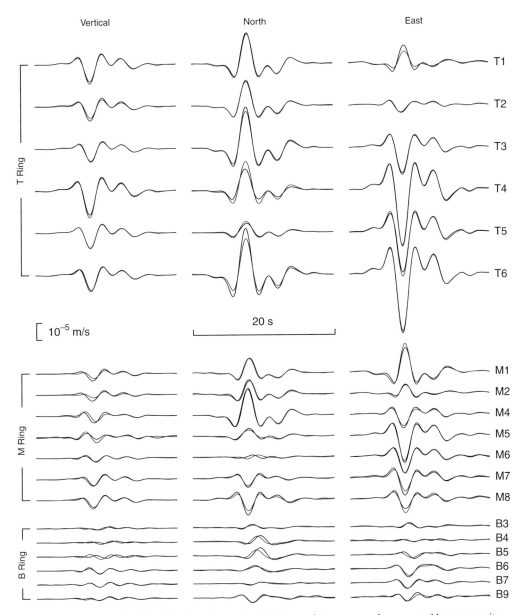

Fig. 10. Waveform match obtained for Type-1 event in which two point sources, each represented by a composite of two intersecting cracks and three single-force components, are assumed to represent the data recorded on the entire network. Black lines represent observed velocity waveforms, and red lines indicate synthetics. The station codes are indicated at the right and the components of motion are indicated at the top of the figure, respectively.

the main sequence observed in the dominant crack. This feature is consistent with a process initiating in the dominant crack and propagating into the subdominant crack. The dominant and subdominant cracks in the lower source show maximum volume changes of 135 m³ and 58 m³, respectively, representing 70% and 30% of the overall volume change in the lower source.

The traces of the four dyke elements imaged in our model for the Type-1 event are shown in map view in Figure 13. Grey swaths represent the uncertainty in position of the surface traces of the two dyke segments in the upper source. Yellow swaths mark the uncertainty in orientation of the hinge connecting the two dyke segments in the lower source, and uncertainty in position of the

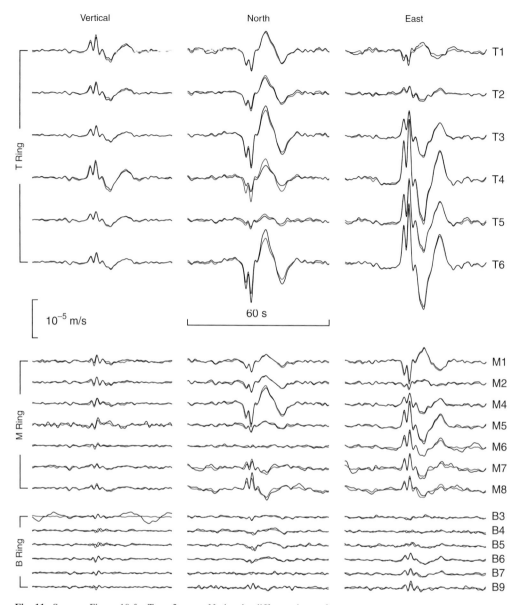

Vertical   North   East

Fig. 11. Same as Figure 10 for Type-2 event. Notice the different timescale.

main dyke segment in the lower source observed at the elevation of the upper source. Both grey and yellow swaths span the range of solutions corresponding to residual errors that are within 0.5% of the minimum error, thus confirming the robustness of our resolution of conduit orientation. The uncertainties in conduit orientations associated with this error range are given in Table 2. The source-time histories corresponding to these

solutions remain virtually identical to those shown in Figure 12.

The best-fitting model in Figure 12 shows that the dips of the dominant cracks in the two point sources are a mere 3° apart, with associated strikes differing by only 8°. The hypocentral positions of the two sources (Fig. 9) and dip of the dominant crack in the lower source are such that the extension of the latter crack to 440 m elevation

Table 1. *Residual error calculated with equation (2) and corresponding AIC calculated with equation (3) for the source mechanisms considered in our inversions of seismic data for Stromboli Volcano*

Source mechanism	Type-1 event		Type-2 event	
	Error, %	AIC	Error, %	AIC
Original point source model[†]	25.899	−32518	31.386	−121019
4 cracks, 6 forces	10.635	−74133	19.489	−192909
4 cracks, 3 forces at upper source	18.396	−58276	31.714	−135653
4 cracks, 3 forces at lower source	16.696	−63258	34.698	−120542
4 cracks only	23.329	−58364	35.487	−141340

Errors and AIC are based on data from the entire network.
[†]Model of Chouet *et al.* (2003).

intersects the position of the upper source nearly perfectly. The closely matching dips of the two dominant cracks point to a conduit that extends essentially straight from 80 m below sea level to the crater floor, 760 m above sea level. At the depth of 80 m below sea level, the conduit features a sharp corner leading into a dyke segment dipping 40° to the southeast. The upper dominant dyke segment, and deep segment below the abrupt corner both strike roughly northeast–southwest, parallel to the elongation of the edifice. At the level of the upper source, the conduit bifurcates. The surface trace of the main conduit segment intersects the northern vent area, while that of the subsidiary segment extends down the Sciara Del Fuoco and intersects the main dyke trace *c.* 170 m north of the northern vent area. The dip of the secondary dyke is shallower than the dip of the dominant dyke; thus, a gas slug rising along the upper wall of the dominant dyke is unlikely to be diverted from its path. Considering this fact, together with the position of the main dyke trace, we conclude that the latter dyke represents the main path of gas slug ascent in the upper conduit structure associated with Type-1 events.

The source-time history of the main dyke segment in the lower source starts with a slight deflation followed by a sequence of inflation, deflation, inflation that is similar in shape, but delayed by *c.* 1.14 s compared to the sequence observed in the dominant dyke segment at the upper source (Fig. 12). Assuming a straight conduit with length of about 560 m between the two sources, this implies a propagation speed of *c.* 490 m/s. Such low speed is consistent with the expected phase velocity of the dispersive crack wave associated with lower modes of crack oscillation (Chouet 1986; Ferrazzini & Aki 1987; Chouet 1988). In contrast, no delay is apparent in the onset of the forces at the lower source compared to the upper source. The vertical forces show roughly synchronous emergent onsets at the two sources, suggesting a

propagation speed between the two sources much faster than the crack wave associated with the volumetric source components. A time delay of 0.16 s between the two sources is expected for a pressure disturbance propagating in a rock with compressional wave velocity $V_p = 3.5$ km/s. Although not readily apparent in the emergent onsets of the vertical forces in Figure 12, such a small delay is certainly not inconsistent with our data. We may therefore conclude that the transmission of the force between the two sources occurs via the rock matrix itself.

The best-fitting source mechanisms obtained for Type-2 event are illustrated in Figure 14, and a map view of the dyke traces for this event is shown in Figure 15. The grey and yellow swaths in the latter figure mark uncertainties in crack orientations and represent all solutions producing residual errors that are within 0.5% of the absolute minimum error. The lower point source is positioned 800 m below and 170 m northwest of the upper point source. The volumetric components of the source mechanism at this location are represented by an inclined crack intersecting a sub-vertical crack whose upward extension crosses the upper source (Fig. 15). At 520 m elevation, the conduit includes a corner leading into two shallower-dipping dyke segments. Our best solution for the volumetric components representing the upper source points to a set of intersecting cracks whose orientations are similar to those seen in Figure 8. These cracks are represented by the two slanted planes in the right panel in Figure 14a. The marked corner linking the subvertical dyke segment in the lower source to the two shallower-dipping dykes at the upper source suggests that a third dyke segment may be required to refine the geometry defining the upper source. Analyses of residual errors and AIC for models including this third dyke segment at the upper source indicate that the inclusion of this segment yields slightly better fits, but only marginally

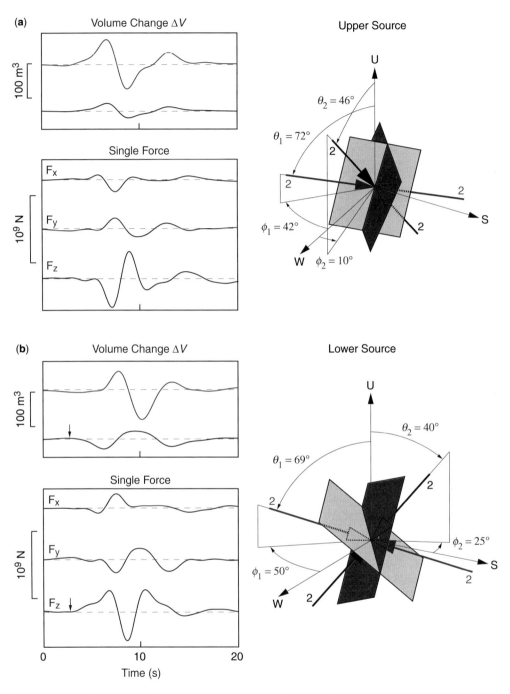

Fig. 12. Sources of Type-1 event. The two point sources are positioned at different depths in the volcanic edifice (see text for details) and each source consists of two intersecting cracks and three single-force components. Volume changes are colour-coded with the colours of the cracks they represent in each point source. (**a**) Upper source. (**b**) Lower source. Arrows mark the onset of deflation in the lower dyke (grey dyke in right panel) and synchronous start of the upward force.

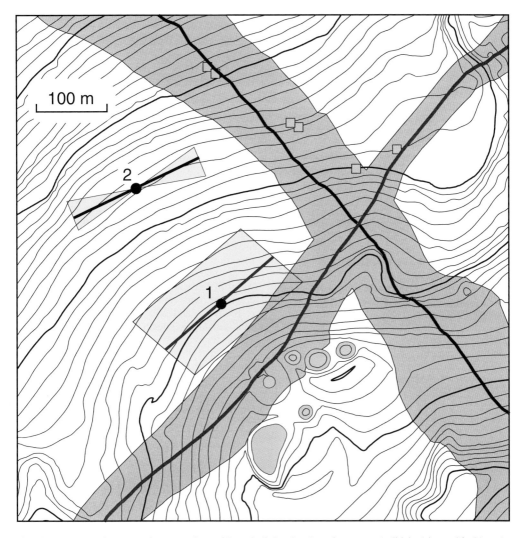

Fig. 13. Map view of upper northwest quadrant of Stromboli showing the point sources (solid dots) imaged for Type-1 event, and traces of individual dyke segments (coloured lines and associated yellow and grey swaths) composing each point source (upper source labelled 1, lower source labelled 2) in relation to the summit vents identified by contours filled in with green. Green squares show the positions of vents that were active in the Sciara Del Fuoco during the flank eruption in 2002–2003 (reproduced from Acocella *et al.* 2006). The blue line segment crossing the lower point source is the horizontal projection of the intersection of the two dykes composing the lower source. The red line segment crossing the upper point source represents the trace at 440 m elevation of the dominant dyke segment in the lower source. The red and blue lines, respectively, mark the surface traces of the dominant and subdominant dyke segments in the upper source. The grey and yellow swaths show the ranges of solutions producing residual errors in fitted waveforms that are within 0.5% of the minimum residual error.

lower AIC. Although such an element would seem to be required to provide a complete picture of the conduit geometry at the upper source, our data do not clearly require this element.

Compared to the solutions shown in Figure 13 for Type-1 event, the uncertainties in crack orientations are similar at the upper source, but significantly larger at the lower source in the Type-2 event (Fig. 15 and Table 2). The larger area swept at 520 m elevation by all the possible traces of the dominant dyke segment in the lower source (see wide yellow swath at level of upper source in Figure 15) reflects an increased sensitivity to changes in the dip angle of this dyke, stemming from a larger distance between the two sources in this event compared to Type-1 event. The

Table 2. *Crack orientations obtained for the best-fitting model and uncertainty (in parentheses) associated with models producing misfits that are within 0.5% of the best-fitting model*

	θ_1, °	ϕ_1, °	θ_2, °	ϕ_2, °
Type-1 event				
Upper source	72 (66–80)	138 (135–139)	46 (38–56)	190 (180–204)
Lower source	69 (64–75)	130 (126–133)	40 (20–50)	295 (280–305)
Type-2 event				
Upper source	51 (46–62)	142 (140–148)	58 (52–64)	109 (98–116)
Lower source	78 (73–86)	130 (115–140)	50 (20–70)	265 (240–300)

Crack orientation is defined by the vector normal to the crack plane with direction fixed by the polar angle θ and azimuth ϕ measured counterclockwise from east.

orientation of the subdominant dyke in the lower source is also less well defined in Type-2 compared to Type-1 event, mainly reflecting the weaker signals recorded on the B ring for Type-2 event (compare data in Figs 10 and 11). Possible azimuths for this lower dyke segment range over c. 60° from northwest–southeast to northeast–southwest, and associated dips range from 20 to 70° to the south (Table 2). Within the angular ranges listed in Table 2, the source-time histories display only minor variations from those shown in Figure 14.

The swaths encompassed by the surface traces of the two slanted dyke segments in the upper source (Fig. 15) show the northeast-trending dominant dyke offset 50–100 m east of the main southern vent and smaller secondary vents to the northeast, and eastwest-trending subdominant dyke bisecting the southern vent. The slanted hinge linking these two shallow segments of conduits intersects the surface 120 m east-northeast of the southern vent. The solutions represented by grey swaths display waveform matches that are indistinguishable from those associated with the minimum error represented by the coloured traces in Figure 15. Considering the areas swept by these swaths, we may envisage the hinge linking the two dykes as a preferred pathway for gas slugs ascending toward the southern vent, and we may further infer the main dyke as feeding not only the southern vent but also the adjacent vents to the northeast.

Interestingly, the main dyke trace imaged at the surface for the solution representing the minimum error (red line in Fig. 15) does not intersect the southern vent, nor does it actually intersect adjacent vents to the northeast. Rather, as mentioned above, the trace of this dyke is offset 50–100 m east of the line of vents along the main crater axis. This apparent incompatibility of our best solution with the vent positions may be easily resolved if one assumes that the conduit becomes steeper at shallow depth. Although near-surface conduit curvature cannot be assessed with our data, it would be consistent with a magma pathway following the steepening dip of a pre-existing sliding surface

activated during a past sector collapse in the Sciara Del Fuoco (Tibaldi 2001). Analogue models of volcano collapse certainly seem to support this idea (see Fig. 4c in Acocella 2005).

The maximum volume changes in Figure 14 are $\Delta V = 219 \text{ m}^3$ and $\Delta V = 68 \text{ m}^3$ in the shallow dominant and subdominant conduit segments, respectively, and $\Delta V = 84 \text{ m}^3$ in the lower dominant conduit segment, and $\Delta V = 19 \text{ m}^3$ in the lower subdominant segment. Assuming a straight conduit and distance of 818 m between the upper and lower sources, the delay of about 4.4 s in the peak amplitude of volume change in the subvertical dyke at the lower source compared to the dominant dyke at the upper source yields a propagation speed of c. 186 m/s, 2.6 times slower than the speed inferred between the two sources in Type-1 event. The slower speed in the Type-2 event is compatible with a slower wave speed of the crack wave expected for the longer period of the signal observed in Type-2 event compared to that of Type-1 event (Chouet 1986, 1988; Ferrazzini & Aki 1987). As in the Type-1 event, a dominantly vertical force accompanies the volumetric source components in both sources. The two forces share similar waveforms with opposite polarities, in accord with a conservation of linear momentum in the overall source volume. As in the Type-1 event, the synchronicity of these forces is consistent with a propagation path in the rock. Unlike the Type-1 event, however, the vertical forces in the Type-2 event start with a clear upward polarity in the upper source, and downward polarity in the lower source. Although only hinted at in the original solution obtained by Chouet *et al.* (2003) (Fig. 4a), and only marginally more apparent in the single-point source reconstruction (Fig. 8), this feature is unequivocal in our final model (Fig. 14).

The surface traces of the upper dyke segments in Type-1 and Type-2 events are shown superimposed on a map of the island of Stromboli in Figures 16a and 16b, respectively. The northeastern track of the dominant dyke and western track of the subdominant dyke in Type-2 event closely parallel the

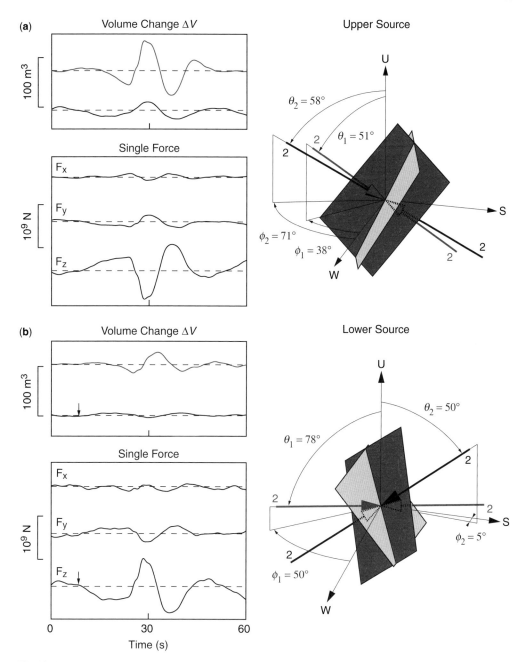

Fig. 14. Sources of Type-2 event. The two point sources are located at different depths in the volcanic edifice (see text for details), and each source is composed of two intersecting cracks and three single-force components. Volume changes are colour-coded with the colours of the cracks they represent in each point source. (**a**) Upper source. (**b**) Lower source. Arrows mark the onset of inflation in the lower dyke (grey dyke in right panel) and synchronous start of the downward force.

collapse scarps flanking the Sciara Del Fuoco to the north and west (Fig. 16b). Taken together, the dips of the two dykes (Fig. 14a) and close fit of the dyke traces to the Sciara boundaries both suggest that these feeder conduits may have become localized along the sliding surfaces activated during one of the major sector collapses affecting the northwest flank of Stromboli. Further support is obtained for

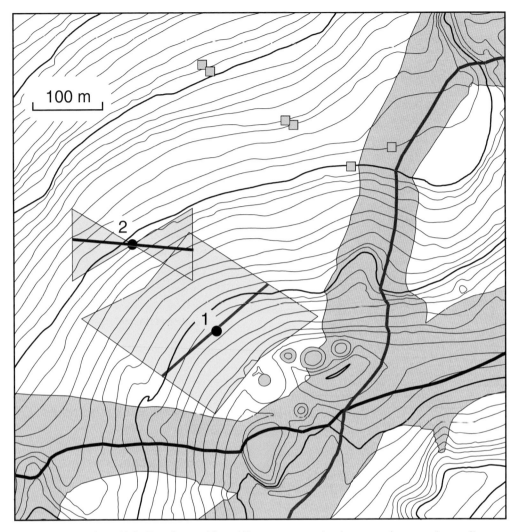

Fig. 15. Same as Figure 13 for Type-2 event. The red line segment crossing the upper point source (source 1) represents the trace at 520 m elevation of the dominant dyke segment in the lower source (source 2).

this idea if one assumes a steepening dip of these fracture planes near the surface as hypothesized above (Acocella 2005). In the Type-1 event, the main conduit appears to have developed along an en echelon fracture sub-parallel to the dominant fracture in Type-2 event (Fig. 16a). The northwest-striking subsidiary dyke imaged in this event is roughly orthogonal to the primary dyke and follows the anticipated trajectories of the maximum gravitational stress in the upper edifice (Gudmundsson 2002; Acocella & Tibaldi 2005). The surface trace of this secondary dyke lines up well with the positions of vents observed in December 2002 in the Sciara Del Fuoco. The surface traces of the primary dykes in Type-1 and Type-2 events

intersect roughly 300 m northeast of the northern perimeter of the summit crater, near the position of a vent active during summer 2003 (Acocella *et al.* 2006). The overlapping solutions obtained for the two dyke traces extending northeastward from the northern crater perimeter also coincide well with a fissure that opened in this area on 28 December 2002 (Acocella *et al.* 2006).

Implications of the model for Strombolian dynamics

Our analyses demonstrate the central role played by conduit geometry in controlling flow disturbances

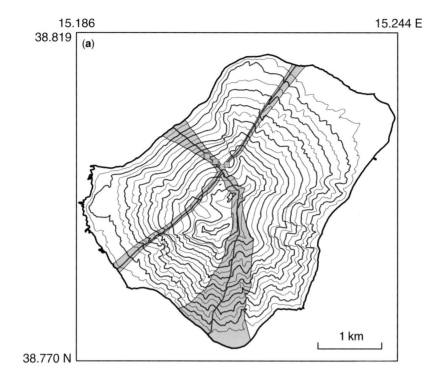

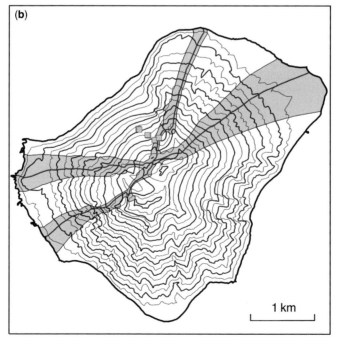

Fig. 16. Surface traces of upper dyke segments in the two types of events analysed in this paper. Vents observed in December 2002 and July 2003 are indicated by green squares. (**a**) Type-1 event. The surface trace of the subsidiary dyke is represented without accounting for the change in surface morphology resulting from the landslide on 30 December 2002. (**b**) Type-2 event. Note the close match between the dyke traces and collapse scarps flanking the Sciara Del Fuoco.

and also providing specific sites where pressure and momentum changes in the fluid can be effectively coupled to the Earth, or where elastic disturbances can feed back into pressure and momentum changes in the fluid. Linear momentum is exchanged between the source and the Earth mainly through vertical forces applied at the top and bottom boundaries of the overall source volume.

Two distinct conduit structures are identified, each representative of explosive eruptions from distinct vents located near the northern and southern perimeters of the summit crater. The main branch of conduit activated during eruptions at the northern vent is composed of a northeast-striking dyke dipping 72° northwest, and extending from the crater floor at 760 m elevation down to a depth of 80 m below sea level, where the conduit features a sharp corner. Below this depth, the imaged conduit consists of a northeast-striking dyke dipping 40° southeast. A subsidiary dyke segment branches off the main conduit at elevations near 440 m. The latter segment strikes approximately northwest and dips 46° west. The surface trace of the main dyke bisects two vents near the north-northwest edge of the crater, while that of the subsidiary segment extends down the northwest flank and intersects the main dyke trace 170 m northeast of the northern vent. A similar analysis carried out for eruptions at the southern vent images an uppermost conduit geometry composed of a northeast-striking dyke dipping 51° northwest, intersecting a west-striking dyke dipping 58° north. The main dyke intersects the surface slightly east of the alignment of the southern vent and adjacent orifices to the northeast, and the subsidiary dyke crosses the southern vent. At 520 m elevation, the two dykes merge into a sub-vertical dyke striking northeast, and at depth of 280 m below sea level the conduit features a second, more abrupt corner leading into a fracture dipping 50° south.

In the Type-1 event, the sharp corner in the conduit at depth of 80 m below sea level marks one site, and a bifurcation in the conduit 440 m above sea level marks another site where disruption in the liquid flow associated with a rising gas slug are strongly coupled to the encasing solid. There is no significant change in the main conduit direction at the upper bifurcation site; however, a marked increase in the main conduit aperture at this location, possibly associated with this bifurcation, may provide the geometry necessary to explain the observed pressure and momentum changes imaged at the upper source. In the Type-2 event, the corner in conduit at depth of 280 m below sea level, and the corner and bifurcation at elevation of 520 m, represent similar flow disruption sites.

The upper seismic source in the Type-1 event represents the intersection of the main dyke with a dyke remnant inclined 46° from the horizontal. This subsidiary dyke appears to be a permanent feature of the upper conduit that was momentarily forced open to the surface in response to excess pressurization during the vigorous activity of 2002–2003. The upper seismic source in the Type-2 event represents a bifurcation of the main conduit coupled with a marked change in conduit dip. This geometry suggests that a fluid disruption mechanism similar to that affecting Type-1 events may be operative at this location. The lower seismic sources imaged for Type-1 and Type-2 events both suggest the possible existence of an additional dyke segment representing the upward extension of the deepest segment of feeder dyke past this corner in the conduit (upper segment of grey dyke in the lower sources in Figs 12b and 14b). Were it filled with magma, this other segment would be thermally unstable and seal; however, if filled with supercritical water it may remain open. Therefore, if present this segment is likely to be filled with water since it will trap any water that enters it.

To gain insights into the origin of the initial pressurization and downward force observed in the source mechanism of the upper source in Type-1 events, we may turn to the laboratory simulations carried out by James *et al.* (2006). These experiments investigate the ascent of a slug of gas in a vertical liquid-filled tube featuring a flare that sharply doubles the cross-sectional area. The tube is instrumented with five pressure transducers mounted flush with the inner tube wall, and one accelerometer mounted on the exterior of the tube, and the whole assembly is free to move in the vertical direction. Detailed measurements of flow transients obtained by James *et al.* (2006) show that the transit of a gas slug through the tube flare involves complex changes in flow pattern. Specifically, the pressure and acceleration records point to strong transients observed at the time the slug clears the flare. A characteristic pinching of the slug tail is observed to occur synchronously with the pressure and acceleration transients, a picture consistent with the downward and inward motion of a liquid piston formed by the thickening film of liquid falling past the slug expanding in the wider tube. The sudden deceleration of the liquid annulus as it impinges the narrow inlet to the lower tube segment generates a pressure pulse in the liquid below the flare and also induces a downward force on the apparatus. These observations are consistent with the pressurization phase and initial downward force seen in the upper source imaged at Stromboli, and we infer that a similar funneling mechanism may be operative there.

The magnitude of the force at the upper source in the Type-1 event is *c.* 4×10^8 N, suggesting a pressure change on the order of 10 MPa (Chouet *et al.* 2003). It was previously demonstrated that forces and pressures of such magnitudes cannot be generated by the steady ascent of gas slugs up vertical or inclined conduits (James *et al.* 2004). Simple calculations also indicate that pressure changes due to foam collapse cannot simply account for the observed signals (Chouet *et al.* 2003). Indeed, a rough analysis of bubble dynamics by Chouet *et al.* (2003) suggests that only liquid inertia during a period of rapid gas-phase expansion is capable of producing superstatic pressures of such magnitudes. This conclusion is confirmed by the laboratory results of James *et al.* (2006), which indicate that the pressure changes induced by transient liquid motions during the passage of the slug through the conduit flare can be more than one order of magnitude larger than those recorded during quasi-steady state flow. As demonstrated by James *et al.* (2006), the resulting force observed under laboratory conditions satisfactorily scales to the magnitude of force seen in the volcanic environment. The repeatability of recorded pressure data and dependence of the magnitude of the pressure transient on slug size seen in these experiments are also in agreement with the observed spatio-temporal properties of VLP signals at Stromboli. Collectively, these laboratory results suggest that such transients represent strong candidates for VLP sources at Stromboli.

The magnitude of the downward force at the upper source in the Type-2 event is comparable to that seen in the Type-1 event, suggesting a similar liquid-piston mechanism underlying the Type-2 event. There is, however, a marked difference in the waveform of the vertical force in the Type-2 compared to the Type-1 event. In the Type-2 event, the onset of the signal shows a clear upward force while no such signal is observed in the Type-1 event. If the source of the downward force at the upper source in both events represents a rapid deceleration of liquid impinging the narrow neck of a flare in the conduit as assumed above, then one might expect that this transient ought to be preceded by a downward acceleration of the liquid mass. A weak upward acceleration was indeed observed to precede the dominant downward force in some of the laboratory results of James *et al.* (2006) (see Fig. 11e in James *et al.* 2006). This upward acceleration of the apparatus was attributed by these authors to a reduced vertical component of force exerted by the liquid on the flare shoulder resulting from the dynamical removal of mass and related pressure decrease as more liquid became wall supported. The synchronism of the initial upward force with conduit deflation in the Type-2 event (Fig. 14a) is consistent with these observations. Therefore, we may view

this force as the result of the downward acceleration of liquid flowing past the expanding slug, while the associated conduit deflation may be attributed to the pressure decrease caused by the dynamical removal of mass (i.e. more liquid becoming wall supported).

The viscous drag force applied by the falling liquid on the conduit walls may offer a possible explanation for this notable difference in behaviour in Type-1 and Type-2 events. Viscous coupling to the wall induces a downward force on the wall, and the net force on the wall is a combination of this force with the upward force resulting from the downward acceleration of the slumping liquid. This net force, however, is much smaller than the force generated once the liquid piston forms (see Fig. 14a; see also Fig. 11e in James *et al.* 2006). Viscous coupling is expected to become more important with long, rapidly expanding gas slugs where liquid velocities are high and significant volumes are held in the falling film. This idea may apply in the case of the Type-1 event where there is no detectable pressure or volume change prior to the downward force. Thus, viscous drag in such an event may have been large enough to cancel out the force induced by the downward acceleration of the liquid, while a lesser contribution from drag in the Type-2 event may have resulted in a net upward force.

In both events, the dominant downward force at the upper source is followed by an upward force of roughly equal magnitude, trailed by a decaying oscillation, all of which occur synchronously with the volumetric oscillations of the source. These dynamics reflect the coupled fluid-solid system's response, which is damped and eventually terminated by changing flow conditions. There are a variety of possible origins for the source rebound, including the elastic response of the conduit walls, compressibility of the bubbly liquid in the conduit, compressibility of a trapped steam pocket in the lower seismic source as hypothesized above, or perhaps even a remnant section of slug below the flare, if the slug happens to be segmented by the piston-like action of the falling liquid (James *et al.* 2006). The shapes of oscillations in Type-1 and Type-2 events do suggest that more than one mechanism may be operating at the upper source.

At the lower source in the Type-1 event, the start of the vertical force signal is synchronous with the onset of deflation of the lower dyke segment (see arrows above F_z trace and above volume change signature of grey dyke in left panels in Figure 12b). As the amplitude of the upward force increases, the lower dyke segment continuously deflates. During the same interval, the dominant dyke (red-coloured trace in Fig. 12b) remains in a slightly deflated state. The lower dyke reaches maximum deflation at the time the upper dyke segment goes through a transition from weak contraction to expansion, and the upward force reaches its peak amplitude *c.* 0.5 s later. This picture is consistent with a compression of the lower dyke synchronous with a downward acceleration of the liquid mass, both of which may be interpreted as reflecting increasing external pressure on the conduit wall induced by the elastic radiation from the downward vertical force at the upper source. Compression of the lower dyke segment proceeds unimpeded until this process is overprinted by the arrival of the much slower volumetric expansion signal from the upper source.

Although more subtle, a similar process is noted at the lower source in the Type-2 event, where a gentle expansion of the lower dyke segment is observed to occur simultaneously with the gradually increasing amplitude of the downward force (see arrows in left panels in Fig. 14b), while the upper dyke segment (red-coloured trace in Fig. 14b) remains essentially stable. The maximum expansion of the lower dyke coincides roughly with the onset of deflation in the dominant dyke. The lower dyke then slowly deflates while the downward force decays slightly. This is followed by a sharp transition leading into a sequence similar to that observed in the Type-1 event. The onset of the signal in the lower source in the Type-2 event thus appears consistent with a downward deceleration of the liquid mass and expansion of the bottom dyke in response to external forcing originating in the upward force at the upper source. The following sequence proceeds as in the Type-1 event and is similarly affected by the arrival of the volumetric signal from the upper source.

In both events, the early response of the lower source relative to the volumetric disturbance arriving from the upper source can, therefore, be interpreted as the passive response of the liquid to the movement of the conduit wall induced by elastic radiation from the force acting at the upper source. Accordingly, the different magnitudes observed in the volumetric responses of the lower conduit in the two events may simply reflect differing conduit lengths and distinct orientations of the bottom dykes in the two conduits. The volumetric response of the bottom dyke is smaller in the Type-2 event because of the greater distance separating the two sources in this conduit, and also because the steeper orientation of the bottom dyke makes it less sensitive to dilation or compression from the applied external force.

The picture emerging from these dynamics is that of an upper source representing an active fluid phase and passive solid phase, and a lower source representing an active solid phase and passive fluid phase. The overall source process may then be summarized as follows. A slug of supercritical water rises through the lower conduit corner (the lower seismic source). This slug is most likely only a few metres long and traverses this corner aseismically; past this corner, the slug then expands on its way to the upper conduit bifurcation (the upper seismic source). The transit time from the lower to the upper seismic source is probably in the range of 5–15 minutes (James *et al.* 2006), so that related changes in magmatic head are well beyond the capability of the broadband seismometers to detect. As it traverses the upper seismic source, the slug length has expanded to tens of metres and this slug is now seismically noisy. Gravitational slumping of the liquid occurs as the slug expands through a flare in the conduit at this location. The slumping liquid rapidly decelerates into the narrowing dyke neck, increasing the liquid pressure and inducing a volume expansion in the main conduit and its subsidiary branch. The rapid deceleration and associated pressurization of the liquid couple to the conduit walls via the flare shoulder and induce a downward vertical force on the Earth. The volumetric signal propagates along the conduit at the slow speed of the crack wave, the phase velocity of which is a function of wavelength, crack stiffness, and gas-volume fraction of the fluid (Chouet 1992). The force signal itself propagates in the solid at the much higher speed of the compressional wave in the rock matrix and arrives at the lower conduit corner well before the crack wave. At this corner, the downward displacement of the rock induced by the force acting at the upper source impinges the more horizontally oriented bottom dyke and acts to decrease the volume of this conduit segment. This segment, possibly including a steam-filled extension above the corner, essentially acts like a spring that absorbs the downward motion of the rock. This scenario is consistent with both the small volume change of the lower dyke segment, as well as its apparent early response. The subsequent conduit response then represents the combined effects of volumetric and force oscillations of this liquid/gas/solid system. Both event types generally fit this picture with some nuances that may reflect differences in viscous drag effects in the two events.

Conclusions

The four point sources imaged in our study provide an optimal description of Strombolian explosion dynamics in the VLP band and yield the elements for a detailed description of the uppermost 1 km of the plumbing system at Stromboli. Our analyses also illuminate the subterranean processes driving eruptions and clearly point to the key role played by the conduit geometry in controlling fluid motion and resultant processes. The observed volumetric changes reflect a sequence of pressurization–depressurization–repressurization, which may be interpreted as the result of a piston-like action of the liquid associated with the disruption of a gas slug transiting through discontinuities in the conduit direction and/or conduit aperture. Each discontinuity in the conduit provides a site where pressure and momentum changes resulting from flow processes associated with the transit of a gas slug through the discontinuity are coupled to the Earth, or where the elastic response of the conduit can couple back into pressure and momentum changes in the fluid.

Seismology alone cannot directly see into the conduit and resolve details of the actual fluid dynamics at the origin of the seismic source mechanism revealed by our analyses. To develop a better understanding of fluid behaviour responsible for VLP events, laboratory experiments are required to explore the links between known flow processes and the resulting pressure and momentum changes. The pressure and momentum changes generated under laboratory conditions may then be compared with the pressure and momentum changes estimated from the time-varying moment-tensor and single-force components imaged from seismic data, yielding clues about the physical flow processes linked to the seismic-source mechanism. The results obtained by James et al. (2006) demonstrate that direct links between the moment-tensor and single-force seismic-source mechanism and fluid-flow processes are possible and could potentially provide a wealth of information not available from seismic data alone.

We are grateful to Stephen Lane and Michael James for many enlightening discussions on two-phase flow dynamics. We also thank Anthony Finizola for providing his DEM of the Stromboli crater terrace. We are indebted to Robert Tilling, Gregory Waite, Minoru Takeo, and Rick Aster for helpful suggestions. This research was supported in part through a grant from the Consiglio Nazionale delle Ricerche, Gruppo Nazionale di Vulcanologia.

References

ACOCELLA, V. 2005. Modes of sector collapse of volcanic cones: insights from analogue experiments. Journal of Geophysical Research, 110, B02205, doi:10.1029/2004JB003166.

ACOCELLA, V. & TIBALDI, A. 2005. Dike propagation driven by volcano collapse: a general model tested at Stromboli, Italy. Geophysical Research Letters, 32, L08308, doi:10.1020/2004GL022248.

ACOCELLA, V., NERI, M. & SCARLATO, P. 2006. Understanding shallow magma emplacement at volcanoes: orthogonal feeder dykes during the 2002–2003 Stromboli (Italy) eruption. Geophysical Research Letters, 33, L17310, doi:10.1029/2006GL026862.

AKAIKE, H. 1974. A new look at the statistical model identification. IEEE Transactions on Automatic Control, AC-9, 716–723.

AKI, K. & RICHARDS, P. G. 1980. Quantitative Seismology. W. H. Freeman, New York.

AUGER, E., D'AURIA, L., MARTINI, M., CHOUET, B. & DAWSON, P. 2006. Real-time monitoring and massive inversion of source parameters of very long period seismic signals: an application to Stromboli Volcano, Italy. Geophysical Research Letters, 33, L04301, doi:10.1029/2005GL024703.

CALVARI, S., SPAMPINATO, L., LODATO, L. ET AL. 2005. Chronology and complex volcanic processes during the 2002–2003 flank eruption of Stromboli volcano (Italy) reconstructed from direct observations and surveys with a handheld thermal camera. Journal of Geophysical Research, 110, B02201, doi:10.1029/2004JB003129.

CHOUET, B. 1986. Dynamics of a fluid-driven crack in three dimensions by the finite difference method. Journal of Geophysical Research, 91, 13967–13992.

CHOUET, B. 1988. Resonance of a fluid-driven crack: radiation properties and implications for the source of long-period events and harmonic tremor. Journal of Geophysical Research, 93, 4375–4400.

CHOUET, B. 1992. A seismic source model for the source of long-period events and harmonic tremor. In: GASPARINI, P., SCARPA, R. & AKI, K. (eds) IAVCEI Proceedings in Volcanology, 3. Springer-Verlag, New York, 133–156.

CHOUET, B. 1996. New methods and future trends in seismological volcano monitoring. In: SCARPA, R. & TILLING, R. I. (eds) Monitoring and Mitigation of Volcano Hazards. Springer-Verlag, New York, 23–97.

CHOUET, B., HAMISEVICZ, N. & MCGETCHIN, T. R. 1974. Photoballistics of volcanic jet activity at Stromboli, Italy. Journal of Geophysical Research, 79, 4961–4976.

CHOUET, B., DAWSON, P., OHMINATO, T. ET AL. 2003. Source mechanisms of explosions at Stromboli Volcano, Italy, determined from moment-tensor inversions of very-long-period data. Journal of Geophysical Research, 108, 2019, doi:10.1029/2002JB001919.

CHOUET, B., DAWSON, P. & ARCINIEGA-CEBALLOS, A. 2005. Source mechanism of Vulcanian degassing at Popocatepetl Volcano, Mexico, determined from waveform inversions of very long period signals. Journal of Geophysical Research, 110, B07301, doi:10.1029/2004JB003524.

EKSTRÖM, G., NETTLES, M. & ABERS, G. A. 2003. Glacial earthquakes. Science, 302, 622–624, doi:10.1126/science.1088057.

FERRAZZINI, V. & AKI, K. 1987. Slow waves trapped in a fluid-filled infinite crack: Implication for volcanic tremor. Journal of Geophysical Research, 92, 9215–9223.

GUDMUNDSSON, A. 2002. Emplacement and arrest of sheets and dykes in central volcanoes. Journal of Volcanology and Geothermal Research, 116, 279–298.

JAMES, M. R., LANE, S. J., CHOUET, B. & GILBERT, J. S. 2004. Pressure changes associated with the ascent and bursting of gas slugs in liquid-filled vertical and inclined conduits. *Journal of Volcanology and Geothermal Research*, **129**, 61–82.

JAMES, M. R., LANE, S. J. & CHOUET, B. A. 2006. Gas slug ascent through changes in conduit diameter: laboratory insights into a volcano-seismic source process in low-viscosity magmas. *Journal of Geophysical Research*, **111**, B05201, doi:10.1029/2005JB003718.

KANAMORI, H. & GIVEN, J. W. 1982. Analysis of long-period waves excited by the May 18, 1980, eruption of Mount St. Helens — a terrestrial monopole? *Journal of Geophysical Research*, **87**, 5422–5432.

KANAMORI, H., GIVEN, J. W. & LAY, T. 1984. Analysis of seismic body waves excited by the Mount St. Helens eruption of May 18, 1980. *Journal of Geophysical Research*, **89**, 1856–1866.

KANESHIMA, S., KAWAKATSU, H., MATSUBAYASHI, H. *ET AL.* 1996. Mechanism of phreatic eruptions at Aso Volcano inferred from near-field broadband seismic observations. *Science*, **273**, 642–645.

KAWAKATSU, H. 1989. Centroid single force inversion of seismic waves generated by landslides. *Journal of Geophysical Research*, **94**, 12363–12374.

KAWAKATSU, H., KANESHIMA, S., MATSUBAYASHI, H. *ET AL.* 2000. Aso94: Aso seismic observation with broadband instruments. *Journal of Volcanology and Geothermal Research*, **101**, 129–154.

KUMAGAI, H., OHMINATO, T., NAKANO, M., OOI, M., KUBO, A., INOUE, H. & OIKAWA, J. 2001. Very-long-period seismic signals and caldera formation at Miyake Island, Japan. *Science*, **293**, 687–690.

KUMAGAI, H., MIYAKAWA, K., NEGISHI, H., INOUE, H., OBARA, K. & SUETSUGU, D. 2003. Magmatic dyke resonances inferred from very-long period seismic signals. *Science*, **299**, 2058–2061.

LEGRAND, D., KANESHIMA, S. & KAWAKATSU, H. 2000. Moment tensor analysis of near-field broadband waveforms observed at Aso Volcano, Japan. *Journal of Volcanology and Geothermal Research*, **101**, 155–169.

NISHIMURA, T., NAKAMICHI, H., TANAKA, S. *ET AL.* 2000. Source process of very long period seismic events associated with the 1998 activity of Iwate Volcano, northeastern Japan. *Journal of Geophysical Research*, **105**, 19135–19147.

OHMINATO, T. & CHOUET, B. A. 1997. A free-surface boundary condition for including 3D topography in the finite difference method. *Bulletin of the Seismological Society of America*, **87**, 494–515.

OHMINATO, T., CHOUET, B. A., DAWSON, P. B. & KEDAR, S. 1998. Waveform inversion of very-long-period impulsive signals associated with magmatic injection beneath Kilauea volcano, Hawaii. *Journal of Geophysical Research*, **103**, 23839–23862.

OHMINATO, T., TAKEO, M., KUMAGAI, H. *ET AL.* 2006. Vulcanian eruptions with dominant single force components observed during the Asama 2004 volcanic activity in Japan. *Earth, Planets and Space*, **58**, 583–593.

OHMINATO, T. 2006. Characteristics and source modeling of broadband seismic signals associated with the hydrothermal system at Satsuma–Iwo Jima volcano, Japan. *Journal of Volcanology and Geothermal Research*, **158**, 467–490.

PASQUARÉ, G., FRANCALANCI, L., GARDUÑO, V. H. & TIBALDI, A. 1993. Structural and geological evolution of the Stromboli Volcano, Aeolian Islands, Italy. *Acta Vulcanologica*, **3**, 79–89.

TAKEI, Y. & KUMAZAWA, M. 1994. Why have the single force and torque been excluded from seismic source models? *Geophysical Journal International*, **118**, 20–30.

TIBALDI, A. 2001. Multiple sector collapses at Stromboli Volcano, Italy: how they work. *Bulletin of Volcanology*, **63**, 112–125.

UHIRA, K. & TAKEO, M. 1994. The source of explosive eruptions at Sakurajima Volcano, Japan. *Journal of Geophysical Research*, **99**, 17775–17789.

UHIRA, K., YAMASATO, H. & TAKEO, M. 1994. Source mechanism of seismic waves excited by pyroclastic flows observed at Unzen Volcano, Japan. *Journal of Geophysical Research*, **99**, 17757–17773.

Trends in activity at Pu'u 'O'o during 2001–2003: insights from the continuous thermal record

EMANUELE MARCHETTI[1,2] & ANDREW J. L. HARRIS[2]

[1]*Dipartimento di Scienze della Terra, Università di Firenze, via G. La Pira, 4 – 50121 Firenze, Italy (e-mail: marchetti@geo.unifi.it)*

[2]*Hawaii Institute of Geophysics and Planetology, School of Oceanography and Earth Science Technology, University of Hawaii, 1680 East-West Road, Honolulu, Hawaii 96822, USA*

Abstract: A permanent thermal monitoring system deployed on the north rim of Pu'u 'O'o crater (Kilauea, Hawaii) provided an 811-day-long data-set spanning March 2001–December 2003. These data allowed us to characterize three emission styles from vents on the crater floor: lava flows, sustained degassing and gas-piston events. Lava flows were recorded as sudden increases in temperature followed by smooth and relatively long-lasting decreases as the lava cooled. Sustained degassing was associated with persistently high levels of thermal signal and was the most common signal type. Finally, gas-piston events were all preceded by marked reductions in temperature (due to diminished degassing) and were marked by abrupt increases (due to the arrival of a gas jet) followed by 50–300 second waning phases. Lava flow occurrence, maximum temperature recorded during degassing, gas-piston thermal amplitude, occurrence and waveform all showed coupled, systematic changes through time. This implies modification of a common source process, and may be a result of a slight change in the magma level beneath the crater so as to modify the conduit geometry/primary degassing pathways, and hence gas collection and release processes, as well as slug ascent dynamics.

Pu'u 'O'o is a cinder-and-spatter cone on the East Rift Zone of Kilauea Volcano (Hawaii), which formed during the Pu'u 'O'o-Kupaianaha eruption (Wolfe *et al.* 1988; Heliker *et al.* 2003). This long-lasting eruption began in 1983 with the opening phases being characterized by a series of fire-fountain events at Pu'u 'O'o (Wolfe *et al.* 1988). Since late 1992, the eruption has been mainly effusive with channel and tube-fed lava flow activity from flank vents on Pu'u 'O'o, creating a lava shield surrounding Pu'u 'O'o and an extensive tube-fed lava-flow-field (Heliker & Mattox 2003). At the same time, activity within the Pu'u 'O'o crater has been characterized by lava flows and lava pond formation (e.g. Barker *et al.* 2003), as well as degassing from a number of open vents located on the crater floor (Heliker *et al.* 2003).

Degassing at Kilauea has been monitored throughout the eruption (e.g. Greenland 1986; Casadevall *et al.* 1987; Sutton *et al.* 2001). Between 1979 and 1997, Kilauea released 9.7×10^6 t of SO_2 (Sutton *et al.* 2001). Most of this emission (8×10^6 t) was accounted for by emissions from Pu'u 'O'o (Sutton *et al.* 2001). Gas monitoring using COSPEC-style instruments allows a precise quantification of degassing rates, but does not allow investigation of the short-term (seconds-to-minute-long) variations in emission during individual degassing events at particular vents. Degassing proceeds at Pu'u 'O'o both as continuous, sustained degassing, and as discrete gas-jetting or gas-piston events. Both styles of emission contribute to the total gas-flux, and their identification and measurement allow for tracking of changes in the dynamics and behaviour of the shallow system (e.g. Barker *et al.* 2003; Johnson *et al.* 2005).

Gas-piston activity was first described at Kilauea during the 1969–1974 Mauna Ulu eruption, where gas-piston events consisted of a sudden release of over-pressurized gas lasting 1–2 minutes (Swanson *et al.* 1979; Tilling *et al.* 1987). In the Mauna Ulu case, each event was preceded by an increase in the magma level in the conduit, over a period of 15–20 minutes, followed by lava overflow from the vent onto the crater floor (Swanson *et al.* 1979). Activity culminated in vigorous bubbling and gas-jetting, after which lava typically drained back into the conduit (Swanson *et al.* 1979). An array of seismometers deployed close to Pu'u 'O'o in 1988 lead to the first seismic investigation of gas-piston events (Chouet & Shaw 1991; Ferrazzini *et al.* 1991). Seismic data showed that gas-piston events at that time appeared to repeat regularly every 10–20 minutes and were characterized by an emergent seismic signal with a spindle-shaped amplitude envelope (Ferrazzini

From: LANE, S. J. & GILBERT, J. S. (eds) *Fluid Motions in Volcanic Conduits: A Source of Seismic and Acoustic Signals.* Geological Society, London, Special Publications, **307**, 85–101.
DOI: 10.1144/SP307.6   0305-8719/08/$15.00 © The Geological Society of London 2008.

et al. 1991). Seismically, while the onset of a gas-piston event corresponded to the bursting of gas bubbles at the surface of a pre-formed lava pond, the maximum seismic amplitude corresponded to the paroxysmal phase, which was followed immediately by lava drain-back (Chouet & Shaw 1991; Ferrazzini *et al.* 1991).

Two possible source mechanisms have been proposed for gas-piston events. On the basis of events observed at Mauna Ulu, gas-piston activity was first ascribed to gases accumulating under a relatively impermeable magma layer (Swanson *et al.* 1979). This first caused the rise of the column and, once the strength of the overburden was exceeded, culminated in gas release by gas-jetting (Swanson *et al.* 1979; Barker *et al.* 2003; Johnson *et al.* 2005). Several years later, Vergniolle & Jaupart (1990) suggested the growth and collapse of foam as a possible source mechanism for gas-piston events at Kilauea. Within this model, exsolved gas accumulates at the roof of a shallow magma body and, once a critical height of the foam is reached, the foam collapses and generates a gas-slug, which ascends the conduit and bursts at the magma free-surface releasing a jet of over-pressurized gas. Such a model has been applied to

several basaltic volcanoes to explain intermittent explosive/degassing activity (e.g. Vergniolle & Jaupart 1990; Vergniolle & Brandeis 1996; Vergniolle *et al.* 2004).

We analyse here activity at Pu'u 'O'o between March 2001 and November 2003, as recorded by three thermal infrared thermometers installed on the north rim of Pu'u 'O'o (Harris *et al.* 2003, 2005). The temporal resolution (0.5 Hz sampling rate) and length (*c.* three years) of the data-set allows detailed tracking of the thermal expression of, as well as temporal trends in, activity at targeted vents within Pu'u 'O'o. We thus use these data to investigate the thermal signature from, and occurrence of, lava flows and gas-piston events through time, as well as their relationships with the thermal signature associated with sustained degassing.

The Pu'u 'O'o permanent thermal system

Thermal data used here were provided by three infrared thermometers deployed during October 2000 on the north rim of Pu'u 'O'o crater (Fig. 1). The instruments consist of three Omega™

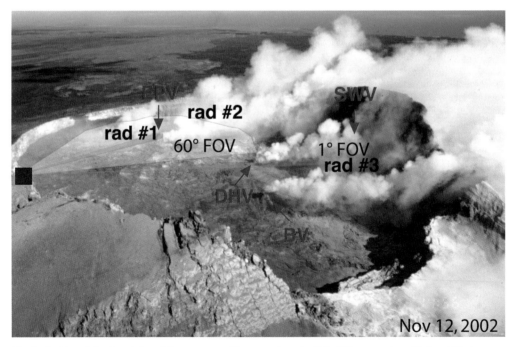

Fig. 1. Position of the three infrared thermometers (blue square) on the Pu'u 'O'o crater rim with 1 and 60 degree FOVs highlighted and marked rad #1, #2 and #3. The picture was taken on November 12, 2002 and shows an oblique aerial view of Pu'u 'O'o from the east. The FOV is representative for the target of the three thermal sensors in May 2003. Some of open degassing vents are marked in the figure (DHV: Drain Hole vent, BV: Beehive vent, SWV: South Wall vent, EPV: East Pond vent).

(OS550 series) infrared thermometers sensitive in the 8 to 14 micron waveband. The instrument reading has a sensitivity of $1 \, mV/degree$ in the temperature range -13 to $1371 \, °C$, with an accuracy of 1% and a response time of 0.25 s. A 16-bit A/D converter samples the continuous analog thermal data provided by the three instruments at 0.5 Hz (Harris *et al.* 2005).

As described by Harris *et al.* (2005), the deployment of two sensors with a field of view (FOV) of $1°$ and a third sensor with a FOV of $60°$ allows a complete and detailed thermal analysis of a large sector of the crater floor, as well as at individual vents targeted by the $1°$ FOV instruments. During the investigation period, Pu'u 'O'o contained a series of open vents, whose morphology, position and activity style changed with time. Accordingly, the sensors were retargeted several times (Table 1) in order to record activity at the most active vents (Harris *et al.* 2005). The two $1°$ sensors, having a *c.* 2.5-m-wide FOV at a distance of 150 m, can target individual vents, typically 4 m in diameter, and are able to detect degassing, gas-piston events and lava flows from the vents as elevated thermal signals (Harris *et al.* 2003, 2005; Johnson *et al.* 2005). The $60°$ instrument, whose FOV corresponds to an area *c.* 170 m in diameter at a distance of 150 m, targets a large part of the crater floor. It thus provides an overview of thermal activity across a large section of the crater (Fig. 1), and is able to detect the thermal expression of significant eruptive events from all vents within its FOV.

Temperature data provided by the infrared thermometers will not be absolute surface temperatures, but is the integrated temperature for an instrument FOV that may contain several thermal components. The recorded temperature is thus a function of four parameters, these being (i) the temperature and fractional coverage of the FOV by each thermal component (active lava, lapilli, vapour, ambient surfaces) inside the FOV; (ii) emissivity of the different objects; (iii) path length effects (plume condensation, atmospheric absorption, emission and transmission); and, in our case, (iv) a window effect. Only the latter effect can be corrected for, where the window fitted to the front of the protective box within which the instrument is housed absorbs up to 30% of the signal. Thus although the data cannot provide a precise quantitative measure of the volcanic activity, they do allow us to detect and define thermal transients related to discrete events (such as lava flows and gas-piston events) and to track the general thermal trends through time.

The permanent thermal system provided a near-continuous record of activity within Pu'u 'O'o between October 2000 and December 2003. Problems with the telemetry caused the system to be taken off-line during January 2004 (Harris *et al.* 2005).

Damage and transmission problems also caused occasional (1–60 day-long) data gaps during the operational lifespan of the system. However, the system functioned for *c.* 70% of the 1188-day-long period, providing an 811-day-long data-set capable of tracking the temporal evolution of thermal activity at Pu'u 'O'o. Johnson *et al.* (2005) consider data for gas-piston events for a 35-day-long period covering June and July 2002. In this paper, we define and consider all event types recorded in the full data-set, focusing on the period spanning March 2001 to December 2003. During this period, there are three major data gaps (November–December 2001, July–August 2002 and February–March 2003). Data provided by the $60°$ FOV instrument were also not available between June and November 2002 because the instrument was damaged and required removal for repair at that time.

Thermal signals recorded at Pu'u 'O'o

Three types of thermal signal were recorded at Pu'u 'O'o during the study period, each of which could be associated with a specific style of activity: lava flow events, sustained degassing and gas-piston events.

Lava flows

Lava flows within Pu'u 'O'o were recorded by the infrared thermometers as a sudden increase in temperature when the high temperature flow entered the instrument FOV. This was followed by a smooth and relatively long-lasting decrease as the flow cooled (Fig. 2a). Peak thermal values recorded by the $60°$ FOV instrument during lava flow events ranged between 100 and $300 \, °C$ (Table 2). The duration of the elevated signal was a function of flow duration. The thermal decay, which lasted generally 1–18 hours, often departed from the expected (steady, logarithmically decaying) cooling curve for a lava flow (Hon *et al.* 1994; Keszthelyi & Denlinger 1996), being complicated by multiple lava pulses or degassing events at the effusive vent. These secondary events overprinted the cooling trend from the initial lava flow event (Fig. 2a). The different amplitude and waveforms for the $1°$ FOV and the $60°$ FOV thermometers may be ascribed to the different targeting of the sensors (Table 1), where the $60°$ FOV sensor, covering a large portion of the crater floor, better describes the lava flow activity and detects the onset of flow emission.

Sustained degassing

Over the three-year-long observation period, a persistently high level of thermal signal in the $1°$ FOV sensor data (Fig. 2b) was the most

Table 1. Targeting of the three thermal infrared thermometers at Pu'u 'O'o following re-installation on 22 February 2001

Date	Sensor	Status	Target	Thermal signal notes
22/02/01	Rad #1	Installed	Central Vent	Rad #3 targeted to sky in West Gap and hence is recording ambient (sky) temperature
	Rad #2 (60°)	Installed	Central Vent	
	Rad #3	Installed	West Gap	
02/05/01	Rad #1	Retarget	July Pit	Rad #1 retargeting explains small increase in Rad #1 signal at this time (Fig. 4)
	Rad #2 (60°)	No Change	Central Vent	
	Rad #3	No Change	West Gap	
14/07/01	Rad #1	Retarget	Beehive Vent	While retargeting of Rad #1 explains small increase in Rad #1 signal at this time, retargeting of Rad #3 away from the ambient (W. Gap) target onto a hot vent in the July Pit is apparent from a marked increase in Rad #3 signal (Fig. 4)
	Rad #2 (60°)	No Change	Central Vent	
	Rad #3	Retarget	July Pit (W. Vent)	
12/10/01	Rad #1	No Change	Beehive Vent	Collapse of the July pit floor removed the hot W vent target from the Rad #3 FOV; the crash in Rad #3 signal during mid-July times this event. Retargeting to the new hot vent is apparent from a signal increase at this time (Fig. 4)
	Rad #2 (60°)	No Change	Central Vent	
	Rad #3	Retarget	July Pit (E. Vent)	
24/03/02	Rad #1	No Change	Beehive Vent	Rad #3 retargeted onto S. Wall vent during lava flow emission. Rad #3 signal before and after re-target indicates that the prior target (July Pit) and new target (S. Wall Vent) were active with lava flows. Rad #1 signal also shows lava flow activity from Beehive Vent at this time (Fig. 4)
	Rad #2 (60°)	No Change	Central Vent	
	Rad #3	Retarget	S. Wall Vent	
18/06/02	Rad #1	Retarget	Drain Hole Vent	That both 1° sensors are now targeting the same vent causes both signals (Rad #1 and #3) to suddenly increase in an identical fashion at this time (Fig. 4). We now have the configuration and gas-piston activity described in Johnson et al. (2005) for June–July 2002
	Rad #2 (60°)	Removed	Water Damage	
	Rad #3	Retarget	Drain Hole Vent	
12/11/02	Rad #1	No Change	Drain Hole Vent	Rad #3 retarget apparent as slight increase in Rad #3 signal. Signal levels, though, similar to those recorded at previous target (Drain Hole Vent) and significantly higher than during previous S. Wall targeting by Rad #3 during 24/03/02–18/06/02 (Fig. 4)
	Rad #2 (60°)	Installed	East Pond Vent	
	Rad #3	Retarget	S. Wall Vent	
08/03/03	Rad #1	No Change	Drain Hole Vent	Collapse of a portion of Pu'u 'O'o's south wall causes S. Wall Vent to become buried and blocked. Crash in Rad #3 in mid-January 2003 times the collapse and burial of this vent. Retargeting to the new hot vent is apparent from post-March Rad #3 signal increase (Fig. 4)
	Rad #2 (60°)	No Change	East Pond Vent	
	Rad #3	Retarget	East Pond Vent	
22/07/03	Rad #1	No Change	Drain Hole Vent	
	Rad #2 (60°)	Retarget	Drain Hole Vent	
	Rad #3	No Change	East Pond Vent	

These targeting notes correct a number of errors in those given by Harris et al. (2005), where Rad #1, #2 and #3 relate to the up-rift, middle and down-rift sensor naming convention of Harris et al. (2005), Rad #2 being the 60° FOV sensor.

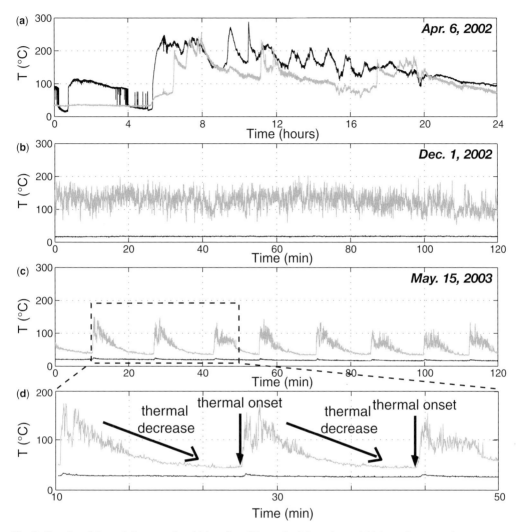

Fig. 2. Samples of thermal signatures for: (**a**) lava flow (**b**) sustained degassing and (**c**) intermittent gas piston events recorded at Pu'u 'O'o crater by 1° (grey) and 60° (black) FOV IR-thermometers. (**d**) Detail of three gas-piston events showing the decrease of temperature to its baseline value before each thermal onset. The thermal records of the 60° FOV thermometer for subplots b–d have been magnified by a factor of five.

common thermal signal recorded. Sustained apparent temperature of up to 250 °C is mostly the result of a sustained degassing process, as confirmed by visual observations during many field visits to Pu'u 'O'o, where thermal radiation from the hot gas and vapour emitted and/or from rock heated by degassing might explain the recorded thermal radiation. The 60° FOV was too large to detect the thermal signal associated with degassing, where such a small and localized thermal source, being a hot vent of just a few square metres in area, is spatially too small to have an effect on the integrated temperature for

such a large FOV (10's of thousands of square metres); the signal being dominated by the cold ambient background.

Gas-piston events

A peculiar feature of degassing activity at Pu'u 'O'o are the gas-piston events (Fig. 2c). Field observations revealed that these events typically involved emissions, from *c*. 4-m-wide vents, of *c*. 10-m-high incandescent (blue) gas jets with rare ballistic material which, when present, consisted of incandescent lapilli. Gas-piston events at Pu'u 'O'o

Table 2. *Lava flows recorded by the thermal system at Pu'u 'O'o 2001–2003, specifying the date (day, month, year), onset time, duration, maximum apparent temperature provided by the infrared thermometers and sensor #, which best detected the events*

Date	Onset time	Duration (hours)	Max. temp. (°C)	Rad #
3-03-2001	12:21	8	100	2
26-04-2001	11:32	3.8	85	2
12-05-2001	10:20	8.5	257	2
3-01-2002	13:02	2.58	128	2
10-02-2002	11:39	6.3	144	2
11-02-2002	13:40	2	111	2
20-02-2002	14:34	2.9	60	2
5-03-2002	9:32	5	282	3
20-03-2002	10:42	7	314	3
20-03-2002	20:51	2.6	224	3
21-03-2002	5:44	4.2	276	3
30-03-2002	5:01	1.2	120	3
31-03-2002	9:40	4.7	243	3
31-03-2002	15:23	1.2	197	3
31-03-2002	18:20	1.45	210	3
1-04-2002	10:58	2.26	219	1
6-04-2002*	5:20	18	288	2
7-04-2002	4:26	1	197	3
7-04-2002	6:03	1.6	227	3
10-04-2002	13:11	4.4	252	1
11-04-2002	9:16	2.9	282	1
14-04-2002	12:54	2.24	221	1
24-04-2002	7:57	6.05	245	1
25-04-2002	5:52	0.9	238	1
11-07-2002	13:06	1.44	375	1
11-07-2002	15:21	1.28	341	1
12-07-2002	10:24	1.7	296	1
23-01-2003	11·49	3.44	158	1
5-11-2003	13:07	5.7	186	2
Min		0.9	60	
Max		18	300	
Mean		3.92	182	
Std		3.47	70	

*This is the most significant lava flow inside Pu'u 'O'o crater during our investigation period, and results from superposition of at least 10 sequential lava pulses (Fig. 2a).

were recorded as abrupt increases in temperature, which then commonly waned over 50–300 second periods (Fig. 2c). Given the small dimension of the process, the gas-piston activity is best described by 1° FOV thermometers (with peaks of 50–150 °C), whereas it is barely detected by 60° FOV sensors (Fig. 2c,d), the contribution to the cooler surrounding material within the instrument FOV leads to a smaller thermal anomaly. Thermal observations reveal that gas-piston events are usually arranged as repeating events, thus resulting in a pulsed regime of degassing (see Johnson *et al.* (2005) for detailed classification of gas-piston events using thermal data). What deserves special attention is the decrease in the thermal level before each gas piston event, pointing to a decline in activity, probably related to a reduction of degassing, before the onset (Fig. 2d).

This behaviour seems to be confirmed by simultaneously collected seismic and thermal records during gas-piston events (Fig. 3), where a reduction of the envelope of seismic tremor is recorded before the seismic transients. Seismic transients of gas-piston events were recorded using a Guralp CMG-40T broadband seismometer during a short, temporary deployment in November 2002, with the instrument being co-located with the three infrared thermometers on the north rim of Pu'u 'O'o (Fig. 1). Recorded transients were similar to those previously recorded at Pu'u 'O'o by Ferrazzini *et al.* (1991), with the seismic transient being simultaneous with smooth and long-lasting thermal transients, suggesting that the gas-jetting activity, coincident with the maximum seismic amplitude (Ferrazzini *et al.* 1991), emits hot gas and fragments that then entered the infrared thermometer FOV.

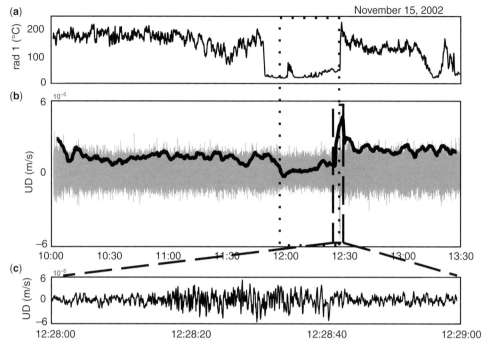

Fig. 3. 3.5-hour-long sample of (**a**) thermal and (**b**) seismic data of a gas-piston event at Pu'u 'O'o. In accordance with the reduction of temperature before the thermal onset, reduced amplitude of seismic tremor (dotted box) is recorded for *c.* 30 minutes before the seismic transient (b). This is better shown by the root mean squared amplitude of the seismic record (black line, magnified here by a factor of five) calculated every five seconds on 60-second-long sliding windows. (**c**) The 60-second-long detail of seismic transient (dashed box in subplot b) points to an emergent onset and a cigar-shaped envelope, in accordance with previous seismic observations of gas-piston activity at Pu'u 'O'o (Ferrazzini *et al.* 1991).

General thermal levels and lava flows

Clouds/fog and condensation of the gas plume during poor weather will reduce visibility and dampen the thermal signal. Thus, given the effect of viewing conditions on the recorded thermal data, we do not attempt to retrieve absolute temperatures. Instead, we use the maximum temperature recorded by the IR thermometers over a 10-hour-long sliding window, with 90% (nine hours) superposition, (T_{max10} to assess any systematic changes in thermal behaviour (Fig. 4). Maximum temperatures selected in this way should reflect measurements made during the best possible weather and observation conditions within each time window, thus reducing the weather and viewing effects. T_{max10} values were then smoothed over a two-day-long period (i.e. using a five-point running mean) to further reduce the effect of external (weather-related) variations and to isolate the thermal fluctuations due to volcanic activity.

Analysis the 60° FOV data revealed generally low levels of T_{max10} interrupted by a few high-amplitude, but short-lived, periods (Fig. 4b). Because the integrated temperature from the large area targeted by the 60° FOV instrument was dominated by the ambient background, trends in T_{max10} from this sensor were not related to volcanic activity. Instead, they mainly reflected non-volcanic influences, such as diurnal solar heating and seasonal heating effects. This trend was, however, interrupted by a number of spikes caused by episodes of lava effusion onto the crater floor (Fig. 4b). Lava flows (due to their extent) can be detected by the 60° FOV sensor and are thus apparent as spikes in the 60° FOV thermal records. Twenty-nine lava flow events, of between 0.9 and 18 hours in duration, were detected thermally during the study period (Table 2). They were particularly common in March–April 2002 (Fig. 4b), when at least 17 different lava flows (i.e. 60% of all lava flow events recorded) were detected by the thermal system. Among them a major lava flow on April 6, 2002 had a duration of *c.* 18 hours and comprised at least 10 sequential lava pulses (Fig. 2a).

In contrast, data from the two 1° FOV ther- mometers allowed detailed analysis of thermal activity at single vents within Pu'u 'O'o, revealing differential thermal levels for different vents and prolonged phases characterized by distinct (low, moderate, high or variable) T_{max10} levels (Fig. 4a,c). Figure 4 shows the variations of T_{max10} during the three-year-long investigation period. Here, it is evident that changes of the recorded T_{max10} are often related to retargeting of the sensors, with thermal levels changing as the sensor is moved to a new, hotter vent. Some of these retargeting events were forced by collapse or burial of an open vent by a crater wall collapse (Table 1). This caused the recorded temperature to drop abruptly (Fig. 4), as the vent dropped out of the bottom of the field of view (during collapse) or was replaced by a pile of blocks at ambient

temperature (in the case of burial). However, at the same time Figure 4 reveals long-term fluctu- ations of T_{max10} at single vents and across multiple vents. This points to a complex degassing process acting at Pu'u 'O'o crater, where changing thermal levels in the longer term point to a variation in the location of preferential degassing pathways within the crater (as some vents die out and new vents start up), while thermal fluctuations over shorter time periods indicate time-varying gas fluxes at single vents.

Despite the evidence that the thermal activity at Pu'u 'O'o is affected by spatial complexities and temporal fluctuations, the three-year-long investi- gation presented here leads to an identification of three general thermal periods at this multivent system, characterized by a change from a low level (c. 50 °C) during March 2001–May 2002,

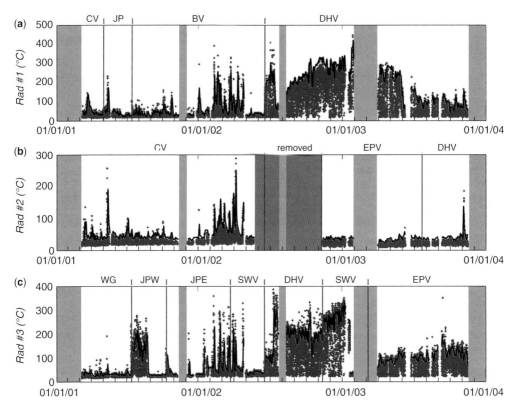

Fig. 4. Thermal activity at Pu'u 'O'o during March 2001–November 2003. Each point (blue) corresponds to the maximum value in the thermal records for one-minute-long sequential windows, while the black line corresponds to the maximum value of the thermal records over a 10-hour-long sliding window with 90% superposition (T_{max10}) and smoothed over two-day-long windows. Grey areas represent gaps in the data stream, and the dark grey area (b) represents the failure of the 60° FOV instrument. Vertical red lines indicate days when re-orientation of the sensors occurred as listed in Table I. Targeted vents by each sensor are specified above each trace (DHV: Drain Hole vent, BV: Beehive vent, JP: July pit, JPW: July pit West, JPE: July pit East, CV: Central vent, WG: West gap, SWV: South Wall vent, EPV: East Pond vent). While the 1° FOV sensors (a, c) provide a detailed record of degassing activity at single vents, the 60° FOV (b) is more useful for recording lava flow activity on the crater floor.

when all targeted vents had a generally low thermal level, to a high level (250–300 °C) between June 2002 and May 2003, when activity at two degassing vents (Drain Hole and South Wall vents) showed a sustained high thermal level. After this period of high levels, T_{max10} decreased again at the targeted vents, to remain stable at c. 100–150 °C until December 2003.

Given the variation in daily visibility and local morphology of the targeted vents, it is necessary to focus on long-term fluctuations of the thermal levels at Pu'u 'O'o crater provided by the three infrared thermometers (Fig. 4). As an example, we focus on the 30-day-long period following June 18 2002, when radiometers #1 and #3 were re-oriented to target the Drain Hole vent. This time period was analysed in detail by Johnson et al. (2005), where data provided by radiometer #1 and radiometer #3 correspond respectively to sensors FOV #2 and FOV #1 in Johnson et al. (2005). Despite being pointed at the target, the FOV of the two instruments was slightly different, with radiometer #3 (FOV #1) pointing slightly above radiometer #1 (FOV #2). Here, an inverse relation of thermal levels at the two sensors, evident from Figure 4 and discussed in Johnson et al. (2005), with thermal level of degassing and gas-piston events increasing at radiometer #3 and decreasing at radiometer #1, has been explained in terms of evolution of the geometry of the targeted vent, where an increase of vent elevation, caused by piling up of pahoehoe lava flows, partially moved the thermal source out of the FOV of radiometer #1 (FOV #2) (Johnson et al. 2005).

Despite the inconsistency of short-term thermal observations described above, thermal data recorded by radiometer #1 and radiometer #3 appear consistent in the long term, where after June 18 2002, both sensors record a general high thermal level (Fig. 4a). Obviously, we cannot state exactly when the increased activity at the Drain Hole vent began simply on the basis of the thermal data presented here, and the timing of the thermal phases evidenced from the permanent infrared thermometers (Fig. 4) is influenced by sensor retargeting. However, we feel confident with the data presented as frequent field work at Pu'u 'O'o, mostly aimed at focusing the thermal monitoring on the most active vents, allowed us to visually confirm the thermal observations. Again, this approach prevents the use of T_{max10} data on short-term (hours–days) analyses, but permits an investigation of long-term (weeks–months) trends.

Gas-piston events: features and trends

Gas-piston events were recorded frequently throughout our investigation period, but tended to cluster into discrete periods of gas-piston dominated activity. During such periods, events with recurrent waveforms repeated in time with a stable frequency (Fig. 2). Most of the events were not associated with the thermal signal expected for a lava flow (i.e. they lacked cooling curves) and were thus not associated with lava effusion, and can therefore be considered non-effusive versions of gas-piston events previously described at Kilauea (Swanson et al. 1979; Chouet & Shaw 1991; Johnson et al. 2005).

To better define the properties of the observed gas-piston activity, we extracted c. 4000 events from thermal data collected by radiometer 1 between March 2001 and November 2003 (Fig. 5). This datastream was chosen as radiometer #1, having been reoriented just three times during the investigation period (Fig. 4a, Table 1), provides a better and more continuous data-set for the description of gas-piston events. In particular, this 1° FOV (rad #1) sensor targeted the Drain Hole vent (Table 1) for c. 18 months (June 2002–Dec 2003) during the long-term fluctuation of T_{max10} described above.

Discrimination and extraction of the thermal waveform of single gas-piston events from the continuous thermal record (0.5 Hz) provided by the 1° FOV infrared thermometer was done automatically, following a procedure based on the difference between the average thermal levels recorded for short-term (sta) and long-term (lta) intervals. The duration of the short-term (10 s) and long-term (10 minute) intervals and the sta-lta threshold value (50°C) have been tuned manually to achieve the best results for this thermal data-set. Although very simple, this procedure proved quite efficient for event discrimination given the simple recurrent thermal waveform commonly associated with gas-piston events, which typically shows an abrupt increase in temperature preceded by a low thermal level and followed by a slow decay (Fig. 2c & 5).

Raw thermal waveforms of all gas-piston events recorded in March 2001, October 2002 and August 2003 (Fig. 5), and extracted following the procedure described above, confirm the efficiency of the discrimination procedure, as no thermal transients, which cannot be related to gas-piston activity, have been extracted automatically. Moreover, comparison of raw thermal data for these three case months (Fig. 5) confirms the general properties of the thermal waveforms ascribed to gas-piston activity. However, thermal records show a large variation in the peak temperature (50–200 °C), in the duration of the thermal transient (100–400 s) and in the thermal waveform through time, with single-peaked thermal transients occurring in March 2001 and August 2003 and double-peaked transients in October 2002 (Fig. 5).

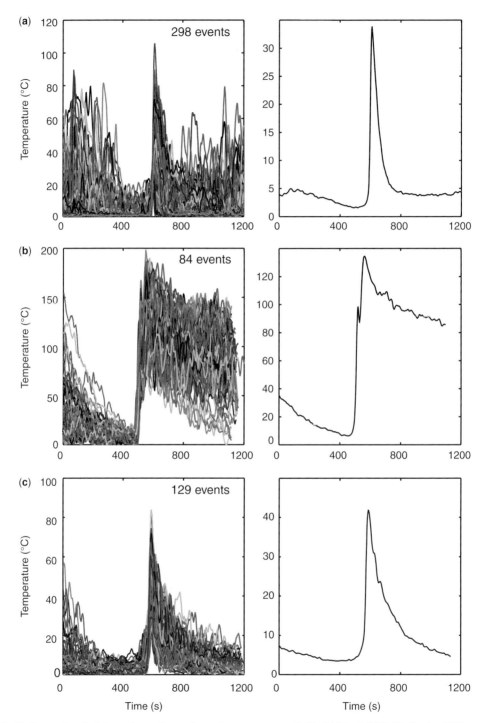

Fig. 5. Raw and stacked thermal waveforms of gas-piston events recorded in (**a**) March 2001 (**b**) October 2002, and (**c**) August 2003. Thermal waveforms shown here have been recorded by a single 1° FOV sensor (rad#1) and describe activity at the Beehive (a) and Drain Hole (b, c) vents. These are taken as case-type examples of gas-piston event types (see text).

The differences in the raw thermal waveforms for gas-piston activity (Fig. 5) cannot be ascribed to sensor characteristics or vent geometry. Effects of sensor response can be ruled out, as all the thermal data have been collected by a single 1° FOV radiometer. In the same way, effects of vent geometry can be neglected, because thermal records of events collected in October 2002 (Fig. 5b) and August 2003 (Fig. 5c), which are characterized by large variations in the amplitude, duration and waveform, are all recorded during gas-piston activity at a single vents (Drain Hole vent), without any reorientation of the sensor. Moreover, these differences in gas-piston activity at the Drain Hole vent (Fig. 5b,c) seem consistent with variations in the T_{max10} values, with long-lasting, double-peaked, high-amplitude events recorded during periods of high T_{max10} (October 2002), and short-lasting, single-peaked, low-amplitude events recorded during medium T_{max10} periods (August 2003). This suggests that a general overview of the properties of gas-piston activity over the entire investigation period is required.

Thermal amplitude of gas-piston events, evaluated from the difference between the peak thermal value and the minimum thermal level for each extracted waveform (Fig. 6a), shows a good fit with the trend apparent in T_{max10} (Fig. 4a). Amplitude of gas-piston events increased from 20–30 °C, during March 2001–February 2002, to 100–170 °C by June 2002–May 2003, when T_{max10} was also at a maximum. It then decreased, over a period of 10 days between May and June 2003, to values of 40–60 °C. Amplitude of gas-piston events then remained stable at this level until November 2003 (Fig. 6a).

In order to obtain information on the occurrence of gas-piston events during the investigation period (March 2001–Dec. 2003), we evaluated the daily number of events (Fig. 6b). Here, the identification and extraction procedure may suffer from low visibility conditions within the crater, and the daily number of events might underestimate the real gas-piston activity occurring at Pu'u 'O'o during single days. However, a general description of the long-term (weeks to months) trends of

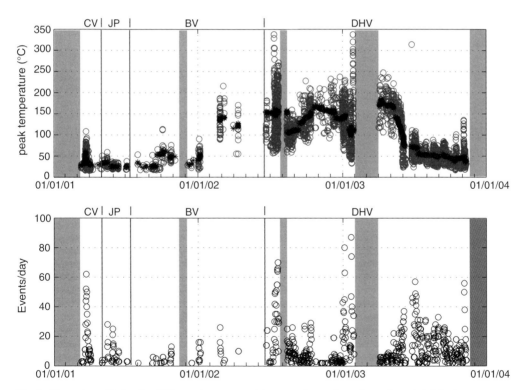

Fig. 6. Thermal amplitude of *c*. 4000 gas-piston events (**a**) extracted over the entire study period (March 2001–November 2003) from thermal data recorded by radiometer #1. Red circles represent the peak temperature for each gas-piston event. The black line results from a smoothing procedure (30 subsequent events) and represents the main thermal trend of gas-piston thermal amplitude. (**b**) Number of gas-piston events over the study period. Grey areas represent gaps in the data stream.

occurrence is likely valid. This analysis points out a relationship between occurrence of gas-piston events (Fig. 6b) and the trends apparent in the gas-piston thermal amplitude (Fig. 6a) and T_{max10} data (Fig. 4a). A low occurrence of gas-piston events was recorded from March 2001 to June 2002 (Fig. 6b), when gas-piston thermal amplitude and T_{max10} were also, generally, low (Figs 4a, 6a). The number of gas-piston events then increased in June 2002, as did the thermal amplitude and T_{max10}. Finally, a moderate number of low temperature events per day was recorded after June 2003 (Fig. 6a) during the final period of moderate T_{max10} (Fig. 4a).

We investigate changes in the thermal waveform for gas-piston events, as suggested by raw thermal data collected for March 2001 (Fig. 5a), October 2002 (Fig. 5b) and August 2003 (Fig. 5c), by stacking all the raw thermal waveforms collected during one-month-long sequential windows (Fig. 7). This points to a systematic change in the piston-related thermal waveform through time. This evolved from short-lasting single-peaked gas jets, as recorded between March 2001 and September 2002, to long-lasting events with multiple peaks, typically recorded during October 2002 to May 2003. The thermal style then reverted to single-peaked emissions in June 2003.

The trends apparent in the thermal waveform and amplitude of gas-piston events lead us to define three types of events occurring during the investigation period (Type I, II, III). As examples, we consider the thermal waveforms for events extracted for March 2001, October 2002 and August 2003 (Fig. 5), to be representative to each type.

Type I gas-piston events were recorded during most of the low T_{max10} phase (March 2001–May 2002) (Fig. 7) and are characterized by single peaked thermal transients with amplitudes of 20–100°C and durations of 100–200 seconds (Figs 5a & 7). Type I events are commonly followed by a rapid decrease in temperature during which the temperature preceding the event is returned to (Fig. 5a). This is then followed by another event. As a result, Type I events show a tendency to cluster into groups each containing 10's of gas-piston events with regular recurrence intervals, as described for the June–July 2002 period of this data-set by Johnson et al. (2005).

Type II events were recorded during most of the high T_{max10} phase (September 2002–May 2003) and were characterized by longer-lasting (200–400 second-long) thermal transients, within which a precursory, but minor, thermal peak was immediately followed by a second major pulse with an amplitude of 50–150 °C (Figs 5b & 7). Such events were usually followed by periods of high, quasi-steady thermal signal associated with

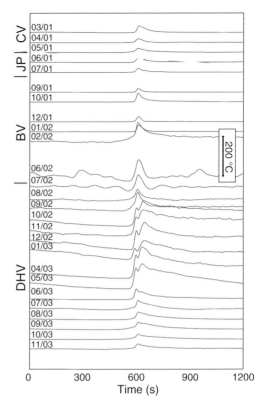

Fig. 7. Thermal waveforms obtained from stacking of all gas-piston events extracted from each month. The mean thermal waveforms of gas-piston events shown here result for the stacking of thermal waveforms recorded by a single 1° FOV infrared thermometer (rad#1), which was re-oriented three times during the investigation period (Table 1), reflecting activity at the Central vent (CV, March–April 2001), at the July pit (JP, May–June 2001), at the Beehive vent (BV, July 2001–June 2002) and at the Drain Hole vent (DHV, June 2002–Dec. 2003). Times of sensor re-orientation are indicated on the left of the figure.

sustained degassing (rather than by further gas-piston events).

Type III events (June–November 2003), like the events recorded before September 2002, appeared as single thermal peaks (Figs 5c & 7). However, unlike the pre-September 2002 events, they were usually followed by high, quasi-steady thermal signal, associated with sustained degassing. This resulted in a longer duration for the waning tail (100–300 sec), as apparent in a comparison of the stacked thermal waveforms (Figs 5, 7).

The time-dependent changes apparent in the gas-piston waveforms (Fig. 7) correlate with periods of different T_{max10} levels (Fig. 4a). Single-peaked, low-amplitude (Type I, Type III) waveforms (Fig. 5a,c) were recorded during periods of low–medium T_{max10}, whereas the long-lasting, high-amplitude, double-peaked (Type II) events (Fig. 5b) were recorded during high T_{max10} periods (Fig. 4a).

Discussion

We see two significant changes in activity at Pu'u 'O'o as recorded by our thermal infrared thermometers during March 2001 to December 2003. We thus define three different phases of activity on the basis of the thermal characteristics of the emissions (Table 3). The first transition occurred during May–June 2002 and involved: (1) increased temperatures at the targeted, persistently degassing vents; (2) an increase in the thermal amplitude and occurrence of gas-piston events; (3) a change in the gas-piston waveform (from single-to double-peaked events); and (4) decreased lava flow activity. The situation changed again during May–June 2003 when activity reverted to a style similar to that recorded during Phase I, but with a continued absence of lava flow activity (Table 3).

Following Stevenson (1993), vent (emission) temperature can vary during sustained degassing as a function of gas flux per unit vent area, depth of the source, conduit radius and/or source temperature. We thus infer that the main long-term trends in the T_{max10} recorded during sustained degassing (Fig. 4) reflect changes in one or all of these parameters. An increase in T_{max10} (as during Phase II) will result, for example, from an increase in gas flux per unit area, an increase in the magma level, a decrease in the conduit radius and/or an increase in the source (magma) temperature (Stevenson 1993). Variation in the last two of these parameters seems unlikely, leaving variation in the gas flux and/or magma level as the two most likely causes of the temperature change.

The change in gas-piston thermal amplitude, occurrence (Fig. 6) and style (Fig. 7) between each of the phases also points to a systematic modification of a common source process. As already

discussed, gas-piston activity at Kilauea is commonly explained in terms of two different source mechanisms: a cyclic effusion–gas-jetting–drain-back model and foam layer collapse mechanism (Fig. 8).

The first model relates gas-piston activity to the build-up of gas beneath a relatively impermeable magma layer (Swanson et al. 1979). This pushes the column upwards so that the free-surface eventually reaches the vent to feed lava flow onto the crater floor (Fig. 8a). Spreading of the lava decreases the overburden of the magma overlying the gas layer, culminating in gas release by gas-jetting (Swanson et al. 1979; Tilling 1987; Barker et al. 2003; Johnson et al. 2005). The lack of lava flows for most of gas-piston activity recorded here led Johnson et al. (2005) to infer that lava spreads inside sub-surface cavities. According to this model, a change in the occurrence and/or thermal amplitude (and hence, we infer, gas flux) during gas-piston events could be ascribed to a change in the gas flux to the conduit and/or a change in conduit geometry. An increase in gas flux, for example, may induce more frequent events of higher thermal amplitude and longer duration, as during Phase II (Table 3). Likewise, a change in the magma level may modify conduit geometry changing, for example, the length of the magma-filled pipe below the crater floor or sub-surface cavity; again causing a change in the occurrence and thermal amplitude.

An alternative explanation for gas-piston events is based on the generation and collapse of foams (Jaupart & Vergniolle 1988, 1989; Vergniolle & Jaupart 1990), where accumulation of gas bubbles at a constriction in the conduit or at chamber roof generates a foam, which, upon attaining a critical thickness, collapses to generate a gas slug. This then enters and ascends the conduit to burst at the magma free-surface, releasing a jet of over-pressurized gas (Fig. 8b). Within this model, a change in gas-piston occurrence, duration, thermal amplitude and/or emission style can be forced by a change in gas flux and/or conduit geometry. Following the formulation proposed by Vergniolle &

Table 3. *Thermal characteristics of lava flow, sustained degassing and gas-piston events at Pu'u 'O'o during March 2001–December 2003*

Phase	I	II	III
Dates	before 06/2002	06/2002–06/2003	after 06/2003
No. Lava Flow Events	24	4	1
Mean T_{max10} (°C)	c. 50	c. 250–300	c. 100–150
Gas Piston Waveform	Type I	Type II	Type III
Mean Gas Piston Thermal Amplitude (°C)	20–100	50–150	20–100
Mean Gas Piston Duration (s)	100–200	200–400	100–300

(a)

(b)

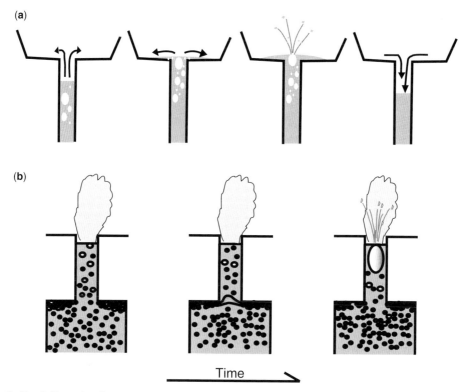

Time

Fig. 8. Sketch illustrating the two source mechanisms for gas-piston activity: based on (**a**) cyclic effusion–gas-jetting–drain back (Swanson *et al.* 1979) and (**b**) on a collapsing foam (Vergniolle & Jaupart 1990).

Jaupart (1990), when all the geometric conditions are stable, the recurrence time of foam collapse decreases when the gas flux increases. Also, movements in the magma level upwards and downwards may result in longer-lasting events when geometries permit larger areas of foam to accumulate. Changes in the pipe geometry may also influence the ascent and break-up of the rising gas slug (James *et al.* 2004, 2006), hence forcing a change in emission style.

The data presented here provide a detailed description of different features of gas-piston activity at Pu'u 'O'o crater in terms of thermal amplitude, occurrence (Fig. 6) and thermal waveform (Figs 5 & 7) of events, but they do not appear to be able to distinguish between the two models proposed (Fig. 8), as both seem able to theoretically explain observed features. However, the thermal data-set described here might be integrated in a wider analysis to investigate changes in the shallow feeding system at the Pu'u 'O'o crater. Gas-piston events extracted from radiometer #1 (Figs 5–7) describe activity at four different vents between March 2001 and December 2003 (Table 1). For all vents, gas-piston activity has been recorded and described, pointing

to a stable thermal waveform for all cases, including gas-piston events at the Drain Hole vent before September 2002 and after June 2003. This suggests that the observed thermal features of gas-piston activity seem most reasonably to be related to the source process rather then to conduit effects, which should result in different thermal waveforms for different vents. At the same time, however, long-term thermal monitoring at the Drain Hole vent reveals a change of thermal waveform of gas-piston events, where the single-peaked transients turn into double-peaked events and then back, after *c.* nine months, to single peaked waveforms. Here, a change in the gas flux regime or magma level below the targeted vent may explain all the different features of the gas-piston activity observed. Longer-lasting, high-amplitude events (Figs 5b & 7), occurring at a higher rate (Fig. 6b), require larger gas volumes, or conduit geometry. Moreover, this is consistent with the matching trend of gas-piston activity (Figs 5–7) and T_{max10} (Fig. 4), which requires an increase of gas flux to, or magma level below the targeted vent to justify the sustained high thermal levels observed.

During January 2002, flow activity increased in the vicinity of Pu'u 'O'o, building a series of shields that, by March 2002, had formed a continuous ridge 2.7 km long and 1.5 km wide in the proximal section of the flow field (hvo.wr.usgs.gov). This near-vent activity was correlated with an increase in activity within Pu'u 'O'o, with several new spatter cones and lava flows forming. Back-up of lava in the system causing higher magma levels at Pu'u 'O'o appears consistent with the increased incidence of lava flow activity recorded by our thermal sensors at this time. However, on 12 May 2002, a new flow (the Mother's Day flow) began to advance down the western edge of the flow field reaching the Chain of Craters Road (c. 10 km from Pu'u 'O'o) during July 2002 (hvo.wr.usgs.gov). This appears to have relieved pressure on Pu'u 'O'o so that observed activity within and around the crater declined (hvo.wr.usgs.gov), as did the incidence of lava flows recorded by our thermal system (Tables 1 & 3). This is consistent with a scenario whereby tapping of the shallow system resulted in a decline in magma levels at Pu'u 'O'o.

These events, being coincident with our move from Phase I to Phase II activity, appear to have triggered the transition in thermal character of the Pu'u 'O'o emissions (as summarized in Table 3). However, some of the thermal changes are difficult to explain by a change in magma level alone. The increase in the temperature recorded during sustained degassing or in the increase in occurrence and/or thermal amplitude during gas-piston events, for example, appear to require shallower levels and/or increased gas fluxes. Aside from short-lived peaks, gas and mass fluxes at Pu'u 'O'o are generally stable (e.g. Sutton et al. 2003). We thus cannot invoke a 12-month-long, system-wide, increase in degassing to explain the Phase II activity. It is, however, possible that the re-organization of the complex shallow system beneath Pu'u 'O'o following the Mother's Day Flow diverted gas to the two conduits tracked by our thermal infrared thermometers at the expense of other conduits. Either way, we have to explain an increase in temperature for a single vent within a multiple vent system that appears to have undergone a decrease in magma level. The only way we can do this is by calling upon increased gas fluxes at certain conduits and/or a change in conduit geometry forced upon the magma resident in the conduit caused by a variation in the magma level.

This hypothesis is confirmed by the thermal investigation presented here. In June 2002 (Table 1), the two 1° FOV radiometers were targeted on the Drain Hole vent, as a result of the higher activity level at this vent at that time. This allowed us to record a five-month-long high thermal period at this vent (Fig. 4a,c). Five months later, in November 2002, radiometer #3 was targeted on the South Wall vent, which was

showing similar thermal levels at that time (Fig. 4a,c). In March 2003, the target of radiometer #3 was changed again towards the East Pond vent following blockage and burial of the South Wall vent by a collapse of the crater wall immediately above it (Table 1). The burial of this vent was marked by a sharp reduction of the thermal level at radiometer #3 in mid-January 2003, while the thermal level at the still-open Drain Hole vent (as detected by radiometer #1) remained high, not decreasing until June 2003. This points to variable thermal level from individual vents within Pu'u 'O'o crater caused by external (collapse, burial, blockage) and internal (changes in magma level) factors, with the vents of the central–south sector (South Wall vent and Drain Hole vent) being particularly active between June 2002 and June 2003.

This simple qualitative discussion on the causes of coupled changes in vent temperature and gas-piston character points to a strong dependence on changes in magma level beneath Pu'u 'O'o, which may influence conduit geometry and/or the location of the preferential degassing pathways: changing the gas flux at one vent at the expense of another. Turning on and off of vents during otherwise persistent degassing is a common feature of the multiple vent degassing at Stromboli (Harris & Ripepe 2007) and points to near-surface complexities in the degassing process at multiple vent systems, which may be controlled by changes in magma level and/or changes in the gas flux to individual conduits, as well as external factors such as vent blockage to shut down certain vents and divert gas to others.

Conclusions

The temporal resolution and length of the thermal record for Pu'u 'O'o has revealed some interesting thermal trends and opened up some associated questions regarding the complexity of degassing at a multi-vent system. It has also showed how a ground-based thermal system can be used to track, document and monitor changes in thermal activity at an active, open, degassing vent system.

Thermal data collected at Pu'u 'O'o over a period of three years (March 2001–November 2003) allow a description of the activity within the crater, which was dominated by sustained degassing and gas-piston activity, with minor lava flow activity. Gas-piston activity is a common feature for Kilauea. However, the absence of lava flow thermal signals during most gas-piston events examined here show that they were typically not associated with effusion, unlike previously reported gas-piston activity. Detailed thermal analysis of c. 4000 gas-piston events recorded during the three-year-long investigation period allows us to define several typical features of this degassing process, such as the existence of typical, repeatable, thermal waveform and a reduction of temperature/degassing before the event onset, as well as systematic changes in the waveform type, waveform duration and recurrence of gas-piston events through time.

The long-term analysis of the full data-set reveals that thermal levels, as well as gas-piston style and occurrence, changed with time (Table 3) and experienced a one-year-long high level period, between June 2002 and June 2003. Such a modification of the system occurred without any apparent fluctuation in the total gas flux and appeared to be triggered by a decline in the magma level at Pu'u 'O'o, as suggested by the decline of lava flow activity inside the crater. This leads us to infer that the fluctuations in thermal and degassing levels and styles described here may reflect changes in the gas collection and release dynamics in the shallow system. This, in turn, likely results from changes in the conduit geometries and preferential pathways produced by slight changes of the magma level beneath the crater, as well as blockage and destruction of certain vents.

This work was funded by NSF grant EAR-0106349. We acknowledge Mike James and Sonia Calvari for their careful reviews. We are extremely grateful for the help and support of the Hawaiian Volcano Observatory, especially Tamar Elias, Christina Heliker, Rick Hoblitt, Jim Kauahikaua, Dave Sherrod, Jeff Sutton, Don Swanson and Carl Thornber, as well as David Okita (Volcano Helicopters). Permanent system instrumentation was developed, purchased and installed using funds from HIGP.

References

BARKER, S. R., SHERROD, D. R., LISOWSKI, M., HELIKER, C. & NAKATA, J. S. 2003. Correlation between lava-pond drainback, seismicity, and ground deformation at Pu'u 'O'o. The Pu'u 'O'o-Kupaianaha eruption of Kilauea Volcano, Hawai'i: the first 20 years. *United States Geological Survey Professional Paper*, **1676**, 53–62.

CASADEVALL, T. J., STOKES, J. B., GREENLAND, L., MALINCONICO, L. L., CASADEVALL, J. R. & FURUKAWA, B. T. 1987. SO₂ and CO₂ emission rates at Kilauea volcano, 1979–1984. Volcanism in Hawai'i. *United States Geological Survey Professional Paper*, **1350**, 771–780.

CHOUET, B. A. & SHAW, H. R. 1991. Fractal properties of tremor and gas-piston events observed at Kilauea Volcano, Hawaii. *Journal of Geophysical Research*, **96**, 10177–10189.

FERRAZZINI, V., AKI, K. & CHOUET, B. A. 1991. Characteristics of seismic waves composing Hawaiian volcanic tremor and gas-piston events observed by a near source array. *Journal of Geophysical Research*, **96**, 6199–6209.

GREENLAND, L. P. 1986. Gas analyses from the Pu'u 'O'o eruption in 1985, Kilauea volcano, Hawaii. *Bulletin of Volcanology*, **48**, 341–348.

HARRIS, A. J. L., PIRIE, D., GABRIEL, H. *ET AL.* 2005 DUCKS: low cost thermal monitoring units for near-vent deployment. *Journal of Volcanology and Geothermal Research*, **143**, 335–360.

HARRIS, A. J. L., JOHNSON, J. B., HORTON, K. *ET AL.* 2003. Ground-based infrared monitoring provides new tool for remote tracking of volcanic activity. *EOS, Transactions, American Geophysical Union*, **84**, 409,418.

HARRIS, A. J. L. & RIPEPE, M. 2007. Temperature and dynamics of degassing at Stromboli. *Journal of Geophysical Research*, **112**, B03205, doi:19.1029/2006JB004393.

HELIKER, C. & MATTOX, T. N. 2003. The first two decades of the Pu'u 'O'o-Kupaianaha Eruption: chronology and selected bibliography. The Pu'u 'O'o-Kupaianaha eruption of Kilauea Volcano, Hawai'i: the first 20 years. *United States Geological Survey Professional Paper*, **1676**, 1–27.

HELIKER, C., KAUAHIKAUA, J. P., SHERROD, D. R., LISOWSKI, M. & CERVELLI, P. F. 2003. The rise and fall of the Pu'u 'O'o cone, 1983–2002. The Pu'u 'O'o-Kupaianaha eruption of Kilauea Volcano, Hawai'i: the first 20 years. *United States Geological Survey Professional Paper*, **1676**, 29–51.

HON, K. A., KAUAHIKAUA, J. P., DENLINGER, R. P. & MACKAY, K. 1994. Emplacement and inflation of pahoehoe sheet flows: Observations and measurements of active lava flows on Kilauea Volcano, Hawaii. *Geological Society of America Bulletin*, **106**, 351–370.

JAMES, M. R., LANE, S. J., CHOUET, B. A. & GILBERT, J. S. 2004. Pressure changes associated with the ascent and bursting of gas slugs in liquid-filled vertical and inclined conduits. *Journal of Volcanology Geothermal Research*, **129**, 61–82.

JAMES, M. R., LANE, S. J. & CHOUET, B. A. 2006. Gas slug ascent through changes in conduit diameter: laboratory insights into a volcano-seismic source process in low-viscosity magmas. *Journal of Geophysical Research*, **111**, B05201, doi:10.1029/2005JB003718.

JAUPART, C. & VERGNIOLLE, S. 1988. Laboratory models of Hawaiian and Strombolian eruptions. *Nature*, **331**, 58–60.

JAUPART, C. & VERGNIOLLE, S. 1989. The generation and collapse of a foam layer at the roof of a basaltic magma chamber. *Journal of Fluid Mechanics*, **203**, 347–380.

JOHNSON, J. B., HARRIS, A. J. L. & HOBLITT, R. P. 2005. Thermal observations of gas pistoning at Kilauea volcano. *Journal of Geophysical Research*, **110**, B11201, doi:10.1029/2005JB003944.

KESZTHELYI, L. & DENLINGER, R. 1996. The initial cooling of pahoehoe flow lobes. *Bulletin of Volcanology*, **58**, 5–18.

STEVENSON, D. S. 1993. Physical models of fumarolic flow. *Journal of Volcanology and Geothermal Research*, **57**, 139–156.

SUTTON, J. A., ELIAS, T., GERLACH, T. M. & STOKES, J. B. 2001. Implication for eruptive processes as indicated by sulfur dioxide emissions from Kilauea Volcano, Hawaii, 1979–1997. *Journal of Volcanology and Geothermal Research*, **108**, 283–302.

SUTTON, J. A., ELIAS, T. & KAUAHIKAUA, J. P. 2003. Lava-effusion rates for the Pu'u 'O'o-Kupaianaha Eruption derived from SO$_2$ emissions and very low frequency (VLF) measurements. The Pu'u 'O'o-Kupaianaha eruption of Kilauea Volcano, Hawai'i: the first 20 years. *United States Geological Survey Professional Paper*, **1676**, 137–148.

SWANSON, D. A., DUFFIELD, W. A., JACKSON, D. B. & PETERSON, D. W. 1979. Chronological narrative of the 1969–71 Mauna-Ulu eruption, Kilauea Volcano, Hawaii. *United States Geological Survey Professional Paper*, **1056**, 1–55.

TILLING, R. I. 1987. Fluctuation in surface height of active lava lakes during 1972–1974 Mauna Ulu eruption, Kilauea volcano, Hawaii. *Journal of Geophysical Research*, **92**, 13721–13730.

TILLING, R. I., CHRISTIANSEN, R. L., DUFFIELD, W. A. ET AL. 1987. Fluctuations in surface height of active lava lakes during 1972–1974 Mauna Ulu eruption; an example of quasi-steady-state magma transfer.

Volcanism in Hawai'i. *United States Geological Survey Professional Paper*, **1350**, 405–469.

VERGNIOLLE, S. & BRANDEIS, G. 1996. Strombolian explosions 1, A large bubble breaking at the surface of a lava column as a source of sound. *Journal of Geophysical Research*, **101**, 20433–20449.

VERGNIOLLE, S. & JAUPART, C. 1990. Dynamics of degassing at Kilauea Volcano, Hawaii. *Journal of Geophysical Research*, **95**, 2793–2809.

VERGNIOLLE, S., BOICHU, M. & CAPLAN-AUERBACH, J. 2004. Acoustic measurements of the 1999 basaltic eruption of Shishaldin volcano, Alaska 1. The origin of Strombolian activity. *Journal of Volcanology and Geothermal Research*, **137**, 109–134.

WOLFE, E. W., NEAL, C. A., BANKS, N. G. & DUGGAN, T. J. 1988. Geologic observations and chronology of eruptive events. The Pu'u 'O'o eruption of Kilauea Volcano, Hawaii; episodes 1 through 20, January 3, 1983, through June 8, 1984. *United States Geological Survey Professional Paper*, **1463**, 1–97.

From Strombolian explosions to fire fountains at Etna Volcano (Italy): what do we learn from acoustic measurements?

S. VERGNIOLLE[1] & M. RIPEPE[2]

[1]*Institut de Physique du Globe de Paris, Institut de recherche associé CNRS et Université de Paris 7, 4 Place Jussieu, 75252 Paris Cedex 05, France (e-mail: vergniolle@ipgp.jussieu.fr)*

[2]*Dipartimento di Scienze della Terra, Universita di Firenze, Firenze, Italy*

Abstract: The 2001 eruption of Etna volcano, prior to the flank eruption, was marked by 16 episodes separated from one another by few days of quiescence. Insights into fire fountain formation are provided by a close comparison of the sound produced by an episode solely involving a series of Strombolian explosions (4 July) and one also showing a transition to a fire fountain (12 July). The best fit between measured and synthetic waveforms gives the temporal evolution of the bubble length, 8–100 m, and overpressure, 0.2 MPa. Both episodes result from the coalescence of a foam layer trapped at the top of the reservoir. At the transition towards a fire fountain, the number of explosions and the bubble length increase simultaneously, suggesting that the foam destabilization is more efficient when a fire fountain is produced. Acoustic records give access to the gas volume trapped within the foam, called 'active' degassing, while the height of fire fountains also includes the gas from passive degassing. The small bubbles from passive degassing are carried to the surface via the wake of the slugs, coming from the depth of the reservoir. The proportion between active and total gas volume represents 38–48%.

Geophysical methods, which can be performed remotely and continuously, have been proven to be extremely useful to understand eruption dynamics without danger. Measurements of tremor and long-period events have been used to provide constraints on the plumbing system (Chouet 1985; Crosson & Bame 1985; Falsaperla et al. 2002; Saccorotti et al. 2004) as well as being useful indicators of impending eruption (Fehler 1983; Aki & Koyanagi 1981; McNutt 2000; Thompson et al. 2002). Furthermore, on basaltic volcanoes such as at Etna and Stromboli (Italy), the tremor results from the superimposition of small impulsive sources associated with magma degassing (Montalto et al. 1992; Ripepe et al. 2001).

Acoustic recordings can also give constraints on the eruptive pattern at the vent. However, these measurements were very sparse in the past and solely done in the audible range, i.e. above 20 Hz (Richards 1963; Woulff & MacGetchin 1976). This method was then entirely forgotten for almost 20 years partially due to the frequency band of volcanic sources, mostly in the infrasonic range (Vergniolle & Brandeis 1994). However, acoustic pressure has now been measured on several volcanoes. It is used to estimate the source characteristics (Vergniolle & Brandeis 1994, 1996; Vergniolle et al. 1996, 2004; Hagerty et al. 2000; Vergniolle & Caplan-Auerbach 2004, 2006) or to calculate the propagation of acoustic pressure in the shallow magma and its transport towards the atmosphere (Buckingham & Garcés 1996; Garcés & McNutt 1997; LePichon et al. 2005; Matoza et al. 2007; Antier et al. 2007). The sound produced by Vulcanian explosions has been associated with the disruption of a plug at the top of the magma column, which is characteristic of a magma with intermediate viscosity, such as those seen at Karymsky (Kamchatka, Russia), Sangay (Ecuador) and Arenal volcanoes (Costa Rica) (Johnson et al. 1998; Johnson & Lees 2000).

Vent pressure has also been estimated from microbarographs (Morrissey & Chouet 1997). Sakurajima (Japan) exhibits a range of vent pressures from 0.5–5 MPa (Morrissey & Chouet 1997), while Shishaldin has a maximum of 1.4 MPa (Vergniolle et al. 2004). Vulcanian explosions at Popocatepetl (Mexico) (Arciniega-Ceballos et al. 1999) are roughly of similar size to Sakurajima explosions (Japan), which in turn are ten to one hundred times larger than the 0.1 MPa explosions recorded at Stromboli (Italy) (Vergniolle & Brandeis 1996). Microbarographs recording explosive volcanoes give vent pressure estimates over 5.0 MPa at Mount Pinatubo (Philippines) in 1991 (Morrissey & Chouet 1997) and 7.5 MPa at the start of the 1980 eruption of Mount St Helens (USA) (Kieffer 1981). In contrast to Mount St Helens, activity at Mount Pinatubo built up gradually with smaller explosions and continuous gas emissions for several hours before leading up to the climactic eruption (Kanamori & Mori 1992).

From: LANE, S. J. & GILBERT, J. S. (eds) *Fluid Motions in Volcanic Conduits: A Source of Seismic and Acoustic Signals.* Geological Society, London, Special Publications, **307**, 103–124.
DOI: 10.1144/SP307.7   0305-8719/08/$15.00 © The Geological Society of London 2008.

Fig. 1. Huge bubble breaking at Etna volcano on 25 July, 2001. Crater at 2500 m elevation and close to Montagnola. Photo taken at 500–700 m from crater with a 90 mm lens (courtesy of T. Pfeiffer; www. decadevolcano.net). This is very similar to what was observed for example at 21 hr 16 min, 21 hr 30 min, 21 hr 35 min, 21 hr 43 min on 4 July (very approximate watch time).

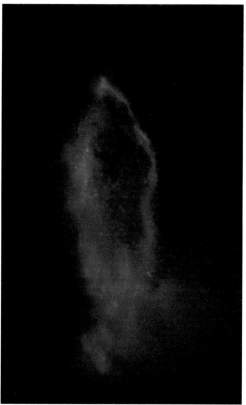

Fig. 2. Fire fountain activity on September 26, 1989 at 20 hr 30 min at the southeast crater (adapted from Bertagnigni *et al.* 1990 and courtesy of A. Bertagnigni). At that moment, the height of fire fountain increased very suddenly to reach 600 m, and the top part immediately began to break up, giving rise to dark ash clouds. It was followed by 10 minutes of sustained activity (Bertagnigni *et al.* 1990). The shape of the fire fountains clearly shows an inner central gas pocket (dark zone) surrounded by a layer of magma. This photo gives unambiguous evidence that fire fountains are driven by a long inner gas jet. Although the intensity on 26 September 1989 largely exceeds what is observed during the 2001 eruption, photos taken during the 1989 eruption (Bertagnigni *et al.* 1990) closely ressemble what was observed visually on 12 July 2001.

Basaltic volcanoes classically display a range of volcanic activity, which includes lava flows, Strombolian explosions and sometimes fire fountains. In very rare occasions, a purely magmatic Subplinian plume can develop at the vent, such as during the 122 BC eruption of Etna and the 1999 eruption of Shishaldin (Alaska) (Coltelli *et al.* 1998*a*; Nye *et al.* 2002; Sable *et al.* 2006). A Strombolian explosion at Etna, such as on 11 September 1989 (p. 13 in Bertagnini *et al.* 1990) or on 25 July 2001 (Fig. 1), clearly corresponds to the bursting of a bubble as large as the conduit, hence to a slug flow (Wallis 1969). The fire fountain at Etna on 26 September 1989 gives superb evidence that a fire fountain was driven by a long inner gas jet (Fig. 2). At that time, the length of the gas core had largely exceeded its radius (Fig. 2), showing that a transition towards an annular flow had occurred (Wallis 1969).

While an increase in the gas volume fraction in the conduit can transform a slug flow into an annular flow (Wallis 1969; Vergniolle & Jaupart 1986), a massive foam coalescence at the depth of the reservoir can explain the regular alternation between phases rich in gas and those relatively poor in gas, as suggested for Kilauea and Stromboli (Jaupart & Vergniolle 1988, 1989; Vergniolle &

Jaupart 1990). Alternatively, an increase in the magma rising rate and volatile flux could trigger the formation of a fire fountain by a progressive coalescence of bubbles in the conduit (Parfitt & Wilson 1995; Parfitt *et al.* 1995; Parfitt 2004). However, recent chemical FTIR measurements at Etna show that the formation of a fire fountain during the 2000 eruption resulted from a foam disruption at a structural barrier at 1.5–1.8 km depth (Allard *et al.* 2005; Spilliaert *et al.* 2006*a*).

Furthermore, several independent data, based on magma ascent rate (Behncke *et al.* 2006), seismic waveforms (Gresta *et al.* 2004), ejecta vesicularity (Polacci *et al.* 2006) or exit pressure (Vergniolle 2008), do not favour Parfitt's model (2004) at Etna. The depth at which the foam is accumulating, 1.5–1.8 km below the surface (Allard *et al.* 2005; Spilliaert *et al.* 2006*a*), is compatible with that of a reservoir, located between the surface and 3 km depth from the position of tremor sources (Saccorotti *et al.* 2004) and tilt modelling (Bonaccorso 2006), and between 3 km and 4 km from tomographic images (Laigle *et al.* 2000; Patanè *et al.* 2002, 2006). Hence, the fire fountains at Etna result from the massive coalescence of a foam trapped in the magma reservoir. This model (Jaupart & Vergniolle 1988, 1989; Vergniolle & Jaupart 1990) implies that the gas content is sufficient for the gas to be exsolved at the depth of the reservoir and to form small bubbles. This is indeed the case at Kilauea, with a total CO_2 content of 0.7 wt% (Gerlach *et al.* 2002) and at Etna with a H_2O content of 3.4 wt% and CO_2 content of ≥ 0.7 wt% (Métrich *et al.* 2004; Spilliaert *et al.* 2006*b*).

On one hand, all the bubbles produced during one episode at Etna are expelled as a train of bubbles of various length, from a slug (Fig. 1) to an annular flow (Fig. 2). On the other hand, the close series of Strombolian explosions or fire fountains correspond to a phase rich in gas. While prior to the paroxysm the activity corresponds to a series of isolated Strombolian explosions, the paroxysm is marked by either a very closely spaced series of Strombolian explosions, a regime called transitional by Parfitt & Wilson (1995) or a sustained Hawaiian fire fountain (Behncke *et al.* 2006; Polacci *et al.* 2006). The magma column is pulsating at a relatively low level during a transitional activity whereas it stays at a fairly constant level of a few hundreds of metres during a fire fountain (Behncke *et al.* 2006). Here, we refer to episodes without a transition towards a fire fountain as quasi-fire fountains because of their richness in gas. In rare occasions, a third regime, also very rich in gas and similar to a small ash-rich column, can exist. A few days of relative quiescence separate each phase rich in gas. Sixteen and 64 episodes of fully developed fire fountains, described as Hawaiian-style, have been reported in 1989 (Bertagnini *et al.* 1990; Privitera *et al.* 2003) and in 2000 (Behncke & Neri 2003; Alparone *et al.* 2003; Behncke *et al.* 2006; Polacci *et al.* 2006), respectively, whereas 22 and 16 episodes of quasi-fire fountains are described in 1998–1999 (La Delfa *et al.* 2001) and in 2001 (Behncke *et al.* 2006; Allard *et al.* 2006), respectively. This behaviour is qualitatively very similar to that of an Hawaiian volcano (Behncke *et al.* 2006), especially during the summit eruption of Kilauea Iki (1959–1960) where 16 fire fountains occurred with a repose time varying between seven hours to four days (Richter *et al.* 1970). However, the repose time at Kilauea can also be one month for the two other large eruptions occurring on the East rift zone, with 12 and 47 episodes of fire fountains for Mauna Ulu (1969–1970) and Pu'u 'O'o eruptions (1983–1986), respectively (Swanson *et al.* 1979; Wolfe *et al.* 1987, 1988; Heliker & Mattox 2003).

Therefore, most of the recent eruptions of Etna, which are at the transition between two classical types of eruption, Strombolian and Hawaiian, can shed light on the links between the two regimes. In this paper, acoustic measurements made between 4 and 13 July, 2001 will be first used to quantify the bubble volume and overpressure during a strong Strombolian episode and one leading to a fire fountain. Then the discrepancy in gas volume obtained from acoustic measurements and from the height of fire fountains will be explained. Lastly, the time evolution of the source will be discussed, together with an interpretation of the origin of fire fountains.

Description of the 2001 eruption

The 2001 eruption of Etna volcano started on 9 May with a mild Strombolian activity and 15 episodes, and lava flows were reported until 13 July (BGVN 2001). Activity occurred at the southeast crater, a scoriae-built cone formed in 1971, at 1 km from the summit craters (Calvari *et al.* 2001, 2002). Lava effusion had been at an output rate of 1 to 10 m^3 s^{-1} since 20 January and there was a progressive build-up of seismic energy with two strong increases in February and in the first half of April (Alparone *et al.* 2004). The eruption consisted of an alternation between eruptive episodes, some with a transition to a fully developed fire fountain, and quiet periods at most limited to degassing at all the summit vents (BGVN 2001; Lautze *et al.* 2004; Behncke *et al.* 2006). Volcanic tremor increased several hours before surface activity started (Behncke *et al.* 2006). It usually began with a lava flow, located on the side of the southeast cone and lasting several hours, before ejecta were propelled at the southeast crater. Each episode started with five to 11 hours of build-up with a series of Strombolian explosions and lava flows, followed by more intense events lasting one hour (BGVN 2001; Lautze *et al.* 2004; Behncke *et al.* 2006). At the beginning of an eruptive episode, explosions were infrequent (i.e. every few minutes), but became noisy and frequent (i.e. every several seconds), after one or two hours

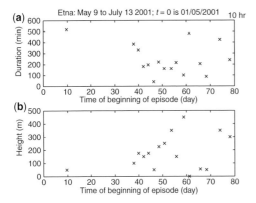

Fig. 3. Fire fountain characteristics during the 2001 eruption as a function of time since May 1 2001 (BGVN 2001): (**a**) duration (min), (**b**) maximum height (m). Note that height and duration of fire fountains have been put equal to 0 if there were no observations. The duration of fire fountains mostly decreases from 520 minutes to 40 minutes, while their height increases until June 30, from 10 m to 450 m. The minimum duration of 40 minutes (15 June, episode 6, day 46) is due to using the duration of the paroxysm rather than the duration of the entire episode. The episode before last, on 12 July, represents one of the largest and most violent fire fountains during the 2001 eruption (Behncke *et al.* 2006). The eruption turns to a flank eruption during the last episode, on 17 July.

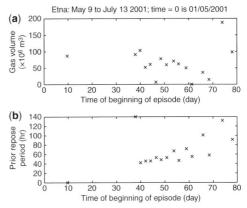

Fig. 4. The 2001 eruption: (**a**) Gas volume released at atmospheric pressure (10^6 m^3) as a function of time since May 1 2001. (**b**) Repose periods prior to fire fountains (h) (BGVN 2001). Note that the gas volume and repose time have been put equal to 0 if there were no observations of fire fountains and that the repose time at 140 hr on day 38 is off scale at 670 hr. Gas volumes decrease from 9.0×10^7 m^3 to 1.4×10^7 m^3, with a peak value of 1.9×10^8 m^3 on 12 July. Gas volume on 9 May is very small due to a very low level of surface activity compared with other episodes. Combining that with the extremely large repose period, 670 hr (June 6, episode 2, day 38), suggests discarding the episode on 9 May as being a proper quasi-fire fountain and it is solely the mark of an open vent. The lowest gas volume is obtained when using the duration of the paroxysm instead of that of the entire episode (15 June, episode 6, day 46). Repose periods prior to fire fountains increase from 42 hr to 132 hr. Gas volumes should be multiplied by the degree of unsteadiness-0.39 or 0.11 for a purely Strombolian episode or an episode leading to a fire fountain, respectively.

(Behncke *et al.* 2006). During its paroxysm, the very closely spaced series of Strombolian explosions, i.e. a transitional activity, sometimes gave way to a well-developed fire fountain (Behncke *et al.* 2006). During the last hour, explosions became sparse and relatively quiet. Each episode lasted from one hour and 30 minutes (7 July) to 10 hours followed by a repose period of two to five days (BGVN 2001; Behncke & Neri 2003; Behncke *et al.* 2006; Fig. 3). Activity was almost steady for the duration of the eruption with a slight and regular increase of the repose time between episodes to a maximum of five days and 12 hours before the very strong activity of 12 July 2001 (Fig. 4). Although episodes generally followed the same evolutionary scheme, they gradually became stronger and significant lava volumes were erupted (Behncke *et al.* 2006). On 17 July, the activity totally changed and the first flank eruption in nearly 10 years of activity started: new fractures opened up, strong lava flows developed and fire fountains did not form anymore at the southeast crater, except early on 17 July. The eruption finished on 9 August (Calvari *et al.* 2001; Behncke & Neri 2003; Behncke *et al.* 2006). In contrast with the flank eruption, the number of studies on the period prior to 17 July is very limited

(Métrich *et al.* 2004; Spilliaert *et al.* 2006*a,b*; Behncke *et al.* 2006; Allard *et al.* 2006), and this period is the scope of our paper.

Results from raw acoustic pressure

Acoustic measurements

Acoustic pressure was successfully recorded for two episodes (4 and 12 July; Fig. 5) among a series of three (4, 7 and 12 July). The surface activity, which consisted of a close series of Strombolian explosions, shares some similarity for the three episodes (4, 7, 12 July). If the duration of the episode was estimated on our visual observations close to the vent, it varied between 1 hr 30 min (7 July), 3 hr 25 min (4 July) and 4 hr 16 min (12 July). However, the start was often very sluggish, so the exact beginning was probably roughly one hour earlier than reported. The continuous recording on 12 July lasted seven hours with Strombolian explosions and a transition towards a fire fountain (04 hr 07 min) before the episode stopped (05 hr 04 min). On 4 July, the activity at the vent stayed purely Strombolian for the entire duration of the episode. On 7 July, an ash-rich eruptive cloud containing lapilli rose to 500–600 m placing the start of the episode at 7 hr 45 min (BGVN 2001). The plume continued its ascent up to 1.5–2 km above the crater area and one of the largest outbursts produced widespread ashfalls (BGVN 2001). The ash-rich eruptive cloud was very well collimated, relatively slow rising, and did not produce large sound waves (C. Jaupart pers. comm. 2007) in contrast with Strombolian explosions, such as on 4 July. Pulsations within the ash-rich eruptive cloud were observed on 7 July, as well as mild Strombolian explosions. In this paper, times are given in UTC (coordinated universal time).

The acoustic sensor, set at a distance of 950 m from the vent, was an infrasonic microphone (Bruel Kjaer 4193) amplified by Nexus (Bruel Kjaer), and had a wide frequency range, from 0.1 Hz to several kHz. Recording on 4 July was performed on a digital acoustic station (Vibra4, Tad) at a sampling frequency of 1200 Hz and a dynamical range of 8

bits. On 12 July, the acoustic pressure was recorded at Tore del Filisofo, at 1070 m from the vent, with a 16 bits accuracy and a sampling frequency of 64 Hz. The second set-up, which consisted of an electret microphone, with a sensitivity of 4.6 mV/Pa and pre-amplified ten times before A/D conversion, has a flat frequency response between 4 and 15 Hz with an attenuation of -3 dB at 2.2 Hz. Acoustic pressure was corrected for the smallest frequencies, as low as 1 Hz. Both microphones were installed very close to the ground to minimize the wind effect, measured between 4 and 6 m s^{-1} at the sensor site. The absence of correlation between wind speed and acoustic pressure shows that acoustic pressure radiated by eruptive activity largely exceeded that of the wind.

Acoustic pressure on 4 and 12 July

Acoustic pressure shows a series of pulses, whose intensity can reach 100 Pa (Fig. 5). The intensity and occurence of these pulses change in time. Explosions are the strongest and most frequent during the third hour of activity on 4 July (Fig. 6) whereas the paroxysm on 12 July is reached after 6 hours and 16 minutes (Fig. 5). The duration of the paroxysm, from 21 hr 24 min to 22 hr 07 min on 4 July and from 4 hr 07 min to 4 hr 47 min on 13 July, is similar for both episodes.

Strombolian explosions at Etna have a waveform (Fig. 7) very similar to explosions at Stromboli and Shishaldin volcanoes, with a first positive peak of less intensity than the negative peak

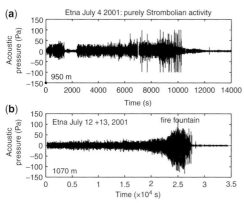

Fig. 5. Acoustic pressure recorded at (**a**) 950 m in the range 0.1 Hz–600 Hz and at a time $t = 0$ equal to 19 hr 07 min 35.2 s on 4 July and at (**b**) 1070 m in the sole infrasonic range, flat between 4 Hz–15 Hz with an attenuation of -3 dB at 2.2 Hz at a sampling frequency of 64 Hz. $t = 0$ is equal to 22 hr 04 min 00 s on 12 July. Acoustic pressure for an episode with purely Strombolian activity, such as on 4 July (a), is fairly constant despite dropping suddenly for a few tens of minutes during the first hour in agreement with visual observations. Acoustic pressure on 12 July (b) undergoes a significant increase towards the transition with a fire fountain. The duration of the paroxysm, from 21 hr 24 to 22 hr 07 min on 4 July (a) and from 4 hr 07 to 4 hr 47 min on July 13 (b), is similar for both episodes.

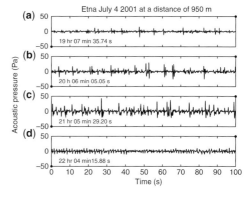

Fig. 6. Time evolution of acoustic pressure on 4 July 2001 for the first 100 s at the beginning of every hour, starting at (**a**) 19 hr 07 min 35 s (**b**) 20 hr 06 min 05 s (**c**) 21 hr 05 min 29 s (**d**) 22 hr 04 min 15 s. Acoustic pressure shows a close series of Strombolian explosions, whose intensity reaches its paroxysm during the third hour before dying off, after 3 hours and 25 minutes of recording. Recording was not done for the early beginning, hence probably the first hour is missing.

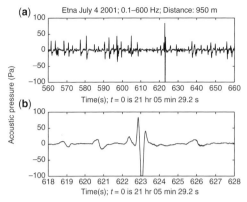

Fig. 7. Acoustic pressure measured at 950 m from the southeast crater, three hours after the start of episode. Time $t = 0$ s corresponds to 21 hr 05 min 29.2 s on 4 July 2001 and explosion occurs at 21 hr 15 min 52 s. Sampling frequency is 1200 Hz. (**a**) 10 s of acoustic measurements. (**b**) enlargement of an explosion. Simultaneous visual observations of the vent showed that large bubbles, similar to those photographed on 11 September 1989 (Fig. 2; Bertagnigni *et al.* 1990) or on 25 July 2001 (Fig. 1), were frequently breaking at the vent.

(Vergniolle & Brandeis 1994, 1996; Vergniolle *et al.* 2004). Furthermore, the acoustic pressure of an explosion has an amplitude which decreases inversely proportional to the distance, suggesting that the source is a monopole. Hence, every strong pulse at Etna is characteristic of a bubble breaking at the top of the magma column (Vergniolle & Brandeis 1996; Vergniolle *et al.* 2004). Simultaneous visual observations show that ejecta frequently form a hemisphere above the vent. This is indeed characteristic of breaking a radially expanding bubble at the surface of a liquid, and is extremely similar to an explosion on 25 July (Fig. 1). Similar explosions, associated with the arrival of large bubbles at the vent, also occurred in November and December 1995 at the northeast crater (Coltelli *et al.* 1998b).

The strong explosion at 21 hr 15 min 52 s (Fig. 7b) shows three precursory peaks in acoustic pressure, separated by 2 s (equivalent frequency of 0.5 Hz) and each with a duration of 0.76 s (equivalent frequency of 1.3 Hz). This feature also exists at Stromboli and has been explained by surface waves, called sloshing waves, induced at the top of the magma column by the bubble approaching the surface (Vergniolle *et al.* 1996). The frequency of sloshing waves depends on the radius of the conduit. It is around 1 Hz for a conduit radius of 1 m as for Stromboli. At Etna, the conduit appears to be 7–8 m in diameter at Bocca Nuova in 1968 and 2–4 m at the northeast crater in 1995

(LeGuern *et al.* 1982; Coltelli *et al.* 1998b) whereas at the southeast crater it is equal to 5 m in 1986 and 10–15 m in 1998 and 2000 (Calvari *et al.* 1994; La Delfa *et al.* 2001; Bonaccorso 2006). The direct observation of the vent, one hour after the end of the episode on 7 July 2001, shows that the vent diameter is 10 m and is located within a small crater of a diameter 20 m (C. Jaupart, pers. comm. 2007). Therefore, the frequency of the sloshing waves is expected between 0.5 Hz and 1 Hz (Vergniolle *et al.* 1996). Explosion frequency is lower at Etna (2 Hz) than at Stromboli (9 Hz) implying that the source at Etna is larger than at Stromboli (Vergniolle & Brandeis 1996; Vergniolle *et al.* 1996). Hence, the conduit is also wider and the frequency of sloshing waves at Etna is lower than 1 Hz.

Spectral content of explosions on 4 and 12 July

Acoustic pressure has been analysed with fast Fourier transforms for a duration of 100 s selected at the beginning of every hour of activity on 4 July (Fig. 8). Several peaks are observed and the frequency between 1 Hz and 2 Hz can be associated with bubble breaking (Fig. 7). Its intensity increases from 0.35 to 0.65 Pa, reaches a maximum of 1 Pa before disappearing during the last hour (0.04 Pa in Fig. 8d). The frequency at 0.5 Hz is compatible with sloshing waves at the surface of the magma, implying a bubble radius of 5 m and in agreement

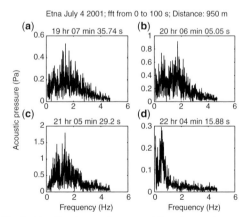

Fig. 8. Fast Fourier transforms on 100 s of acoustic pressure selected at the beginning of every hour of activity on 4 July, starting at (**a**) 19 hr 07 min 35 s (**b**) 20 hr 06 min 05 s (**c**) 21 hr 05 min 29 s (**d**) 22 hr 04 min 15 s. Explosions have a frequency between 1 and 2 Hz. The approach of large bubbles towards the surface triggers sloshing waves, with a frequency *c.* 0.5 Hz, compatible with a bubble radius of 5 m.

with visual observations of conduit radius (LeGuern et al. 1982; Calvari et al. 1994; Coltelli et al. 1998b; La Delfa et al. 2001; C. Jaupart pers. comm. 2007). The frequency of 0.5 Hz has an amplitude which increases as the 1–2 Hz mode becomes stronger, suggesting that the frequency of 0.5 Hz is also induced by the behaviour of large bubbles. This is indeed compatible with sloshing waves at the top of the magma column, suggesting that the bubble radius is close to 5 m. There are very few events corresponding to bubble breaking during the last hour of activity (Fig. 8d) and the sound, of lesser intensity, is probably produced by the residual motion of magma into the conduit.

The bubble is almost as large as the conduit (Fig. 16) and the thickness of the lateral film around the slug, δ_∞, depends on the magma viscosity, μ_{liq}, as (Batchelor 1967; Vergniolle et al. 1996, 2004; Vergniolle & Caplan-Auerbach 2006):

$$\delta_\infty = 0.9 R_{cond}\left(\frac{\mu_{liq}^2}{\rho_{liq}^2 R_{cond}^3 g}\right)^{1/6} \qquad (1)$$

where ρ_{liq}, g and R_{cond} are the magma density, 2700 kg m^{-3} (Williams & McBirney 1979), the acceleration due to gravity, 9.81 m s^{-2} and the conduit radius, assumed at first order to be the bubble radius R_0. The high dissolved water content at Etna, 3.4 wt%, combined with a temperature of 1140 °C, suggests that the magma viscosity could be as low as 1 Pa s (Giordano & Dingwell 2003; Métrich et al. 2004). However, the magma may have a phenocryst content similar to that of scoriae, from 19% to 40% (Polacci et al. 2006), suggesting that a value of 10 Pa s is probably the best estimate of the in-situ viscosity (Dingwell 1998). For a magma viscosity equal to 10 Pa s, the lateral film has a thickness of 0.22 m and the conduit radius is equal to 5.2 m (equation 1).

On 12 July, the reduced recording frequency band, compared with 4 July, only allows observation of explosions but not sloshing waves. These occur at the same frequency (Fig. 9) and with similar waveforms to those on 4 July. The increase in acoustic pressure corresponds to a widening in the frequency range, both towards the lowest and the highest frequency (Fig. 9).

Results from the acoustic synthetic waveforms

Model of bubble vibration

The similarity between acoustic waveforms at Stromboli and Etna suggests that the source is the same (Fig. 7b). Hence, each explosion at Etna is modelled by the vibration for a large bubble, induced by a residual overpressure left in the bubble approaching the top of the magma column (Vergniolle & Brandeis 1996; Vergniolle et al. 2004). Similarly to Stromboli, the low frequency content of the explosion suggests discarding a mechanism solely based on bubble bursting like a balloon (Vergniolle & Brandeis 1994). After less than one cycle, the bubble may break while in contraction by the growth of instabilities at the bubble surface (Vergniolle & Brandeis 1996; Vergniolle et al. 2004). Equations for bubble vibration and the related acoustic pressure can be found in Vergniolle & Brandeis (1996) and Vergniolle et al. (2004). The mass of the oscillator is related to the thickness of the magma layer above the bubble. This parameter, difficult to constrain, is assumed to be equal to the mean size of ejecta (Vergniolle & Brandeis 1996; Vergniolle et al. 2004). Photoballistic studies for one explosion of the northeast crater in June 1969 shows that 50% of ejecta have a size between 10 and 40 cm, with a median value at 15 cm (McGetchin et al. 1974), whereas the mean diameter is only 2 cm at Stromboli (Chouet et al. 1974). Scoriae blocks produced by Strombolian activity on 25–31 July 2001 also had a mean size between 10 and 40 cm (Métrich et al. 2004), making these dimensions characteristic of ejecta at Etna.

The best fit between a synthetic waveform based on the bubble vibration and each pulse recorded during the two episodes of 4 and 12 July provides an estimate of bubble radius, length and overpressure (Vergniolle & Brandeis 1996; Vergniolle et al. 2004). Although the signal of acoustic pressure is saturated into its negative part, the fit is very good (Fig. 10) for a bubble with a radius of 5 m, a length of 8 m and an overpressure of 0.39 MPa if the thickness of the magma layer above the bubble is 10 cm (Fig. 10). If the thickness of magma layer is equal to 15 cm, bubble radius and length are equal to 5 m and 2 m, respectively, for an overpressure of 0.85 MPa. Our determination of bubble radius is in perfect agreement with visual observations and the interpretation of the frequency content (see previous section). Simultaneous visual observations show that both the bubble hemispherical cap and the bubble bottom can sometimes be observed at the vent in quick succession. The delay of a few seconds between the appearance of both sections of the bubble gives an estimate of the bubble length if we assume that the bubble bottom is still rising at its equilibrium velocity, the one calculated far from the top of the magma column. A bubble as large as the conduit, i.e. a slug, rises at a constant velocity U_{gslug}, which only depends on the conduit radius R_{cond} (Wallis

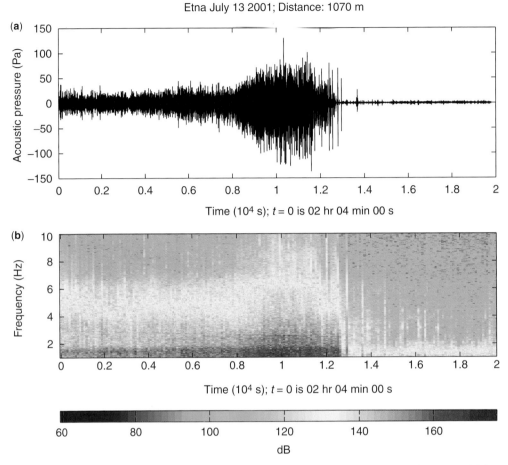

Fig. 9. (**a**) Acoustic pressure during the second half of July 12 episode, with $t = 0$ is 02 hr 04 min 00 s on 13 July (**b**) Spectrogram in decibel scale (dB). Intensity in dB, I_{dB}, is equal to $20 \times \log_{10}(P_{ac}/P_{ref})$, where P_{ac} and P_{ref} are acoustic pressure and reference pressure, equal to 2×10^{-5} Pa, respectively. Note that the increase in acoustic pressure corresponds to a widening in the frequency range, at the time when the fire fountains are observed. The episode clearly ends at 05 hr 04 min 00 s.

1969; Vergniolle & Jaupart 1986; Vergniolle & Caplan-Auerbach 2006):

$$U_{gslug} = 0.345\sqrt{2gR_{cond}} \qquad (2)$$

Because a bubble of 5 m in radius rises at an upwards velocity of 3.5 m s^{-1}, a delay of 2–3 seconds between the bubble top and bottom corresponds to a bubble length between 7 and 11 m. Indirect measurements of the bubble length can also be provided by a radar. This instrument, whose Doppler shift gives estimate on velocity, receives a reflected power, which is proportional to the mass of the ejecta. During a Strombolian explosion, ejecta mainly come from top and

bottom of the bubble, by analogy from a small bubble breaking at the surface of the ocean (Resch et al. 1986). Hence, the two bursts of energy, observed on 11–12 October 1998 one hour after a fully developed fire fountain (Dubosclard et al. 2004), and separated by 6 s, correspond to a bubble length of 21 m. Although the bubble seen on 25 July (Fig. 1) has already decompressed and burst as shown by numerous discrete ejecta, its volume is compatible with our estimates from acoustic measurements. Therefore, the estimate of 0.10 m for the thickness of the magma layer above a bubble is the most appropriate. It is most likely that this parameter does not fluctuate significantly between explosions (Vergniolle & Brandeis 1996; Vergniolle et al. 2004), hence a thickness

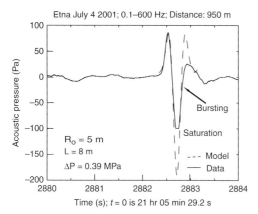

Fig. 10. Comparison between a synthetic waveform and data recorded on 4 July 2001. $t = 0$ is 21 hr 05 min 29.2 s, so the explosion occurs at 21 hr 53 min 31 s, when where large bubbles were frequently observed breaking at the vent. The constant acoustic pressure at -100 Pa results from the saturation of the instruments. Bubble radius, length and overpressure are equal to 5 m, 8 m and 0.39 MPa, respectively. Results from the bubble vibration model depend on the thickness of the magma layer above the bubble, h_{eq} assumed to be equal to 0.10 m, according with observations (see text for details).

of 0.10 m is the value chosen for each explosion. At Etna, the equilibrium thickness of the magma layer above the bubble is 1% of its diameter, in the same way as for Stromboli and Shishaldin (Vergniolle & Brandeis 1996; Vergniolle *et al.* 2004). Our

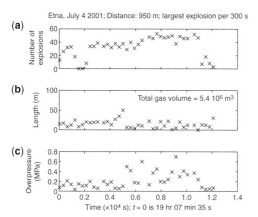

Fig. 11. Results from synthetic waveforms done manually on the largest explosion per five minutes occurring on 4 July $t = 0$ is 19 hr 07 min 35 s. (**a**) Number of explosions. (**b**) Bubble length (m). (**c**) Bubble overpressure (MPa). Eruption stopped at 22 hr 40 min, after 3 hours and 25 minutes of recording. The total gas volume when calculated at atmospheric pressure is *c.* 5.4×10^6 m^3 for the all duration of the episode (Table 1).

determination of the gas volume is very robust because the bubble vibration model uses the sound wave frequency and our results are probably valid within a 20% accuracy (Vergniolle *et al.* 2004). Furthermore, both the estimates on bubble radius and length are in excellent agreement with visual observations, reinforcing the validity of our determination of gas volume.

Time evolution of the source of Strombolian explosions

Due to the large number of explosions, we have only selected the largest explosions per five minutes. The best fit between measured and synthetic acoustic pressure is done manually, which gives the maximum accuracy. Bubble length and overpressure are 15 ± 9.9 m and 0.22 ± 0.16 MPa and 8.6 ± 15.7 m and 0.20 ± 0.083 MPa on 4 and 12 July, respectively (Figs 11 & 12).

The number of explosions per five minutes and the bubble length stay roughly constant on 4 July, at 33 ± 15.5 and 15 ± 9.9 m, respectively (Fig. 11; Table 1). Despite the increase in the bubble overpressure, from 0.12 MPa to 0.29 MPa

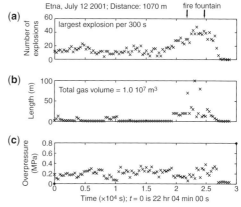

Fig. 12. Results from synthetic waveforms done manually on the largest explosion per five minutes occurring on 12 July. $t = 0$ is 22 hr 04 min 00 s. (**a**) Number of explosions. (**b**) Bubble length (m). (**c**) Bubble overpressure (MPa). Simultaneous visual observation, marked by the vertical arrows, shows that a fire fountain starts to develop at 04 hr 07 min, which is in perfect agreement with a sudden increase in bubble length, from 20 m to 70 m between 04 hr 05 min and 04 hr 10 min, and with the number of explosions, 30 per five minutes. The paroxysm is reached at 04 hr 20 min 00 s, with a bubble length of *c.* 100 m and more than 50 explosions per five minutes. The eruption stopped at 05 hr 04 min, after seven hours. The total gas volume when calculated at atmospheric pressure is *c.* 1.0×10^7 m^3 for the whole duration of the episode (Table 1).

Table 1. *Physical parameters at the vent of Etna*

Etna 2001 (SEC) Surface activity	4 July Str	7 July Th/Str	12 July ff
Acoustic			
$n_{totslug}$	1284	unknown	1614
n_{slug}	33 ± 16	unknown	17 ± 11
L_{slug} (m)	15 ± 9.9	unknown	8.6 ± 16
ΔP_{slug} (MPa)	0.22 ± 0.16	unknown	0.20 ± 0.083
V_{mean} ($\times 10^5$ m^3)	1.4 ± 1.0	unknown	1.0 ± 1.7
V_{max} ($\times 10^5$ m^3)	3.6	unknown	9.6
C_{unst}	0.39	0.39?	0.11
V_{gac} ($\times 10^7$ m^3)	0.54	0.22?	1.0
Fire fountain			
V_{ff} ($\times 10^7$ m^3)	1.4	1.2	2.1
V_{gact}/V_{gtot}	0.38	0.38?	0.48
α_{wake} (%)	65 (54–82)	unknown	44 (36–55)
Ejecta			
V_{mag} ($\times 10^5$ m^3)	1.5	unknown	1.7
$V_{magwake}$ ($\times 10^6$ m^3)	4.7	unknown	14
V_{totmag} ($\times 10^6$ m^3)	0.15– 4.7	unknown	0.17–14
V_{ejecta} ($\times 10^6$ m^3)	0.73	unknown	0.77

SEC, Str, th and ff stand for the SE crater, Strombolian activity, thermal and fire fountain, respectively. $n_{totslug}$, n_{slug} L_{slug}, ΔP_{slug} are the total number of bubbles, the number of bubbles per five minutes, the bubble length and overpressure, respectively. V_{mean}, V_{max} and C_{unst} correspond to the average gas volume expelled per five minutes, its peak value and their ratio, called the degree of unsteadiness, respectively. V_{gac}, V_{ff} are the total gas volume based on acoustic records and based on the height of fire fountains once corrected by the degree of unsteadiness, respectively. V_{gact}/V_{gtot} and α_{wake} correspond to the proportion between the volume of the slugs coming from the reservoir (active) and the total gas volume and gas volume fraction within the slug wake, respectively (Fig. 16). V_{mag}, $V_{magwake}$, V_{totmag} and V_{ejecta} are the magma volume around the slug, the magma volume within the slug wake, the total magma volume and the measured ejecta volume (Behncke et al. 2006).

after 6000 s, the vent is displaying purely Strombolian activity, without significant change over the 3 hour 25 minutes duration. However, on 12 July, the number of explosions, initially around 10, shows a regular increase towards the last third of the episode and reaches a peak value at 50 explosions per five minutes (Fig. 12; Table 1). The bubble length, initially fairly constant at 2.6 ± 3.2 m, increases towards the end of the episode to a peak value of 100 m (Fig. 12). Our visual observations show that a transition to a fully developed fire fountain starts at 4 hr 07 min and lasts 40 minutes. During that period, between 04 hr 05 min and 04 hr 45 min, the bubble length exceeds 50 m (Fig. 12), suggesting that the transition between a Strombolian explosion and a fire fountain occurs when the bubble length exceeds ten times the bubble radius, here 5 m. This criteria is in perfect agreement with a fire fountain being driven by a long inner gas jet (Fig. 2) and corresponding to an annular fow (Wallis 1969). This period also has the maximum number of explosions. However, the gas overpressure does not change over the 7 hours' duration of the episode

(Fig. 12), and the visual trend in increasing acoustic pressure during the transition with a fire fountain (Fig. 5) is due to both the regular increase in the number of explosions and in the bubble length (Fig. 12). Therefore, a fire fountain at Etna simply results from a close series of very long Strombolian explosions (Figs 11 & 12). A fire fountain was also observed for a period of 40 minutes on 14 June, 2000 (Allard et al. 2005), with a period varying between 10 to 60 minutes for other episodes (Alparone et al. 2003), suggesting that 40 minutes is typical for the duration of fire fountains at Etna.

The large overpressure measured in Strombolian bubbles, ≥ 0.1 MPa (Figs 11 & 12), largely exceeds the pressure obtained by the coalescence of two bubbles in the conduit, at most several thousands of Pa (Vergniolle & Caplan-Auerbach 2004). This implies that a fire fountain cannot solely result from a progressive coalescence of bubbles in the conduit as suggested by Parfitt & Wilson (1995), Parfitt et al. (1995) and Parfitt (2004).

The number of bubbles during the paroxysm, 50–60 explosions per five minutes (Figs 11 & 12), is so high that this activity should be called transitional, as defined by Parfitt & Wilson (1995), even during a fully developed fire fountain. However, a fire fountain at Etna is described as a continuously sustained jet of liquid magma and gas, hence clearly corresponds to an Hawaiian-style fire fountain (Benhcke *et al.* 2006; Polacci *et al.* 2006). At Kilauea, pulsating fire fountains were also often reported (Richter *et al.* 1970; Swanson *et al.* 1979; Wolfe *et al.* 1987), suggesting that indeed fire fountains at Etna and Kilauea have the same origin.

Comparison with other Strombolian activity

Strombolian activity recorded in June 1998 at Voragine crater (summit area) has shown that pressure pulses were only 1 to 7 Pa at a distance of 200 m (Ripepe *et al.* 2001). If we assume that the source is approximatively of the same size, because frequency is similar in 1998 and 2001, acoustic pressure of 7 Pa is 50 times smaller than on 4 July 2001, giving a bubble overpressure of 0.08 MPa.

The bubble radius, length and overpressure at Etna in 2001 (5 m, 10–100 m, 0.2 MPa) are similar to those at Stromboli (0.9 m, 7 m and 0.2 MPa; Vergniolle & Brandeis 1996) and Shishaldin (5 m, 24 m, 0.15 MPa; Vergniolle *et al.* 2004). The Strombolian activity of Heimaey (Iceland) in February 1973 showed large bubbles of several metres in diameter updoming at the surface of the lava and having an excess pressure of 0.025 MPa (Blackburn *et al.* 1976). This is much smaller than explosions measured at other Strombolian volcanoes, suggesting that the bubbles had already expanded.

Gas volume

Gas volume from acoustic measurements

Visual observations (BGVN 2001) can also be used to deduce a gas volume. However, this gas volume corresponds to the value obtained after decompression at atmospheric pressure. The strength of explosions, described from visual observations (BGVN 2001), is directly related to the bubble overpressure. The gas volume from acoustic measurements is obtained by multiplying the number of explosions recorded per five minutes with the volume of a typical bubble. We choose the most overpressurized bubble as characteristic to ensure the best ratio between signal and noise and also because its acoustic waveform is very similar to others explosions. Its gas volume only exceeds by 22% the gas volume of the average explosion obtained from stacking all explosions per five minutes. Therefore, our determination of the gas

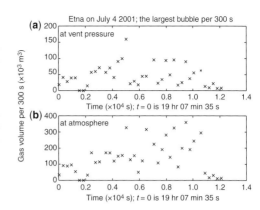

Fig. 13. Gas volume expelled per five minutes on less 4 July 2001 (**a**) at the pressure of the bubble at the vent prior to bursting (**b**) at atmospheric pressure. The average gas volume at atmospheric pressure, $1.4 \times 10^5 \pm 1.0 \times 10^5$ m³, is lower by a factor of 2.5 from its peak value, 3.6×10^5 m³, as a result of the quasi-steady activity (Fig. 11). That factor is extremely useful to quantify the average gas volume from its maximum, as done when using the maximum height of fire fountains.

volume from the most overpressurized bubble is reliable.

The gas volume expelled in the atmosphere can then be deduced from the gas volume at the vent by using the perfect gas law and the overpressure estimated from acoustic measurements (Figs 11 & 12). The gas volume expelled in the atmosphere, $1.4 \times 10^5 \pm 1.0 \times 10^5$ m³, exceeds its value at the vent, $4.7 \times 10^4 \pm 3.4 \times 10^4$ m³, by a factor of 3.0 on 4 July due to the bubble overpressure (Fig. 13). On 12 July, the gas volume at atmospheric pressure, $1.0 \times 10^5 \pm 1.7 \times 10^5$ m³ exceeds the one at the vent, $2.5 \times 10^4 \pm 4.8 \times 10^4$ m³, by a factor of 4.3 (Fig. 14).

The average gas volume released at atmospheric pressure per five minutes, $1.4 \times 10^5 \pm 1.0 \times 10^5$ m³, differs by a factor of 2.5 from its peak value, 3.6×10^5 m³ on 4 July (Fig. 13; Table 1). This difference is much larger on 12 July with a factor of 9.0 between average gas volume per five minutes, $1.0 \times 10^5 \pm 1.7 \times 10^5$ m³, and maximum gas volume, 9.6×10^5 m³ (Fig. 14; Table 1). The difference betwen average and maximum gas volume represents the steadiness of the surface activity and a large value is obtained for a very unsteady episode. The difference is indeed large for an episode leading to a fire fountain, which is visually observed as very unsteady (Benhcke *et al.* 2006; Polacci *et al.* 2006) whereas an episode with solely Strombolian explosions is relatively steady and has a smaller value. Hence, we define the degree of unsteadiness, C_{unst}, by the ratio

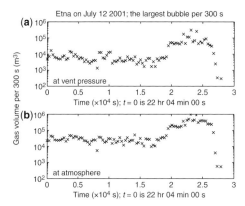

Fig. 14. Gas volume expelled per five minutes on 12 July 2001 (a) at the pressure of the bubble at the vent prior to bursting and (b) at atmospheric pressure. The average gas volume, $1.0 \times 10^5 \pm 1.7 \times 10^5$ m^3, at atmospheric pressure is lower by a factor of 9.0 from its peak value, 9.6×10^5 m^3. This large factor is related to the unsteadiness of a fire fountain episode at Etna. It is extremely useful to quantify the average gas volume from its maximum, as done when using the maximum height of fire fountains. Note that there is a sudden increase in gas volume at the transition to fire fountain, as already seen on the number of explosions and the bubble length (Fig. 12).

between the average gas volume and the maximum gas volume and this varies between 0.39 and 0.11 for a Strombolian activity and a fire fountain, respectively.

The total gas volume expelled in the atmosphere can be estimated from the number of explosions and the gas volume present in each explosion, leading to a gas volume of 5.4×10^6 m^3 on 4 July and 1.0×10^7 m^3 on 12 July when calculated at atmospheric pressure. However, the large gas volume on 12 July compared with 4 July partially reflects that the continuous recording on 12 July gives the early start of the episode. If we compare the 4 hours and 16 minutes before 05 hr 04 min on 12 July as being the equivalent of 3 hours and 25 minutes on 4 July there is a gas volume of 9.0×10^6 m^3 expelled on 12 July. It is still above the value on 4 July even when considering the 20% accuracy in the results obtained from the best-fit method on the bubble vibration model (Vergniolle *et al.* 2004). Hence, the gas volume on 12 July is 1.7 times the one obtained on 4 July. The quantitative analysis of the foam coalescence is given in the companion paper (Vergniolle 2008).

Gas volume from surface activity

The gas volume can be estimated by analogy with Kilauea from the height of fire fountains, its

duration τ_{ff} and the area of the vent S_{vent} (Wilson 1980; Vergniolle & Jaupart 1990; Vergniolle 2008). The fire fountain height H_{ff} is related to the gas velocity, U_{ff}, at the vent (Wilson 1980; Vergniolle & Jaupart 1990; Sparks *et al.* 1997) by:

$$U_{ff} = \sqrt{2gH_{ff}} \tag{3}$$

Equation 3, which results from a balance between kinetic and potential energy, is valid for both a steady fire fountain and a single Strombolian explosion. It neglects the entrainment of air from atmosphere into the fire fountain. However, when it exists, the radius increases with height, as for plumes (Turner 1973). This is not observed (Fig. 2), hence equation 3 is very appropriate to calculate the gas volume. Furthermore, a fire fountain of 300 m high, pushed by a vertical velocity of $\cong 77$ m s^{-1} if equation 3 is valid, has been monitored by a radar measuring an average ejecta velocity of $\cong 86$ m s^{-1}, in very good agreement with equation 3 (Dubosclard *et al.* 1999). At Stromboli, the average ejecta velocity measured by a Sodar, 40–50 m s^{-1}, is also in very good agreement with that deduced from the explosion height, 44–54 m s^{-1} (Weill *et al.* 1992). However, the gas velocity is twice the average ejecta velocity at Stromboli (Weill *et al.* 1992), suggesting that equation 3, although very accurate for finding the average velocity, might underestimate the gas velocity at most by a factor of 2.

Conservation of the gas flux at the vent leads to the total gas volume expelled by a fire fountain, V_{ff}, equal to (Vergniolle & Jaupart 1990):

$$V_{ff} = U_{ff} S_{ff} t_{ff} \tag{4}$$

A height of 50–60 m as on 4 July leads to a gas velocity of 33 m s^{-1}, whereas on 12 July a height of 300–400 m for the fire fountain corresponds to a gas velocity of 83 m s^{-1} (Table 2). The respective duration of each episode, 3 hours 25 minutes and 7 hours, leads to a gas volume at atmospheric pressure of 3.8×10^7 m^3 and 2.0×10^8 m^3. This greatly exceeds the values estimated from acoustic measurements, 5.4×10^6 m^3 and 1.0×10^7 m^3, respectively (Table 1).

The discrepancy results from the assumption that the maximum height, the only one reported, is close to the average height. But the time evolution of the bubble characteristics shows that the average gas volume is less than the maximum gas volume and should be corrected by the degree of unsteadiness, 0.39 and 0.11 on 4 and 12 July, respectively (Figs 13 & 14; Table 1). Although both gas volumes are estimated from synthetic acoustic waveforms, the ratio between them is

Table 2. *Data from surface activity during the 2001 eruption of Etna (BGVN 2001)*

Ep.	Start	End	Paroxysm	Height (m)	Duration	Repose
1	May 9 15 hr 20	May 10 0?	May 9 15 hr 20–15 hr 40	modest 50?*	8 hr 40 min	unknown
2	June 6 21 hr 36	June 7 04 hr 00	unknown unknown	Stromb 50–150	6 hr 24 min	670 hr ?
3	June 8 >22 hr	June 9 03 hr 30	unknown unknown	Stromb 150–200	< 5 hr 30 min	>42 hr
4	June 11 01 hr 00	June 11 04 hr 00	June 11 2 hr 30–?	Stromb 150	3 hr	45 hr 30 min
5	June 13 01 hr 45	June 13 05 hr 00	unknown unknown	ff 150–200	3 hr 15 min	45 hr 45 min
6	June 15 <9 hr 50	June 15 >10 hr 30	June 15 9 hr 50–10 hr 50	minor ff 50?	unknown >40 min	<52 hr 50 min
7	June 17 10 hr 50	June 17 14 hr 30	June 17 13 hr 00	Stromb 150–300	3 hr 40 min	<48 hr 20 min
8	June 19 19 hr 20	June 19 <22	June 19 20 hr 40	ff 200–300	<2 hr 40 min	52 hr 40 min
9	June 22 17 hr 00	June 22 19 hr 40	June 22 18 hr 00–19 hr 00	ff 300–400	2 hr 40 min	<67 min
10	June 24 18 hr 15	June 24 21 hr 50	June 24 20 hr 30–?	Stromb 200	3 hr 55 min	46 hr 35 min
11	June 27 <21 hr 20	June 27 23 hr 00	June 27 21 hr 20–22 hr	Stromb 400–500	1 hr 40 min	<72 hr
12	June 30 6 hr 00	June 30 14 hr 00	June 30 11 hr 00–12 hr 00	unknown	8 hr	55 hr
13	July 4 19 hr 00	July 4 22 hr 25	July 4 21 hr 24–22 hr 07	55	3 hr 25 min	101 hr
14	July 7 8 hr 33	July 7 10 hr 00	July 7 unknown	ash cloud 50	1 hr 27 min	58 hr 08 min
15	July 12 22 hr 00	July 13 05 hr 04	July 13 04 hr 07–04 hr 47	strong ff 300–400	7 hr 04 min	132 hr
16	July 17 0 hr 30?	July 17 04 hr 30	July 17 unknown	strong ff 300–400	4 hr	91 hr 26 min

Stromb, ff and ? stand for Strombolian activity, fire fountains and very approximate estimate. The star indicates cases for which the height is only known from that of the average ejecta.

very well constrained as it is corrected for any systematic bias. The difference in the correction factor between 4 and 12 July results from the quasi-steadiness of a purely Strombolian episode compared with the unsteadiness of an episode leading towards a fire fountain. If we multiplied the gas volume estimated from the maximum height, $3.6 \times 10^7 \text{ m}^3$ and $1.9 \times 10^8 \text{ m}^3$ on 4 and 12 July, respectively, by these factors, 0.39 and 0.11 (Figs 13 & 14), we obtain $1.4 \times 10^7 \text{ m}^3$ and $2.1 \times 10^7 \text{ m}^3$, respectively (Table 1). This is still larger than the gas volumes estimated from acoustic measurements, $5.4 \times 10^6 \text{ m}^3$ and $1.0 \times 10^7 \text{ m}^3$ on 4 and 12 July, respectively, by a factor of 2.6 and 2.1.

Passive and active degassing

Acoustic measurements can only detect bubbles with an inner overpressure and these bubbles are likely to be the ones coming from depth.

The overpressure, also detected on seismic records on Stromboli, has been associated with pressurization of the conduit induced by the formation and release of a gas slug at a structural barrier (Chouet et al. 2003; James et al. 2006). The seismic source of explosions is located 300 m below the vents (Chouet et al. 2003) and this depth is also where the foam coalescence, responsible for low-frequency seismic signals, is located (Ripepe et al. 2001). The resemblance between the series of Strombolian explosions at Etna (Fig. 7) and those at Stromboli (Vergniolle & Brandeis 1996) suggests that the overpressurized slugs at Etna are also coming from the base of the conduit, estimated at between 1.5 km and 3 km (Laigle et al. 2000; Allard et al. 2005; Spilliaert et al. 2006a; Saccorotti et al. 2004; Bonaccorso 2006; Patanè et al. 2002, 2006).

Surface activity observed at the vent results from both the bubbles coming from depth and those quasi-stagnant in the conduit, silent on acoustic records. Observations of scoriae indeed shows that syn-eruptive volatile exsolution, albeit moderate, exists during each single eruptive episode, together with foam collapse, the latter being the main driving mechanism for fire fountain activity (Polacci *et al.* 2006). The volume of gas coming from passive degassing is calculated by the difference between the total gas volume deduced from the height of fire fountains (equations 3 and 4) and the gas volume detected from acoustic records (Figs 13 & 14). The gas volume associated with passive degassing exceeds the one coming from depth by a factor of 1.6 and 1.1 on 4 and 12 July, respectively (Table 1). The similarity between these two values, despite the difference of a factor of 1.9 in the gas volume expelled on 4 and 12 July, $5.4 \times 10^6 \, \mathrm{m}^3$ and $1.0 \times 10^7 \, \mathrm{m}^3$, respectively, suggests that the volume of gas from passive degassing is proportional to the volume of gas passing from the reservoir to the surface.

The bubble, which is formed by massive coalescence at the top of the reservoir, is sufficiently large to induce a recirculation zone at its back while rising, and this region is called a wake. The existence and dimension of the wake is deduced from the value of the Reynolds number R_{ebub} based on the bubble (Batchelor 1967; Brennen 1995):

$$R_{\mathrm{ebub}} = \frac{\rho_{\mathrm{liq}} \, 2 \, R_0 \, U_b}{\mu_{\mathrm{liq}}} \qquad (5)$$

where U_b is the rise speed of the large bubble (equation 2) and R_0 its radius.

It is experimentally found that the wake, or vortex ring, develops close to the rear stagnation point for a bubble Reynolds number above 30 (Brennen 1995). The wake reaches a diameter comparable to that of the bubble when the bubble Reynolds number is 130 (Brennen 1995). In these conditions, the wake is laminar and closed (Fig. 15; Brennen 1995). The wake detaches from a cylindrical bubble for a Reynolds number equal to 500 (Batchelor 1967). As the Reynolds number increases past 500, the flow develops a fairly steady near-wake and a turbulent far-wake (Fan & Tsuchiya 1990). At 1000, the flow becomes unsteady with a periodic shedding of vortices from the large bubble, leading to a cyclic detachment of the wake (Fan & Tsuchiya 1990). Once the Reynolds number has reached this value, the maximum volume of the wake solely depends on the bubble volume and not on the Reynolds number. Laboratory experiments have shown that the maximum volume of the wake is four to six times that of the bubble (Fan & Tsuchiya 1990).

Laboratory experiments have been performed to mimic physical process at Etna, within the magma reservoir and the conduit, by producing a large

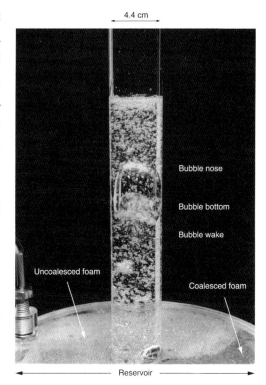

Fig. 15. Laboratory experiment of a slug flow rising in a tube of diameter 4.4 cm and produced from the coalescence of a foam trapped in the reservoir. The fluid is a silicone oil, Rhodorsil 47V100, and the gas flux in the reservoir is $\cong 3 \times 10^{-5} \, \mathrm{m}^3 \, \mathrm{s}^{-1}$. The bubble bottom has the shape of an inward dome. The bubble wake is well visualized by the trajectory of the small bubbles, which are solely in fast motion at the rear of the slug. The stretching of the small bubbles is related to their upwards rise velocity forced to be equal that of the overlying slug. The bubble Reynolds number, $\cong 96$, is much smaller than close to the surface on Etna, and the wake has a size between half and equal to the one of the slug, whereas at Etna, the volume of the wake is four to six times larger than the volume of the slug. The elongation of the small bubbles is very likely to occur within the conduit at Etna. The small bubbles in the slug wake correspond to passive degassing whereas the slug comes directly from the massive coalescence of the foam.

bubble, i.e. a slug, from the coalescence of a foam trapped at the top of a reservoir, and letting it rise through a bubbly liquid in a conduit (Fig. 15). These experiments are designed to understand the fate of small bubbles present in a magma while the slug rises in a conduit. The reservoir and the tube have a diameter of 28 cm and 4.4 cm, respectively. The gas flux, $\cong 3 \times 10^{-5} \, \mathrm{m}^3 \, \mathrm{s}^{-1}$, corresponds to a bubble diameter of $\cong 2 \, \mathrm{mm}$ for the small bubbles produced at the base of the reservoir. The fluid is a silicone oil, Rhodorsil 47V100, with a viscosity equal to 0.097 Pa s. The slug bottom has the shape of an inward dome (Fig. 15). In the wake, the motion of the liquid is convective with an upwards central region and a downwards flow towards the conduit walls (Batchelor 1967; Brennen 1995). That secondary flow is favourable to an accumulation of small bubbles as the slug rises through a bubbly magma once the wake is open (Fig. 16). The slug wake is well visualized by the trajectory of the small bubbles solely elongated within the slug wake. The bubble Reynolds number, $\cong 96$, leads to a wake with a size between half of and equal to the one of the slug, as predicted (Brennen 1995). The stretching of a small bubble within the wake is related to its rise velocity, forced to be that of the overlying slug, 0.23 m s^{-1} (equation 2) instead of its equilibrium velocity, i.e. the Stokes velocity, $2.2 \times 10^{-2} \text{ m s}^{-1}$.

A second dimensionless number, called the Weber number W_{ebub}, is indicative of the extent of bubble deformations (Brennen 1995). That number corresponds to a ratio of pressures,

between destabilization due to bubble rise and stabilization due to surface tension. For small bubbles with low Reynolds numbers, such as in our laboratory experiments, the Weber number is equal to (Brennen 1995):

$$W_{\text{ebub}} = \frac{\rho_{\text{liq}} \, 2 \, R_0 \, U_{\text{b}}^2}{\sigma} \qquad (6)$$

where σ is the surface tension between liquid and gas.

The 47V100 silicone oil has a density and surface tension equal to 965 kg m^{-3} and 2.1×10^{-2} kg s^{-2}, respectively. Hence, the Reynolds and Weber numbers are equal to $\cong 4.7$ and $\cong 5.2$, respectively (equations 5 and 6), when using the velocity of the overlying slug (equation 2). In these conditions, the assumption of a low Reynolds number is valid and the Weber number can be calculated from equation 6. The value of $\cong 5.2$ lies between a Weber number equal to unity, corresponding to small deformations and a Weber number large compared to 1, for which deformations are very strong. A Weber number of $\cong 5.2$, as for our laboratory experiments, is compatible with an aspect ratio between bubble length and diameter between 2 and 3 as experimentally observed (Fig. 15).

The use of dimensionless numbers ensures that processes observed at the scale of laboratory experiments are analogous to those occurring in the volcanic conduit, providing that their values are similar. The magma viscosity at Etna, 10 Pa s (Giordano & Dingwell 2003; Métrich et al. 2004), leads to a Reynolds number equal to 9.4×10^3 (equation 5) for a slug rising at c. 3.5 m s^{-1} in a conduit radius of c. 5.2 m (equation 2). If the amount of quasi-stagnant bubbles in the conduit ahead of the slug is as large as reaching a gas volume fraction of 10%, the bubbly magma density is reduced by the same amount, 2430 kg m^{-3}, while the viscosity is increased by 10% (Wallis 1969; Faber 1995). Since the bubble rise speed is independent on both parameters (equation 2), the Reynolds number (equation 5) is only decreased by 20%. In these conditions, the slug wake undergoes periodic detachments and reaches a maximum volume, equal to four to six times the slug volume. The fast pressure drop induced while breaking the slug at the surface is very likely to be responsible for releasing the gas volume trapped within the slug wake, almost simultaneously to that from the slug.

Most of the bubbles observed in scoria correspond to the small bubbles trapped within the slug wake (Fig. 16) because the syn-eruptive gas exsolution is only moderate (Polacci et al. 2006). Therefore, the bubble diameter and the gas volume

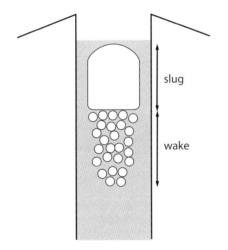

Fig. 16. Sketch for the formation of passive and active degassing. The slug represents the gas volume coming from the reservoir and detected from acoustic measurements, while the small bubbles within the slug wake represent the passive degassing, included in the surface measurements together with the gas directly from the reservoir. The small bubbles present in the volcanic conduit are pushed aside as the slug rises and may be incorporated within the slug wake if the wake is open, due to strong convective motions in that region (Fig. 15) (Fan & Tsuchiya 1990). For the Reynolds number of the slug at Etna, 9.4×10^3, the wake is open and has a maximum volume between four and six times the slug volume (Fan & Tsuchiya 1990). The gas volume fraction within the slug wake is 65% and 44% for 4 and 12 July, respectively (Table 1).

fraction measured in scoria may be indicative of these parameters within the shallowest part of the conduit prior to the slug breaking. If the diameter of these small bubbles, between 0.315 mm and 0.150–0.192 mm for Strombolian activity and fire fountains, respectively (Polacci *et al.* 2006), is taken to be representative of the diameter of bubbles from passive degassing, the Reynolds number lies between 0.30 and 0.14 for these two cases, respectively (equation 5). This corresponds to a low Reynolds number and equation 6 is valid. For a surface tension of 0.36 kg s^{-2} (Proussevitch *et al.* 1993), the Weber number ranges between 26 and 12 for Strombolian activity and fire fountains, respectively (equation 6). Indeed, the small bubbles from passive degassing must be very elongated if there is a sufficient time to reach an equilibrium in shape. The bubble elongation is favourable to bubble coalescence within the slug wake and that process is enhanced by strong convective motions in that area. At Etna, the vigour of convective motions close to the surface is identical for Strombolian activity and fire fountains because the rise velocity of the slug is solely dependent on the conduit radius (equation 2). Hence, the efficiency of bubble coalescence is mainly related to the extent of bubble elongation, i.e. to the bubble diameter. It must be stressed that the bubble coalescence is solely likely to appear if the bubbles are deformable, and this can be directly deduced from the Weber number (equation 6). Therefore, the shape of the vesicles in scoria, which are convoluted for Strombolian activity and spherical for fire fountains (Polacci *et al.* 2006), only results from the bubble diameter prior to coalescence, hence to the time available for bubble growth since nucleation. Direct measurements of the bubble diameter prior to coalescence are therefore impossible from the largest bubbles in the scoria. This suggests that 0.15 mm is probably the maximum diameter for the bubbles resulting from passive degassing and prior to coalescence.

At Etna, the volume of gas from passive degassing is equal to $\cong 1.6$ and $\cong 1.1$ times the volume of the slugs coming from the reservoir, V_{gslug}. The volume of gas V_{gwake}, which can be trapped into the slug wake, depends on the gas volume fraction, α_{wake} and on the wake volume V_{wake} as:

$$V_{gwake} = \alpha_{wake} n_{wake} V_{gslug} \qquad (7)$$

where n_{wake} is the coefficient of proportionality between the slug volume and the wake volume, between 4 and 6 for Reynolds number exceeding 1000 (Fan & Tsuchiya 1990). If the gas volume from passive degassing is trapped into the slug wake, the wake contains a gas volume fraction between 27% and 41% for Strombolian activity

and between 18% and 27% for fire fountains (equation 7; Fig. 16). These values are minimum because the wake is not always present at its maximum volume, due to its periodic detachment and regrowth.

In fact, the wake volume oscillates between a value assumed to be close to zero, right after detachment, to its maximum volume (equation 7). If we further assume that the time spent at each stage of growth is similar, the average wake volume is half its maximum value, leading to a coefficient of proportionality n_{wake}, between 2 and 3. This implies that the wake contains a large amount of small bubbles, with a gas volume fraction between 54% and 82% and 36% and 55% on 4 and 12 July, respectively (equation 7), making the bubbly wake almost a foam (Table 1). Therefore, the gas volume fraction within the slug wake is probably *c.* 65% and *c.* 44% on 4 and 12 July, respectively, obtained for an average n_{wake} equal to 5 (Fig. 16). This is compatible with the high vesicularity measured in scoria, from 52% to 56% and from 53% to 74% for Strombolian activity and fire fountains, respectively (Polacci *et al.* 2006). These large gas volume fractions could be explained partially by gas exsolution resulting from a depressurization induced by magma removal during fire fountains and partially by a local amount of small bubbles in the uppermost conduit. Indeed, scoria at Etna show a moderate amount of syn-eruptive gas exsolution (Polacci *et al.* 2006), suggesting that most of the bubbles correspond to bubbles already present in the uppermost conduit.

Alternatively, some of the small bubbles present in the slug wake may come directly from the base of the conduit (Fig. 17). This implies that the wake

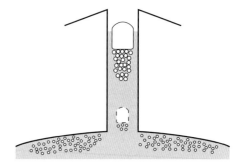

Fig. 17. Sketch for the formation of passive and active degassing. The small bubbles present in the volcanic conduit are pushed aside as the slug rises and can be incorporated within the slug wake as early as the base of the conduit due to a Reynolds number exceeding 1000. The depressurization induced by magma removal during fire fountains can lead to gas exsolution in the conduit. However, the syn-eruptive gas exsolution is only found to be moderate (Polacci *et al.* 2006).

exists at the depth of the reservoir, hence that the Reynolds number exceeds 30 (Brennen 1995). However, the gas compresses with depth following the perfect gas law and may also change shape. While at the surface the bubble fills the entire width of the conduit, a transition might exist in the conduit between a slug in the uppermost part of the conduit, and a spherical cap bubble below. In contrast with a slug, the rise velocity of a spherical cap, U_{gcap}, depends on its equivalent diameter d_e defined as:

$$d_e = \left(\frac{6\, V_{gcap}}{\pi}\right)^{1/3} \qquad (8)$$

where V_{gcap} is the volume of spherical cap bubble (Batchelor 1967).

$$U_{gcap} = \frac{2}{3}(g\, d_e)^{1/2} \qquad (9)$$

The coefficient of $2/3$ is empirical and has been deduced from laboratory experiments (Batchelor 1967). This velocity (equation 9) is the one to be used to estimate the Reynolds number of a spherical cap (equation 5).

The gas volume, once compressed at the depth of the reservoir, $\cong 2$ km (Laigle et al. 2000; Saccorotti et al. 2004; Allard et al. 2005; Spilliaert et al. 2006a; Bonaccorso 2006; Patanè et al. 2002, 2006) from the perfect gas law, varies between 9.2 m³ and 5.7 m³ for a bubble length of $\cong 15$ m and $\cong 8$ m, characteristic of the Strombolian activity for 4 and 12 July, respectively (Figs 11 and 12). This leads to an equivalent bubble diameter equal to c. 2.6 m and c. 2.2 m, respectively. Hence, these bubbles have the shape of a spherical cap for a conduit radius of $\cong 5.2$ m and a Reynolds number equal to 2.4×10^3 and 1.9×10^3, respectively. During the paroxysm of a fire fountain, the bubble, which has a characteristic length exceeding 50 m at the surface, is also a spherical cap at the reservoir depth. Its equivalent diameter, equal to 3.7 m, is less than that of the conduit, 10.4 m, and the bubble has a Reynolds number of 4.0×10^3. Although there is a factor of two between the Reynolds number characteristic of a Strombolian activity and that of a paroxysm during a fire fountain, the maximum volume of the wake is solely dependent on the volume of the bubble ahead. Hence, the proportion of gas present in the spherical caps, resulting from the foam coalescence and previously called slugs, and in the wake is similar for both types of surface activity. Furthermore, it is very likely that most of the small bubbles in the slug wake are coming from the reservoir (Fig. 17). In fact, the driving slug, which already occupies a fourth of the conduit at its base, can aggregate there a volume of bubbly magma at most between four and six times its own volume (Fan & Tsuchiya 1990).

But the proportion between these deep bubbles and those present at very shallow levels and left in the conduit long before the eruptive episode is as yet unknown (Fig. 17). The detailed processes occurring in the conduit, however, are complex due to the wake periodic detachment, hence their study is out of the scope of the present paper.

The existence of a significant gas volume within the wake reinforces the crucial role played by the bubbles coming from the reservoir in driving the surface activity. Although the bubbles resulting from passive degassing in the conduit have a large gas volume at the surface, the eruption is driven by the large bubbles coming from the reservoir. Therefore, acoustic measurements are extremely valuable to deduce the sole gas volume trapped at the top of the magma reservoir and not a gas volume combined with the quasi-stagnant bubbles present in the conduit as measurements on the fire fountains height do.

Ejecta volume

The interpretation of a Strombolian explosion as a slug and its bubbly wake leads to a crude estimate of the volume of magma expelled per episode (Fig. 16). The magma is initially contained above and around the slug, and in a wake of volume V_{wake} (equation 7), leading to a magma volume V_{mag} equal to:

$$V_{mag} = \frac{\pi R_0^2\, h_{eq}}{2} + 2\pi R_0 \delta_\infty L_{slug}$$
$$+ (1 - \alpha_{wake})\, V_{wake} \qquad (10)$$

where L_{slug}, R_0, h_{eq} and δ_∞ are the slug length, the slug radius, the thickness of the film above the slug and the thickness of the lateral film around the slug (equation 1), respectively. The first two terms represent the magma volume around one slug and the third term corresponds to the magma volume present in the slug wake. The total magma volume becomes:

$$V_{mag} = n_{slug}\left(\frac{\pi R_0^2\, h_{eq}}{2} + 2\,\pi R_0 \delta_\infty L_{slug}\right.$$
$$\left. + (1 - \alpha_{wake})\, n_{wake} V_{gslug}\right) \qquad (11)$$

when using n_{slug} for the number of explosions and n_{wake} for the coefficient of proportionality between the wake volume and the slug volume, equal to $2-3$ (see discussion above and equation 7).

The number of explosions, 1284 and 1644 on 4 and 12 July, respectively, leads to a total magma volume above and around the slugs equal to $1.5 \times 10^5 \ m^3$ and $1.7 \times 10^5 \ m^3$, respectively (Table 1). A gas volume fraction, on average between 65% and 44% on 4 July and 12 July, respectively, for a coefficient of proportionality between the wake volume and the slug volume equal to 2.5, leads to a volume of magma in the wake equal to $4.7 \times 10^6 \ m^3$ and $1.4 \times 10^7 \ m^3$, respectively (Table 1). The absence of overpressure within the small bubbles corresponding to passive degassing suggests that only some of the magma trapped within the wake can fall outside of the conduit while all the gas can escape upwards. Therefore, our estimate of ejecta volume, $1.5 \times 10^5 - 4.7 \times 10^6 \ m^3$ and $1.7 \times 10^5 - 1.4 \times 10^7 \ m^3$, is in good agreement with the measurements of the volume of ejected magma, between $7.3 \times 10^5 \ m^3$ and $7.7 \times 10^5 \ m^3$, respectively (Behncke et al. 2006). This suggests that the transport of gas both in the slug and in its wake is an appropriate mechanism to explain the volume of gas and magma expelled during one episode (Fig. 16). However, the ejecta volume, only partly dominated by the average volume of the slug wake, cannot be used directly at Etna as a tool to estimate the gas volume.

Time evolution of the gas volume from surface activity

Nevertheless, observations of surface activity (BGVN 2001) can be used to find the trends in the gas volume at the depth of the reservoir in the absence of acoustic measurements. Here, we include the episode on 17 July, first because its repose period is similar to that of previous episodes and secondly because it also started at the southeast crater before the eruption turned to the flank eruption (Allard et al. 2006; Behncke et al. 2006).

The data presented here are based on visual observations from the local volcanological observatory (BGVN 2001) as well as our own on 4, 7 and 12 July. However, not all reports include the maximum height of the fire fountains or explosions, but solely those reached by the ejecta (BGVN 2001). On the sole occasion where that approximation is needed (BGVN 2001), we assume that the height reached by an average ejecta is equivalent to half the maximum height of a fire fountain based on visual observations (Table 2). We also interpret early morning, morning, late morning, early afternoon, afternoon, late afternoon, evening and late evening (BGVN 2001) as 07 hr 00, 09 hr 00, 11 hr 00, 14 hr 00, 16 hr 00, 18 hr 00, 20 hr 00, and 22 hr 00, respectively. We assume that a few hundred metres equals 200 m and several hundred metres 500 m.

The duration of the fire fountain decreases mostly from 520 minutes to 40 minutes while its height increases until 30 June, from 10 m to 450 m (Fig. 3; Table 2; BGVN 2001). The repose period prior to fire fountain increases from 42 hours to 132 hours (Fig. 4; Table 2; BGVN 2001).

Except on three occasions (9 May, 15 June, 12 July), the gas volume shows a regular decrease between 6 June and 7 July, from $9 \times 10^7 \ m^3$ to $1.4 \times 10^7 \ m^3$ (Fig. 4). The very low values on 15 June result from using the duration of the paroxysm, 40 minutes, instead of the duration of the entire episode (Table 2). The first episode is followed by a very long repose period but its gas volume is similar to other episodes, owing to its long duration (Fig. 3). The very large gas volume estimated on 12 July is very accurate compared with episodes in May and June, due to our continuous observation on 4, 7 and 12 July (Fig. 4). Furthermore, having the largest gas volume on 12 July is in excellent agreement with that episode being visually observed as the most violent of that eruption (Behncke et al. 2006). That episode coincides with a vigourous seismic crisis, which ended up on 17 July with the migration of the vents down on the volcano flank (Allard et al. 2006). Although observations are crude (BGVN 2001), the intermittency and the gas volume have a smooth temporal evolution (Fig. 4), suggesting that our methods are good up to order of magnitude levels.

The largest values of the gas volume on 12 July are indeed the mark of a more active foam coalescence compared with a purely Strombolian episode (Vergniolle 2008). The efficiency of the foam coalescence (Jaupart & Vergniolle 1989) on 12 July is explained by a gas flux larger by a factor 1.7 on that day compared to 4 July (Vergniolle 2008). The quantitative analysis of the foam behaviour at the depth of the reservoir, which led to quantitative estimates of bubble diameter, gas volume fraction, surface area and the height of the degassing reservoir, can be found in the companion paper (Vergniolle 2008).

These gas volumes, which are maximum, should be multiplied by the degree of unsteadiness between 0.39 and 0.11 (Table 1), to deduce the average gas volume. During the 2001 eruption, most of the episodes resembled a purely Strombolian episode, except for 12 and 17 July (Corsaro & Pompilio 2004; Behncke et al. 2006), hence the same degree of unsteadiness could be used throughout the entire eruption, and taken to be equal to 0.39. A further correction of 0.38–0.48 should be then applied if only the gas volume from the reservoir is to be found (Table 1). The remaining 62–52% of the total gas is related to passive degassing.

The interpretation that the small bubbles from passive degassing are carried towards the surface via the wake behind the slug and expelled quasi-simultaneously to the breaking of the slug at the surface, is a process that can be generalized on any volcano. However, finding the gas volume trapped at the top of the reservoir from the total gas volume requires knowing the gas volume fraction within the slug wake, as well as the conduit radius and the magma viscosity (equations 2, 5 and 7). The gas volume fraction, which is the most unknown of the three parameters, could be estimated from the vesicularity of scoria as well as by comparing the surface activity with Etna.

Conclusion

Acoustic measurements are a very powerful tool for remote sensing of volcanoes, as they provide strong constraints on physical mechanisms. Sound waves give quantitative evidence for large volumes of gas being present during Strombolian to quasi-fire fountain episodes of Etna volcano in 2001. Modelling the sound produced on basaltic volcanoes provides estimates on bubble volume and overpressure during explosions.

The best fit between the strongest explosion recorded per five minutes and its synthetic waveform gives the time evolution of the bubble characteristics over the duration of the episode. On 4 July, the bubble length and overpressure stayed constant, at 15 m and 0.22 MPa, respectively, for a bubble radius of 5 m. On 12 July, the bubble length, initially constant, increases from 8 m towards ≥ 50 m, simultaneously to visual observations of fire fountains at the vent. The bubble overpressure stayed constant on 12 July, at 0.2 MPa; however, the number of explosions per five minutes increased from 20 to 40 at the peak of the fire fountain. Therefore, a fire fountain at Etna only results from a closely spaced series of very long Strombolian explosions.

The simultaneous increase in the number of explosions and in the bubble length suggests that the foam coalescence, if at the origin of both Strombolian explosions and fire fountains, is more efficient in the later than in the former case. This is indeed compatible with the larger gas volume expelled on 12 July than 4 July, 1.0×10^7 m^3 and 5.4×10^6 m^3, respectively.

The maximum gas volume can also be calculated by using the maximum height of fire fountains. The average gas volume is then calculated by multiplying by the degree of unsteadiness estimated from continuous acoustic measurements, and equal to 0.39 and 0.11 for Strombolian and fire-fountaining activities, respectively. The gas volume deduced from surface activity results partially from the gas volume coming from massive foam coalescence at depth, with a proportion of 38–48% of the total gas volume, and partially from quasi-stagnant bubbles resulting from passive degassing in the conduit, with 62–52%. This result is independent of whether surface activity consists of purely Strombolian activity or a transition to fire fountains, despite a factor of two difference in their gas volumes.

This suggests that the small bubbles from passive degassing are carried to the surface via the wake of the slugs, coming from the depth of the reservoir (Fig. 17), in agreement with the large Reynolds numbers of the slugs, 9.4×10^3. Indeed, the volume of the gas from passive degassing can be accumulated within the slug wake with a gas volume fraction of $\cong 65\%$ and $\cong 44\%$ for Strombolian activity and fire fountains, respectively. Such an interpretation leads to an ejecta volume between 1.5×10^5–4.7×10^6 m^3 and 1.7×10^5–1.4×10^7 m^3, respectively, in good agreement with observations, 7.3×10^5 m^3 and 7.7×10^5 m^3, respectively (Behncke et al. 2006). This reinforces the crucial role of the bubbles coming from the reservoir in driving fire-fountain activity. The time evolution of the gas volume during the entire eruption shows a general decrease, except for the last two episodes during which it doubles its value while turning to a flank eruption.

Thorough reviews by L. Wilson and L. Parfitt have greatly improved the manuscript. We also thank the help of C. Tirel, P. Allard, G. Dubosclard, R. Cordesses, C. Hervier, C. Jaupart, M. Burton, A. Hirn, J. C Lepine, A. Nercessian, D. Bréfort, B. Alcoverro and the Comissariat Energie Atomique, P. Briole, Y. Gaudemer, P. Tinard, L. Milelli, C. Zielinski, S. Gresta, A. Bertagnini, T. Pfeiffer, www.decadevolcano.net, and S. Lane and J. Gilbert for careful editing. S. V. thank the CNRS, INSU, IPGP and MAE for their support via PNRN and ACI, BQR (IPGP) and grant numbers 122/2000 and 02000146. This is an INSU contribution number (407) and an IPGP contribution number (2310).

References

AKI, K. & KOYANAGI, R. Y. 1981. Deep volcanic tremor and magma ascent mechanism under Kilauea, Hawai'i. *Journal of Geophysical Research*, **86**, 7095–7110.

ALLARD, P., BURTON, M. & FILIPPO, F. 2005. Spectroscopic evidence for a lava fountain driven by previously accumulated magmatic gas. *Nature*, **433**, 407–410.

ALLARD, P., BEHNCKE, B., D'AMICO, S., NERI, M. & GAMBINO, S. 2006. Mount Etna 1993–2005: anatomy of an evolving eruptive cycle. *Earth Science Reviews*, **78**, 85–114.

ALPARONE, S., ANDRONICO, D., LODATO, L. & SGROI, T. 2003. Relationship between tremor and volcanic activity during the southeast crater eruption on Mount Etna in early 2000. *Journal of Geophysical Research*, **108**, 2241, ESE 6-1-6-13.

ALPARONE, S., ANDRONICO, D., GIAMMANCO, S. & LODATO, L. 2004. A multidisciplinary appoach to detect active pathways for magma migration and eruption at Mt Etna (Sicily, Italy) before the 2001 and 2002–2003 eruptions. *Journal of Volcanology and Geothermal Research*, **136**, 121-6-140.

ANTIER, K., LEPICHON, A., VERGNIOLLE, S., ZIELINSKI, C. & LARDY, M. 2007. Multi-year validation of the NRL-G2S wind fields using infrasound from Yasur. *Journal of Geophysical Research*, **112**, D23110, doi:10.1029/2007JD008462.

ARCINIEGA-CEBALLOS, A., CHOUET, B. & DAWSON, P. 1999. Very long-period signals associated with Vulcanian explosions at Popocatepetl volcano, Mexico. *Geophysical Research Letters*, **26**, 3013–3016.

BATCHELOR, G. K. 1967. *An Introduction to Fluid Dynamics*. Cambridge University Press, Cambridge.

BEHNCKE, B. & NERI, M. 2003. The July–August eruption of Mt Etna (Sicily). *Bulletin of Volcanology*, **65**, 461–476.

BEHNCKE, B., NERI, M., PECORA, E. & ZANON, V. 2006. The exceptional activity and growth of the Southeast crater, Mt Etna (Italy) between 1996 and 2001. *Bulletin of Volcanology*, **69**, 149–173.

BERTAGNINI, A., CALVARI, S., COLTELLI, M., LANDI, P. & POMPILIO, M. 1990. The 1989 eruptive sequence. *In:* BARBERI, F., BERTAGNINI, A. & LANDI, P. (eds) *Mt Etna: The 1989 Eruption*. Gruppo Nationale Vulcanologia, Italy, 10–22.

BLACKBURN, E. A., WILSON, L. & SPARKS, R. S. J. 1976. Mechanics and dynamics of Strombolian activity. *Journal of the Geological Society, London*, **132**, 429–440.

BONACCORSO, A. 2006. Explosive activity at Mt Etna summit craters and source modeling by using high-precision continuous tilt. *Journal of Volcanology and Geothermal Research*, **158**, 221–234.

BRENNEN, C. E. 1995. *Cavitation and Bubble Dynamics*. Oxford Engineering Science series 44, Oxford University Press.

BUCKINGHAM, M. J. & GARCÉS, M. A. 1996. A canonical model of volcano acoustics. *Journal of Geophysical Research*, **101**, 8129–8151.

BULL. GLOBAL VOLCANISM NETWORK (BGVN). 2001. Sistema Posidon, Instituto Nazionale di Geofisica e Vulcanologia-Sezione di Catania, Italy, Etna, *Smithsonian Institution*, Etna activity reports, **26**.

CALVARI, S., COLTELLI, M., MULLER, W., POMPILLIO, M. & SCRIBANO, V. 1994. Eruptive history of the south-eastern crater of Mount Etna, from 1971 to 1994. *Acta Vulcanologica*, **5**, 11–14.

CALVARI, S. Research Staff of the Instituto Nazionale di Geofisica e Vulcanologia- Sezione di Catania, Italy 2001. *EOS, Transactions of AGU*, **82**, 653–656.

CALVARI, S. & PINKERTON, H. 2002. Instabilities in the summit region of Mount Etna during the 1999 eruption. *Bulletin of Volcanology*, **63**, 526–535.

CHOUET, B. 1985. Excitation of a buried magmatic pipe: a seismic source model for volcanic tremor. *Journal of Geophysical Research*, **90**, 1881–1893.

CHOUET, B., HAMISEVICZ, N. & MCGETCHIN, T. R. 1974. Photoballistics of volcanic jet activity at Stromboli, Italy. *Journal of Geophysical Research*, **79**, 4961–4975.

CHOUET, B. A., DAWSON, P. A. & OHMINATO, T. *ET AL.* 2003. Source mechanisms of explosions at Stromboli

volcano, Italy, determined from moment-tensor inversions of very-long-period data. *Journal of Geophysical Research*, **108**, 2019, ESE 7, 1–25.

COLTELLI, M., DEL CARLO, P. & VEZZOLI, L. 1998*a*. Discovery of a Plinian basaltic eruption of Roman age at Etna volcano, Italy. *Geology*, **26**, 1095–1098.

COLTELLI, M., POMPILIO, M., DEL CARLO, P., CALVARI, S., PANNUCCI, S. & SCRIBANO, V. 1998*b*. Etna, 1. Eruptive activity. *Acta Vulcanologica*, **26**, 141–148.

CORSARO, R. A. & POMPILIO, M. 2004. Magma dynamics in the shallow plumbing system of Mt Etna as recorded by compositional variations in volcanics of recent summit activity (1995–1999). *Journal of Volcanology and Geothermal Research*, **137**, 55–71.

CROSSON, R. S. & BAME, D. A. 1985. A spherical source model for low frequency volcanic earthquakes. *Journal of Geophysical Research*, **90**, 10 237–10 247.

DINGWELL, D. B. 1998. Recent experimental progress in the physical description of silicic magma relevant to explosive volcanism. *In:* GILBERT, J. S. & SPARKS, R. S. J. (eds) *The Physics of Explosive Volcanic Eruptions*. Geological Society, London, Special Publications, **145**, 9–26.

DUBOSCLARD, G., CORDESSES, R., ALLARD, P., HERVIER, C., COLTELLI, M. & KORNPROBST, J. 1999. First testing of a volcano Doppler radar (Voldorad) at Mount Etna, Italy. *Geophysical Research Letters*, **26**, 3389–3392.

DUBOSCLARD, G., DONNADIEU, F. & ALLARD, P. *ET AL.* 2004. Doppler radar sounding of volcanic eruption dynamics at Mount Etna. *Bulletin of Volcanology*, **66**, 443–456.

FABER, T. E. 1995. *Fluid Dynamics for Physicists*. Cambridge University Press.

FALSAPERLA, S., PRIVITERA, E., CHOUET, B. & DAWSON, P. 2002. Analysis of long-period events recorded at Mount Etna (Italy) in 1992 and their relationship to eruptive activity. *Journal of Volcanology and Geothermal Research*, **114**, 419–440.

FAN, L.-S. & TSUCHIYA, K. 1990. *Bubble Wake Dynamics in Liquids and Liquid-Solid Suspensions*. Butterworth-Heinemann series in chemical engineering, Boston.

FEHLER, M. 1983. Observations of volcanic tremor at Mount St Helens volcano. *Journal of Geophysical Research*, **88**, 3476–3484.

GARCÉS, M. A. & MCNUTT, S. R. 1997. Theory of the airborne sound field generated in a resonant magma conduit. *Journal of Volcanology and Geothermal Research*, **78**, 155–178.

GERLACH, T. M., MCGEE, K. A., ELIAS, T., SUTTON, A. J. & DOUKAS, M. P. 2002. Carbon dioxide emission rate of Kilauea volcano: implications for primary magma and the summit reservoir. *Journal of Geophysical Research*, **107**, 2189, ESV 3, 1–15.

GIORDANO, D. & DINGWELL, D. B. 2003. Viscosity of hydrous Etna basalt: implications for Plinian-style basaltic eruptions. *Bulletin of Volcanology*, **65**, 8–14.

GRESTA, S., RIPEPE, M., MARCHETTI, E., D'AMICO, S., COLTELLI, M., HARRIS, A. J. L. & PRIVITERA, E. 2004. Seismo-acoustic measurements during July–August 2001 eruption of Mt Etna volcano, Italy. *Journal of Volcanology and Geothermal Research*, **137**, 219–230.

HAGERTY, M. T., SCHWARTZ, S. Y., GARCES, M. A. & PROTTI, M. 2000. Analysis of seismic and acoustic observations at Arenal Volcano, Costa Rica, 1995–1997. *Journal of Volcanology and Geothermal Research*, **101**, 27–65.

HELIKER, C. C. & MATTOX, T. N. 2003. The first two decades of the Pu'u 'O'o-Kupaianaha eruption: chronology and selected bibliography. *US Geological Survey Professional Papers*, **1676**, 1–28.

JAMES, M. R., LANE, S. L. & CHOUET, B. 2006. Gas slug ascent through changes in conduit diameter: laboratory insights into a volcano-seismic source process in low-viscosity magmas. *Journal of Geophysical Research*, **111**, B05201, doi:10.1029/2005JB003718.

JAUPART, C. & VERGNIOLLE, S. 1988. Laboratory models of Hawaiian and Strombolian eruptions. *Nature*, **331**, 58–60.

JAUPART, C. & VERGNIOLLE, S. 1989. The generation and collapse of a foam layer at the roof of a basaltic magma chamber. *Journal of Fluid Mechanics*, **203**, 347–380.

JOHNSON, J. B., LEES, J. M. & GORDEEV, E. I. 1998. Degassing explosions at Karymsky Volcano, Kamchatka. *Geophysical Research Letters*, **25**, 3999–4002.

JOHNSON, J. B. & LEES, J. M. 2000. Plugs and chugs: seismic and acoustic observations of degassing explosions at Karymsky, Russia and Sangay, Ecuador. *Journal of Volcanology and Geothermal Research*, **101**, 67–82.

KANAMORI, H. & MORI, J. 1992. Harmonic excitation of mantle Rayleigh waves by the 1991 eruption of Mount Pinatubo, Philippines. *Geophysical Research Letters*, **19**, 721–724.

KIEFFER, S. W. 1981. Fluid dynamics of the May 18 blast at Mount St. Helens. *US Geological Survey Professional Papers*, **1250**, 379.

LADELFA, S., PATANE, G., CLIOCCHIATTI, R., JORON, J.-L. & TANGUY, J.-C. 2001. Activity of Mount Etna preceding the February 1999 fissure eruption: inferred mechanism from seismological and geochemical data. *Journal of Volcanology and Geothermal Research*, **105**, 121–139.

LAIGLE, M., HIRN, A., SAPIN, M. & LEPINE, J.-C. 2000. Mount Etna dense array local earthquake P and S tomography and implications for volcanic plumbing. *Journal of Geophysical Research*, **105**, 21633–21646.

LAUTZE, N. C., HARRIS, A. J. L. & BAILEY, J. E. *ET AL.* 2004. Pulse lava effusion at Mount Etna during 2001. *Journal of Volcanology and Geothermal Research*, **137**, 231–246.

LEGUERN, F., TAZIEFF, H., VAVASSEUR, C. & ZETTWOOG, P. 1982. Resonance in gas discharge of the Bocca Nuova, Etna (Italy, 1968–1969. *Journal of Volcanology and Geothermal Research*, **12**, 161–166.

LEPICHON, A., BLANC, E. & DROB, D. *ET AL.* 2005. Continuous infrasound monitoring of volcanoes to probe high-altitude winds. *Journal of Geophysical Research*, **110**, D13106, doi:1029/2004JD005587.

MATOZA, R. S., HEDLIN, M. A. H. & GARCES, M. A. 2007. An infrasound array study at Mount St Helens. *Journal of Volcanology and Geothermal Research*, **160**, 249–262.

MCGETCHIN, T. R. SETTLE, M. & CHOUET, B. 1974. Cinder cone growth modeled after northeast crater, Mount Etna, Sicily. *Journal of Geophysical Research*, **79**, 3257–3272.

MCNUTT, S. R. 2000. *Volcanic Seismicity, Encyclopedia of Volcanoes*. Academic Press.

MÉTRICH, N., ALLARD, P., SPILLIAERT, N., ANDRONICO, D. & BURTON, M. 2004. 2001 flank eruption of the alkali and volatile-rich primitive basalt responsible for Mount Etna's evolution in the last three decades. *Earth and Planetary Science Letters*, **228**, 1–17.

MONTALTO, A., DISTEFANO, G. & PATANE, G. 1992. Seismic patterns and fluid-dynamics features preceding and accompanying the January 15, 1990 eruptive paroxysm on Mt Etna. *Journal of Volcanology and Geothermal Research*, **51**, 133–143.

MORRISSEY, M. M. & CHOUET, B. 1997. Burst conditions of explosive volcanic eruptions recorded on microbarographs. *Science*, **275**, 1290–1293.

NYE, C. J., KEITH, T. & EICHELBERGER, J. C. *ET AL.* 2002. The 1999 eruption of Shishaldin volcano, Alaska: monitoring a distant eruption. *Bulletin of Volcanology*, **64**, 507–519.

PARFITT, E. A. & WILSON, L. 1995. Explosive volcanic eruptions: IX the transition between Hawai'ian-style lava fountaining and Strombolian explosive activity. *Geophysical Journal International*, **121**, 226–232.

PARFITT, E. A., WILSON, L. & NEAL, C. A. 1995. Factors influencing the height of Hawai'ian lava fountains: Implications for the use of fountain height as an indicator of magma content. *Bulletin of Volcanology*, **57**, 440–450.

PARFITT, E. A. 2004. A discussion of the mechanisms of explosive basaltic eruptions. *Journal of Volcanology and Geothermal Research*, **134**, 77–107.

PATANÉ, D., CHIARABBA, C., COCINA, O., DE GORI, P., MORETTI, M. & BOSCHI, E. 2002. Tomographic images and 3D earthquake locations of the seismic swarm preceding the 2001 eruption Mt Etna eruption: Evidence for dyke intrusion. *Geophysical Research Letters*, **29/10**, doi:10.1029/2001GL014391.

PATANÉ, D., BARBERI, G., COSINA, O., DE GORI, P. & CHIARABBA, C. 2006. Time-resolved seismic tomographic detects magma intrusion at Mount Etna. *Science*, **313**, 821–823.

POLACCI, M., CORSARO, R. A. & ANDRONICO, D. 2006. Coupled textural and compositional characterization of basaltic scoria: insights into the transition from Strombolian to fire fountain activity at Mount Etna. *Geology*, **34**, 201–204.

PRIVITERA, E., SGROI, T. & GRESTA, S. 2003. Statistical analysis of intermittent volcanic tremor associated with the September 1989 summit explosive eruptions at Mount Etna, Sicily. *Journal of Volcanology and Geothermal Research*, **120**, 235–247.

PROUSSEVITCH, A. A., SAHAGIAN, D. L. & KUTOLIN, V. A. 1993. Stability of foams in silicate melts. *Journal of Volcanology and Geothermal Research*, **59**, 161–178.

RESCH, F. J., DARROZES, J. S. & AFETI, G. M. 1986. Marine liquid aerosol production from bursting of air bubbles. *Journal of Geophysical Research*, **91**, 1019–1029.

RICHARDS, A. F. 1963. Volcanic sounds, investigation and analysis. *Journal of Geophysical Research*, **68**, 919–928.

RICHTER, D. H., EATON, J. P., MURATA, K. J., AULT, W. U. & KRIVOY, H. L. 1970. Chronological narrative of the 1959–60 eruption of Kilauea volcano, Hawaii. *US Geological Survey Professional Papers*, **537**, 1–70.

RIPEPE, M., COLTELLI, C., PRIVITERA, E., GRESTA, S., MORETTI, M. & PICCININI, D. 2001. Seismic and infrasonic evidences for an impulsive source of the shallow volcanic tremor at Mount Etna, Italy. *Geophysical Research Letters*, **28**, 1071–1074.

RIPEPE, M., CILIBERTO, S. & DELLA SCHIAVA, M. 2001. Time constraints for modeling source dynamics of volcanic explosions at Stromboli. *Journal of Geophysical Research*, **106**, 8713–8727.

SABLE, J. E., HOUGHTON, B. F., DEL CARLO, P. & COLTELLI, M. 2006. Changing conditions of magma ascent and fragmentation during the Etna 122 BC basaltic eruption: Evidence from clast microtextures. *Journal of Volcanology and Geothermal Research*, **158**, 333–354.

SACCOROTTI, G., ZUCCARELLO, L., DEL PEZZO, E., IBANEZ, J. & GRESTA, S. 2004. Quantitative analysis of the tremor wavefield at Etna volcano, Italy. *Journal of Volcanology and Geothermal Research*, **136**, 223–245.

SPARKS, R. S. J., BURSIK, M. I., CAREY, S. N., GILBERT, J. S., GLAZE, L. S., SIGURDSSON, H. & WOODS, A. W. 1997. *Volcanic Plumes.* Wiley, Chichester.

SPILLIAERT, N., METRICH, N. & ALLARD, P. 2006a. S-Cl-F degassing pattern of water-rich alkali basalt: modelling and relationship with eruption styles on Mount Etna volcano. *Earth and Planetary Science Letters*, **248**, 772–786.

SPILLIAERT, N., ALLARD, P., METRICH, N. & SOBOLEV, A. 2006b. Melt inclusion record of the conditions of ascent, degassing and extrusion of volatile-rich alkali basalt during the powerful 2002 flank eruption of Mount Etna (Italy). *Journal of Geophysical Research*, **111**, B04203, doi:10.1029/2005JB003934.

SWANSON, D. A., DUFFIELD, D. A., JACKSON, D. B & PETERSON, D. W. 1979. Chronological narrative of the 1969–1971 Mauna-Ulu eruption of Kilauea volcano, Hawaii. *US Geological Survey Professional Papers*, **1056**.

THOMPSON, G., MCNUTT, S. R. & TYTGAT, G. 2002. Three distinct regimes of volcanic tremor associated with the eruption of Shishaldin volcano, Alaska, April 1999. *Bulletin of Volcanology*, **64**, 535–547.

TURNER, J. S. 1973. *Buoyancy Effects in Fluids.* Cambridge University Press.

VERGNIOLLE, S. & JAUPART, C. 1986. Separated two-phase flow and basaltic eruptions. *Journal of Geophysical Research*, **91**, 12842–12860.

VERGNIOLLE, S. & JAUPART, C. 1990. Dynamics of degassing at Kilauea volcano, Hawaii. *Journal of Geophysical Research*, **95**, 2793–2809.

VERGNIOLLE, S. & BRANDEIS, G. 1994. Origin of the sound generated by Strombolian explosions. *Geophysical Research Letters*, **21**, 1959–1962.

VERGNIOLLE, S. & BRANDEIS, G. 1996. Strombolian explosions: 1. A large bubble breaking at the surface of the lava column as a source of sound. *Journal of Geophysical Research*, **101**, 20433–20448.

VERGNIOLLE, S., BRANDEIS, G. & MARESCHAL, J. C. 1996. Strombolian explosions: 2. Eruption dynamics determined from acoustic measurements. *Journal of Geophysical Research*, **101**, 20 449–20 465.

VERGNIOLLE, S., BOICHU, M. & CAPLAN-AUERBACH, J. 2004. Acoustic measurements of the 1999 basaltic eruption of Shishaldin volcano, Alaska: 1) Origin of Strombolian activity. *Journal of Volcanology and Geothermal Research*, **137**, 109–134.

VERGNIOLLE, S. & CAPLAN-AUERBACH, J. 2004. Acoustic measurements of the 1999 basaltic eruption of Shishaldin volcano, Alaska: 2) Precursor to the Subplinian activity. *Journal of Volcanology and Geothermal Research*, **137**, 135–151.

VERGNIOLLE, S. & CAPLAN-AUERBACH, J. 2006. Basaltic thermals and Subplinian plumes: constraints from acoustic measurements at Shishaldin volcano, Alaska. *Bulletin of Volcanology*, **68**, 7–8, 611–630.

VERGNIOLLE, S. 2008. From sound waves to bubbling within a magma reservoir: comparison between eruptions at Etna (2001, Italy) and Kilauea (Hawaii). *In:* GILBERT, J. S. & LANE, S. (eds) *Fluid Motions in Volcanic Conduits: A Source of Seismic and Acoustic Signals.* The Geological Society, London, Special Publications, **307**, 125–146.

WALLIS, G. B. 1969. *One Dimensional Two-Phase Flows.* McGraw-Hill, New York.

WEILL, A., BRANDEIS, G. & VERGNIOLLE, S. *ET AL.* 1992. Acoustic sounder measurements of the vertical velocity of volcanic jets at Stromboli volcano. *Geophysical Research Letters*, **19**, 2357–2360.

WILLIAMS, H. & MCBIRNEY, A. R. 1979. *Volcanology.* Freeman Cooper, San Francisco.

WILSON, L. 1980. Relationships between pressure, volatile content and ejecta velocity. *Journal of Volcanology and Geothermal Research*, **8**, 297–313.

WOLFE, E. W., GARCIA, M. O., JACKSON, D. B., KOYANAGI, R. Y., NEAL, C. A. & OKAMURA, A. T. 1987. The Pu'u 'O'o eruption of Kilauea volcano, episodes 1–20, January 3, 1983 to June 8, 1984, in Volcanism in Hawaii. *US Geological Survey Professional Papers*, **1350**, 471–508.

WOLFE, E. W., NEAL, C. A., BANKS, N. G. & DUGGAN, T. J. 1988. Geologic observations and chronology of eruptive events, in the Pu'u 'O'o eruption of Kilauea volcano, episodes 1–20, January 3, 1983 to June 8, 1984. *In:* WOLFE, E (ed.) *US Geological Survey Professional Papers*, **1463**, 471–508.

WOULFF, G. & MCGETCHIN, T. R. 1976. Acoustic noise from volcanoes: theory and experiments. *Geophysical Journal of the Royal Astronomical Society*, **45**, 601–616.

From sound waves to bubbling within a magma reservoir: comparison between eruptions at Etna (2001, Italy) and Kilauea (Hawaii)

S. VERGNIOLLE

Institut de Physique du Globe de Paris, Institut de recherche associé CNRS et Université de Paris 7, 4 Place Jussieu, 75252 Paris Cedex 05, France (e-mail: vergniolle@ipgp.jussieu.fr)

Abstract: The 2001 eruption of Etna, prior to the flank eruption, has shown an alternation between episodes rich in gas, composed with a series of Strombolian explosions sometimes leading to a fire fountain, and repose periods. The regular alternation results from the coalescence of a foam trapped in the reservoir and periodically rebuilt prior to each episode. The degassing of a magma reservoir depends on bubble diameter, gas volume fraction, surface area and height of the reservoir. These four parameters are deduced from the measured gas flux, the timescale over which the gas flux decreases and the foam dynamics. The dimensionless foam thickness, 0.76 for a purely Strombolian episode, increases to 0.89 for an episode leading to a fire fountain, indicating a more efficient foam coalescence. At Etna, the bubble diameter, gas volume fraction, surface area and height of the reservoir are estimated at 0.50–0.59 mm, 0.25–0.39%, >0.20 km^2 and 97–220 m, respectively. At Kilauea, the excellent agreement between the prediction of the foam model, 0.94–1.2 km^2 and that resulting from deformations, 1 km^2, reinforces the validity of the foam model qualitatively and quantitatively. The thickness of the degassing reservoir, 16–12 m in 1959, is now 1.3–1.1 km.

Basaltic eruptions often begin with a curtain of fire, such as at Kilauea (Hawaii) or in Icelandic fissure eruptions (Scarth 1994). However, after a day, the curtain of fire focalizes into a series of vents, as a result of a thermal instability within a rising magma (Bruce & Huppert 1989), and is very rapidly transformed into a fire fountain at a single vent. The large eruptions of Kilauea (1959–60, 1969–70, 1983–present) display a regular time series of numerous fire fountains, a feature also present at Etna (1989, 1995–96, 1998–99, 2000, 2001) (Richter *et al.* 1970; Swanson *et al.* 1979; Wolfe *et al.* 1987, 1988; Bertagnini *et al.* 1990; Heliker & Mattox 2003; Privitera *et al.* 2003; Alparone *et al.* 2003; Behncke *et al.* 2006; Allard *et al.* 2006).

Fire fountains can also be observed in a rapid erratic succession such as during the 2002 eruption of Etna (Italy). However the 2002–03 eruption is very different from most recent eruptions (1989, 1995–96, 1998–99, 2000 and 2001), both by its products and by the lack of cyclicity of a few days between the repose periods (Clocchiatti *et al.* 2004; Spilliaert *et al.* 2006b; Bertagnini *et al.* 1990; La Delfa *et al.* 2001; Privitera *et al.* 2003; Behncke & Neri 2003; Alparone *et al.* 2003; Behncke *et al.* 2006; Spilliaert *et al.* 2006b; Allard *et al.* 2006). Another difference is related to the simultaeous quiescence at the summit vents,

making the 2002–03 eruption solely a flank eruption. Furthermore that eruption, in contrast with the 'classical' eruptions, results from the rapid rise of a primitive basaltic magma, rich in volatiles, which does not stop in a magma reservoir between its initiation at 7–9 km depth below the summit vents and the surface (Clocchiatti *et al.* 2004; Spilliaert *et al.* 2006a, b). Alternatively, Carbone *et al.* (2006) proposed that an accumulation of a gas cloud, i.e. a concentration of bubbles, at some level in the conduit plexus feeding the new eruptive vent, could have acted as a joint source of gravity and tremor anomalies during the 2002–03 eruption.

The fire fountain on 26 September 1989 at Etna was obviously driven by a long inner gas jet while Strombolian explosions, photographed in 1989 and 2001, show clearly that they correspond to bursting bubbles as large as the conduit (Bertagnini *et al.* 1990; Vergniolle & Ripepe 2008). However, all the bubbles produced during one episode are expelled as a train of bubbles of various lengths, from slug to annular flow (Vergniolle & Ripepe 2008). Hence, an eruptive episode corresponds to a phase very rich in gas, which is separated from other episodes by a repose period of a few days. Although not all eruptive episodes reach a transition to a fully developed fire fountain, any episode without a transition to a fire fountain should be

From: LANE, S. J. & GILBERT, J. S. (eds) *Fluid Motions in Volcanic Conduits: A Source of Seismic and Acoustic Signals.* Geological Society, London, Special Publications, **307**, 125–146.
DOI: 10.1144/SP307.8   0305-8719/8/$15.00 © The Geological Society of London 2008.

called a 'quasi-fire fountain' because of its richness in gas (Vergniolle & Ripepe 2008).

Sixteen and 64 episodes of fully developed fire fountains, described as Hawaiian-style, have been reported in 1989 (Bertagnini et al. 1990; Privitera et al. 2003) and in 2000 (Behncke & Neri 2003; Alparone et al. 2003), respectively, whereas 22 and 16 episodes of quasi-fire fountains are described in 1998–99 (La Delfa et al. 2001) and in 2001 (Behncke et al. 2006; Allard et al. 2006), respectively. This behaviour is qualitatively very similar to that of a Hawaiian volcano (Behncke et al. 2006), especially during the summit eruption of Kilauea Iki (1959–60) where 16 fire fountains occurred with a repose time varying between seven hours to four days (Richter et al. 1970). However, the repose time at Kilauea is one month for the Mauna Ulu and Pu'u 'O'o eruptions, both located on the east rift zone (Swanson et al. 1979; Wolfe et al. 1987, 1988; Heliker & Mattox 2003).

The regular repose time between a phase rich in gas and a phase relatively poor in gas has been explained by the behaviour of a foam layer trapped at the roof of the magma chamber, which can be periodically unstable (Jaupart & Vergniolle 1988, 1989; Vergniolle & Jaupart 1990). When the foam becomes unstable, a large gas volume is released in the volcanic conduit, leading to either fire fountains or explosions, with a repose time of days to a month at Kilauea or one hour at Stromboli. Furthermore, a stable foam in the reservoir can lead to the formation of a stable foam in the conduit, which can break and produce a basaltic Subplinian plume, such as at Shishaldin (Alaska) (Vergniolle & Caplan-Auerbach 2004, 2006, in prep.; Vergniolle in prep.). Recently, the 122 BC eruption of Etna has been recognized as a basaltic Subplinian eruption, a rare event with only 11 of these eruptions being reported since 1500 BC (Coltelli et al. 1998; Sable et al. 2006).

The high vesicularity of scoria ejected by fire fountains at Etna, $\cong 65 \pm 5\%$ (Polacci et al. 2006), is very similar to that produced by Kilauea (Mangan & Cashman 1996). For Polacci et al. (2006), a fire fountain at Etna is mainly driven by the foam coalescence accumulated at a constriction, whereas for Mangan & Cashman (1996), it results at Kilauea from fast ascent and expansion of volatile-saturated magma.

An increase in the magma rising rate and volatile flux, coupled with rapid exsolution during ascent, could also trigger the formation of a lava fountain by a progressive coalescence of bubbles in the conduit (Parfitt & Wilson 1995; Parfitt et al. 1995; Parfitt 2004), although several points do not favour that model. First, the large overpressure measured in Strombolian bubbles, ≥ 0.1 MPa,

largely exceeds the pressure obtained by the coalescence of two bubbles in the conduit, at most several thousands of Pa (Vergniolle & Brandeis 1996; Vergniolle et al. 2004; Vergniolle & Caplan-Auerbach 2004; Vergniolle & Ripepe 2008). Secondly, if both models could explain the episodic fire fountaining at Etna in 1998–99 (Harris & Neri 2002), the repose time between fire fountains was incorrectly used as a characteristic timescale for the entire eruption (Vergniolle 1996), cancelling the main argument against the foam model. Thirdly, recent chemical FTIR measurements show that the formation of a fire fountain during the 2000 eruption resulted from a foam disruption at a structural barrier located at 1.5–1.8 km (Allard et al. 2005; Spilliaert et al. 2006a), favouring a model based on foam accumulation at depth (Jaupart & Vergniolle 1988, 1989; Vergniolle & Jaupart 1990). This is indeed compatible with large gas content both at Etna, with 3.4 wt% of H_2O and 0.41 wt% of CO_2 (Métrich et al. 2004), and at Kilauea, with 0.7 wt% of CO_2 (Gerlach et al. 2002). Fourthly, the transition at Kilauea between the first part of the eruption, marked by cyclic fire fountains for several months, and the second part, without a fire fountain for the last several months, was explained (Parfitt & Wilson 1995; Parfitt et al. 1995; Parfitt 2004) by the decrease in the magma supply and ascent rate. However, such a transition, which exists at Etna, is marked by a nearly three-fold increase in the eruption rate, precluding Parfitt's model (2004) being applicable at Etna (Behncke et al. 2006). Fifthly, seismic waveforms measured during the 2001 eruption of Etna are also in agreement with a model based on a foam collapse at a structural barrier (Gresta et al. 2004). Therefore, observations of eruptive activity at Etna suggest that a fire fountain is driven by a long inner gas jet (Bertagnini et al. 1990; Vergniolle & Ripepe 2008), and is formed at the depth of the reservoir rather than in the conduit.

The variety of surface activity at Etna with Strombolian episodes, fire fountains and small ash-rich eruptive clouds (Allard et al. 2006; Behncke et al. 2006), here referred to as thermals by analogy to Shishaldin (Vergniolle & Caplan-Auerbach 2006) makes this volcano a very good candidate to quantify the foam dynamics in a magma reservoir (Fig. 1). Another remarkable feature of Etna is that it is at the transition between two classical types of eruption, Strombolian and Hawaiian. To understand this characteristic, the 2001 eruption at the southeast crater of Etna will here be compared with three large eruptions of Kilauea (1959–60, 1969–71, 1983–present). The respective role of the reservoir geometry, i.e. the surface area and the height of the degassing layer,

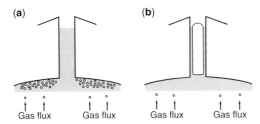

Fig. 1. (**a**) Accumulation of small bubbles at the top of the reservoir during the repose period. (**b**) Formation of a quasi-fire fountain by coalescence of the foam trapped at the top of the reservoir. Foam coalescence at Etna can vary between midly active, such as for purely Strombolian activity (4 July; Fig. 1 in Vergniolle & Ripepe 2008) and strongly active, such as for episodes leading to a fire fountain (12 July; Fig. 2 in Vergniolle & Ripepe 2008).

and of the bubble content in the reservoir, i.e. the bubble diameter and the gas volume fraction, will then be discussed.

Constraints from other studies

Gas volume and gas flux

The gas volume, expelled at the surface, can be estimated from the height of fire fountains H_{ff} and their duration τ_{ff} (Vergniolle & Jaupart 1990; Vergniolle & Ripepe 2008), because the gas velocity at the vent, U_{ff}, depends on the height by (Wilson 1980; Sparks *et al.* 1997):

$$U_{ff} = \sqrt{2gH_{ff}} \qquad (1)$$

where g is the acceleration due to gravity, $9.81\ \mathrm{m\ s^{-2}}$. Conservation of the gas flux at the vent leads to the total gas volume expelled by a fire fountain, V_{ff}, equal to:

$$V_{ff} = U_{ff}\, S_{vent} t_{ff} \qquad (2)$$

where S_{vent} is the vent area.

Recording the sound waves produced by an explosion is also a very powerful tool to remotely monitor the gas volume at the vent (Vergniolle & Brandeis 1994, 1996; Vergniolle *et al.* 2004; Vergniolle & Caplan-Auerbach 2004; Vergniolle & Ripepe 2008). This technique, based on a best fit between a recorded explosion and its synthetic waveform, also provides an estimate for both the average and the maximum gas volume for a given duration (Vergniolle & Ripepe 2008). Using this ratio is a very robust technique to quantify the steadiness of surface activity. In the following, the

degree of unsteadiness, C_{unst}, is defined as the ratio between the average gas volume and the maximum gas volume. At Etna, this coefficient varies between 0.39 and 0.11 for Strombolian activity and a fire fountain, respectively (Table 1; Vergniolle & Ripepe 2008).

At Kilauea, the degree of unsteadiness C_{unst} can be deduced, in the absence of acoustic measurements, from a detailed monitoring of the height of several fire fountains observed during the Pu'u 'O'o eruption. The maximum and average height are related to one another by a factor between 1.5 and 2.9 (Fig. 1.27 in Wolfe *et al.* 1987, 1988), leading to a degree of unsteadiness C_{unst} between 0.83 and 0.59 (Table 2). Visual observations show that fire fountains at Kilauea are very steady compared with Etna, as indeed shown by this coefficient C_{unst} being closer to 1 for the former than the latter.

The gas flux in the reservoir (Fig. 1) is obtained, as previously done (Vergniolle & Jaupart 1990; Vergniolle 1996), from the ratio between the gas volume compressed at the reservoir pressure using the perfect gas law, and the repose period prior to each episode during which the foam rebuilds.

The depth of the magma reservoir

At Etna, bodies of accumulating magma seem to exist below the summit craters from a depth of 3 km to the surface (Chiarabba *et al.* 2000), as also suggested by the location of the tremor source (Saccorotti *et al.* 2004). Tomographic images also show that a low Vp/Vs zone is found at 3–4 km below the vents, which can be interpreted as a region of molten material rich in gas (Laigle *et al.* 2000; Patanè *et al.* 2002, 2006). Two other storage zones are located at 6 km and 13 km below the summit (Murru *et al.* 1999; Chiarabba *et al.* 2000; Caracausi 2003). Tilt modelling showed that the data recorded in June 2000 are explained by a conduit of radius 5–8 m and a depth of 1.5–1.9 km below the southeast crater (Bonaccorso 2006). That depth, 1.5–1.8 km, is also the depth of a constriction, below which the gas accumulated before leading to the fire fountain on 14 June 2000 (Allard *et al.* 2005; Spilliaert *et al.* 2006a). Although seismic studies suggest that the reservoir could be made of several large magma pockets (Aki & Ferrazzini 2001), the magma chamber at Etna can be assumed to be located at a depth of 2 km (Laigle & Hirn 1999; Laigle *et al.* 2000; Calvari *et al.* 2001; Patanè *et al.* 2002, 2006).

Deformations at Kilauea show that a single summit reservoir can explain both inflations and deflations measured between 1970 and 1985 (Yang *et al.* 1992). The summit reservoir shape,

Table 1. *Bubble characteristics within the degassing reservoir of Etna using an estimate for the gas volume based on acoustic records (V_{gac}) and on the height of fire fountains once corrected by the degree of unsteadiness (V_{ff})*

Etna 2001 (SEC)	4 July	7 July	12 July
Surface activity	Str	Th/Str	ff
V_{gac} ($\times 10^7$ m^3)	0.54	0.22 ?	1.0
C_{unst}	0.39	0.39 ?	0.11
V_{ff} ($\times 10^7$ m^3)	1.4	1.2	2.1
V_{gact}/V_{gtot}	0.38	0.38 ?	0.48
Repose (hr)	101	58 hr 08 min	132
Reservoir (acoustic)			
V_{gres} ($\times 10^4$ m^3)	1.0	0.41	1.9
Q_{gres} ($\times 10^{-2}$ m^3 s^{-1})	2.7	1.9	4.0
N_1	0.76	0.65	0.89
d_{gres} (mm)	0.54	0.50	0.59
S_{res} (km^2)	0.10	0.038	0.20
α_{res} (%)	0.62	1.3	0.39
α_{res} (%) ($S_{res} = 0.20$ km^2)	0.31	0.25	0.39
H_{gres} (m)	110 (260)	97 (220)	130 (300)
Reservoir (ff)			
V_{gres} ($\times 10^4$ m^3)	2.7	1.1	3.9
Q_{gres} ($\times 10^{-2}$ m^3 s^{-1})	7.1	5.2	8.3
N_1	0.76	0.65	0.89
d_{gres} (mm)	0.43	0.40	0.44
S_{res} (km^2)	0.21	0.076	0.32
α_{res} (%)	1.3	2.9	0.77
H_{gres} (m)	69 (160)	61 (130)	89 (200)

SEC, Str, th and ff stand for the southeast crater, Strombolian activity, thermal and fire fountain, respectively. C_{unst}, V_{gact}/V_{gtot} and 'Repose' correspond to the degree of unsteadiness, i.e. the ratio between the average gas volume expelled per five minutes and its maximum value, the ratio between the volume of the slugs coming from the reservoir (active) and the total gas volume, respectively (Fig. 3). V_{gres}, Q_{res}, N_1, d_{res}, α_{res}, S_{res} are the foam volume, gas flux, dimensionless foam height, bubble diameter, gas volume fraction and area of the reservoir, respectively (Fig. 2). The height of the reservoir is calculated without the episode on 9 May and values in brackets include that episode.

directly deduced from deflationary periods, is best modelled by a spherical body with a centre of deflation spreading over an area of 1 km^2 and located at 2.6 km below the summit (Yang *et al.* 1992). That depth is compatible with a low-velocity zone observed in seismic tomography (Thurber 1987) and with the high Vp/Vs ratio of seismic waves at 3 km depth (Dawson *et al.* 1999). A depth of a few kilometres (Vergniolle & Jaupart 1990) is also deduced from the time taken by the gas pistons to rise from the reservoir to the surface (Swanson *et al.* 1979; Wolfe *et al.* 1987, 1988). All these results make us consider, in the rest of the paper, that the reservoir depth at Kilauea is 2 km below the summit.

Gas volume fraction in reservoir from fluid inclusions

Gas volume fraction can be estimated at the depth of the reservoir from the total amount of volatiles in the magma. A dissolved gas phase at 1.5 wt%

of water was measured during the 1989 eruption of Etna (Métrich *et al.* 1993). The fluid inclusions, also rich in CO_2, suggested that their minimum trapping depth is 3 km if solely based on water (Métrich *et al.* 1993), and this value could represent the depth of the local shallow reservoir below the southeast crater. Although we have no measurements of fluid inclusions during the 2001 eruption, the first days of the 2002 eruption expelled a magma very similar to the one expelled during the 2001 eruption (Métrich *et al.* 2004; Spillaert *et al.* 2006b), making fluid inclusion measurements on the 2002 eruption a good approximation. However, if the magma is saturated in gas at the depth of the reservoir, the gas content in fluid inclusions does not represent the initial gas content in the magma (Wallace 2005; Spillaert *et al.* 2006b). But the very primitive magma feeding the 2002 eruption is believed to come from a depth of 7–9 km in five hours, without stopping and degassing in a magma reservoir (Métrich *et al.* 2004; Spillaert *et al.* 2006b). Hence, measurements of its fluid inclusions are

Table 2. *Bubble characteristics within the degassing reservoir of Kilauea using an estimate for the gas volume based on the height of fire fountains (V_{ff})*

Kilauea	Kilauea Iki	Mauna Ulu	Pu'u 'O'o
surface activity	ff (summit)	ff (ERZ)	ff (ERZ)
ff duration (days)	37	221	1237
C_{unst}	0.59–0.83	0.59–0.83	0.59–0.83
V_{gact}/V_{gtot}	1 ?	1 ?	1 ?
Reservoir (ff)			
V_{gres} ($\times 10^6$ m^3)	0.25–0.36	17–24	1.0–1.4
Q_{gres} (m^3 s^{-1})	2.0–2.9	2.7–3.9	0.47–0.66
N_1	0.82	0.82	0.82
d_{gres} (mm)	0.20–0.18	0.18–0.17	0.29–0.26
S_{res} (km^2)	0.94–1.2	56–76	5.2–7.1
α_{res} (%)	37–49	0.98–1.3	0.74–0.98
H_{gres} (m)	16–12	93–79	1300–1100
Eruption			
V_{gtot} ($\times 10^{10}$ m^3)	(1.0–1.4)	(8.7–12)	(6.2–8.8)
+ unobserv (10^{10} m^3)	1.0–1.4	8.7–12	7.5–11
End previous eruption	28/02/1955	19/02/1960	15/10/1971
(summit)	(27/06/1952)		
Repose (days)	1718 (2673)	3379	3730
Q_{ginter} (m^3 s^{-1})	(0.084–0.12)	(1.1–1.6)	(0.77–1.1)
+ unobserv (m^3 s^{-1})	0.13–0.18	1.1–1.9	0.95–1.4

ff, summit and ERZ stand for fire fountains, summit eruption and east rift zone, respectively. C_{unst} and V_{gact}/V_{gtot} correspond to the degree of unsteadiness and the ratio between the volume of the slugs coming from the reservoir (active) and the total gas volume, respectively (Fig. 3). V_{gres}, Q_{res}, N_1, d_{res}, S_{res}, α_{res} and H_{res} are the foam volume, minimum gas flux, dimensionless foam height, bubble diameter, surface area of the reservoir, gas volume fraction, and reservoir height, respectively (Fig. 2). V_{gtot} is the total gas expelled for the entire eruption with adding unobserved episodes (values in brackets are without unobserved episodes). 'Repose' and Q_{ginter} are the period of quiescence prior to the eruption and the inter-eruptive gas flux, respectively.

representative of the maximum gas content available in the reservoir at each magma reinjection. This primitive melt contains high contents of H_2O (3.4 wt%), CO_2 (0.11 to 0.41 wt%; and even ≥ 0.7 wt%) together with S (0.32 wt%) and Cl (0.16 wt%) (Métrich *et al.* 2004).

For simplicity, we use Henry's law of solubility (Mysen 1977; Stolper & Holloway 1988) to estimate the dissolved and exsolved gas volume fraction at the depth of the reservoir for either a pure CO_2 or H_2O phase. The mass fraction of exsolved gas x_{exs} at the pressure of the reservoir P_{res} depends on the total amount of gas x_{tot}:

$$x_{tot} = x_{exs} + s\, P_{res}^n \quad (3)$$

where s is the solubility constant, 5×10^{-12} and the exponent coefficient, n equal to 1 for CO_2 (Stolper & Holloway 1988; Dixon *et al.* 1995). For pure water in basalts, the solubility constant is 6.8×10^{-8} for an exponent coefficient of 0.7 (Mysen 1977).

The gas volume fraction α_{exs} exsolved at the depth of the reservoir depends on the mass fraction of exsolved gas x_{exs} if bubbles and liquid have roughly the same rise velocity, so they rise together, giving:

$$x_{tot} = \cfrac{1}{1 + \cfrac{(1-x_{exs})\,\rho_{g\,2\,km}}{x_{exs}\,\rho_{liq}}} \quad (4)$$

where ρ_{liq} and $\rho_{g\,2\,km}$ are the liquid and gas density at 2 km depth, 2700 kg m^{-3} (Williams & McBirney 1979), 87 and 265 kg m^{-3} for H_2O and CO_2, respectively, at 1050 °C (Table 3).

If we asume that CO_2 is the sole volatile phase at Etna, with a total mass content of 0.41 wt% (Métrich *et al.* 2004), the exsolution occurs at a depth of 31 km below the summit, whereas for a pure H_2O phase with 3.4 wt%, the bubbles form at a depth of 5.2 km. It is therefore clear that there would be a significant population of nucleated bubbles at a depth of 2 km (Figs 1 and 2). The gas volume fraction is between 4.4% and 7.7% for a total mass content of CO_2 between 0.41 wt% (Métrich *et al.* 2004) and 0.70 wt% (Spilliaert *et al.* 2006), respectively (equation 4). The gas volume fraction corresponding to 3.4 wt% of water is equal to 11%.

Table 3. *Physical parameters and references used throughout the text*

Physical parameters	Etna	Kilauea	References
ρ_{liq} (kg m^{-3})	2700	2700	W
μ_{liq} (Pa s)	10	10	G ; M ; S
σ (kg s^{-2})	0.36	0.36	P
ε_{res}	0.6	0.6	J
C_{unst}	0.39 (Str)–0.11 (ff)	0.59–0.83	V
V_{gact}/V_{gtot}	0.38 (Str)–0.48 (ff)	1 ?	V
z_{gres} (km)	2	2	
N_1	0.76 (Str)–0.89 (ff)	0.82	
S_{res} (km^2)	>0.20	0.94–1.2	
others S_{res} (km^2)	>0.072	1.0	H ; Y
V_{gtot} ($\times 10^9$ m^3)	0.12–0.42	10–14	
End previous eruption	24/06/2000	28/02/1955 (ERZ)	
Repose (days)	318	1718	
Q_{ginter} (m^3 s^{-1})	0.0025–0.0095	0.13–0.18	

(W), (G; M; S), (P), (J), (V) and (H; Y) stand for Williams & McBirney 1979; Giordiano & Dingwell 2003; Métrich *et al.* 2004; Schmincke 2004; Proussevitch *et al.* 1993; Jaupart & Vergniolle 1988; Vergniolle & Ripepe 2008; Harris & Neri 2002; Yang *et al.* 1992, respectively. ρ_{liq}, μ_{liq}, σ and ε_{res} are liquid density, viscosity, surface tension and gas volume fraction within the foam. C_{unst} and V_{gact}/V_{gtot} correspond to the degree of unsteadiness, the ratio between the volume of the slugs coming from the reservoir (active) and the total gas volume, respectively (Fig. 2). z_{res}, N_1 and S_{res} are the reservoir depth, foam dimensionless height and surface area of the reservoir, respectively. V_{gtot} is the total gas expelled for the entire eruption with adding unobserved episodes. 'Repose', 'end previous', ERZ and Q_{inter} are the period of quiescence prior to the eruption, the time when previous eruption had stopped, the east-rift zone at Kilauea and the inter-eruptive gas flux, respectively.

At Kilauea, the maximum gas volume fraction is equal to 7.7% and 11% (equation 4) when using 0.7 wt% of CO_2 and 3.4 wt% of H_2O (Gerlach *et al.* 2002) and a reservoir depth of 2 km (Yang *et al.* 1992).

Physical processes in the conduit and the reservoir at Etna

Description of the 2001 eruption at the southeast crater

On 9 May 2001, a relatively weak quasi-fire fountain marked a distinct increase in the level of activity, hence the beginning of the eruption (Allard *et al.* 2006; Behncke *et al.* 2006). Unusually vigorous eruptive episodes occurred in early June with the first strong phase on 6 June. During the following six weeks, the vigour of the eruptive episode gradually increased and at times culminated in true lava fountaining from the southeast crater. Fourteen episodes occurred between 7 June and 13 July ending with the most violent of the series (Behncke *et al.* 2006). A vigorous lava fountain occurred at the southeast crater, shortly after midnight on 17 July, in a very similar fashion to the previous episode, while turning into a flank eruption (Behncke *et al.* 2006). The last two episodes coincided with a vigorous seismic crisis that affected the south flank of Etna and indicated that

a significant change in the eruptive dynamics of the volcano was at hand (Allard *et al.* 2006).

Transition between purely Strombolian activity and a fire fountain

The difference between purely Strombolian activity, as on 4 July, and an episode leading to a fire fountain, as on 12 July (Vergniolle & Ripepe 2008), is marked, for the latter, by a simultaneous increase in the number of explosions and in the slug length (Fig. 12 in Vergniolle & Ripepe 2008). The gradual transition between Strombolian explosions and fire fountains also suggests that both activities share a common origin. This reinforces the model of foam coalescence at the depth of the reservoir (Jaupart & Vergniolle 1988, 1989) as being at the origin of both Strombolian explosions (Ripepe & Gordeev 1999; Ripepe *et al.* 2001a,b; Gresta *et al.* 2004) and fire fountains (Allard *et al.* 2005; Spilliaert *et al.* 2006a; Behncke *et al.* 2006; Polacci *et al.* 2006).

Transition to a thermal on 7 July

On 7 July, a thermal, i.e. a small ash-rich eruptive cloud, containing lapilli rose to 500–600 m placing the start of the episode at 07 hr 45 min. It was very well collimated, relatively slow rising and did not produce energetic sound waves

(C. Jaupart pers. comm. 2007) in contrast with Strombolian explosions, such as on 4 July. The plume continued its ascent up to 1.5–2 km above the crater area and one of the largest outbursts produced widespread ashfalls (BGVN 2001). The relatively low level of sound is related to the source being a dipole for an ash and gas plume, which hence attenuates strongly with the distance d_{vent} as $1/d_{vent}^2$ in the near field and $1/d_{vent}$ in the far field, in contrast with the monopolar radiation of Strombolian explosions (Vergniolle & Caplan-Auerbach 2006).

The formation at Etna of a thermal resembles, albeit at a much-reduced scale, what happened at Shishaldin (Alaska) in 1999, where a 16 km high basaltic Subplinian plume was produced (Vergniolle & Caplan-Auerbach 2006; Vergniolle subm.). The analogy with Shishaldin suggests that, also on 7 July 2001 at Etna, the foam in the reservoir was just starting to be unstable. At Etna, the Strombolian activity, the fire fountain and the thermal are hence likely to be driven by the foam dynamics, which can be quantified from the gas flux and the bubble diameter in the reservoir (Jaupart & Vergniolle 1989).

Equations for a degassing reservoir

The physical process at work during the degassing of a magma reservoir depends on the bubble diameter d_{res}, on the gas volume fraction α_{res}, on the surface area of the reservoir S_{res} and on the reservoir height H_{res} (Fig. 2). The first two constraints are given by the measured gas flux and, by analogy to Kilauea (Vergniolle 1996), by the timescale over which the gas flux decreases. The two other equations are given by the foam dynamics (Jaupart & Vergniolle 1989).

Gas flux in reservoir

Rising small bubbles in a degassing magma reservoir (Fig. 2) provide a gas flux Q_{res} at its top (Vergniolle & Jaupart 1990) of:

$$Q_{gres} = \alpha_{res} \frac{d_{res}^2 \, (\rho_{liq} - \rho_{g\,2\,km}) \, g}{18 \, \mu_{liq}} S_{res} \quad (5)$$

where μ_{liq} is the magma viscosity (Table 3). The gas density is assumed to be negligible compared with the liquid density (equation 4).

Foam dynamics in reservoir

The foam behaviour depends on two dimensionless numbers (Jaupart & Vergniolle 1989). The dimensionless foam height, N_1, which is the ratio

between the maximum and the critical foam height, is given by:

$$N_1 = \left[\frac{3 \, Q_{res} \, \mu_{foam}}{\pi \, \varepsilon_{res}^2 \, \rho_{liq} \, g} \right]^{1/4} \Big/ \left[\frac{4 \, \sigma}{\varepsilon_{res} \, \rho_{liq} \, g \, d_{res}} \right] \quad (6)$$

where ε_{res} and σ are the gas volume fraction within the foam in the reservoir and the surface tension (Fig. 2; Table 3). The foam viscosity μ_{foam} depends on the magma viscosity (Jaupart & Vergniolle 1989; Vergniolle et al. 2004):

$$\mu_{foam} = \frac{\mu_{liq}}{(1 - \varepsilon_{res})^{5/2}}. \quad (7)$$

The surface area of the reservoir

When the foam is just above its limit of stability, we can assume that the foam height is close to its critical height for which coalescence can be spontaneously initiated. If we further assume that the foam shape is rectangular, the foam volume V_{foam} is (Jaupart & Vergniolle 1989):

$$V_{foam} = \frac{4 \, \sigma \, S_{res}}{\varepsilon_{res} \, \rho_{liq} \, g \, d_{res}}. \quad (8)$$

The foam volume is proportional to the gas volume V_{gfoam} by:

$$V_{gfoam} = \varepsilon_{res} \, V_{foam}. \quad (9)$$

However, a foam close to its stability is not entirely widthdrawn by its coalescence (Jaupart & Vergniolle 1998). Hence, the surface area of the reservoir is best characterized by using an episode during which the foam coalescence is very efficient.

The height of the degassing reservoir

The decrease of the gas flux within the reservoir, which occurred at Etna and for Kilauea Iki (see next sections) is similar to results already obtained for the Mauna Ulu and Pu'u 'O'o eruptions on Kilauea (Vergniolle & Jaupart 1990; Vergniolle 1996). The large initial value of the gas flux was explained at Kilauea by the rise of the largest bubbles of a population initially existing in the magma chamber, whereas the small late gas flux resulted from the smallest bubbles of the population (Vergniolle 1996). At the transition between the period with cyclic fire fountains and the period with solely effusive activity, the foam trapped in the reservoir had undergone a transition between unstable and stable (Vergniolle & Jaupart 1990).

The duration of an eruption, which is related to the progressive depletion in gas content, can then be calculated from the time taken for a single bubble to rise from the bottom to the top of the reservoir, τ_{stokes} (Vergniolle 1996):

$$\tau_{stokes} = \frac{18\,\mu_{liq}\,H_{res}}{d_{res}^2(\rho_{liq} - \rho_{g\,2\,km})\,g}. \quad (10)$$

A dimensionless time τ_{ff*}, based on the first period of activity with periodic fire fountains, τ_{ff}, is defined as:

$$\tau_{ff*} = \tau_{ff}/\tau_{stokes} \quad (11)$$

and is equal to 1 at Kilauea (Vergniolle 1996). The correct duration for the period with fire fountains is equal to the duration of the entire phase with fire fountains, 7 and 42 months at Mauna Ulu and Pu'u 'O'o eruptions, respectively (Table 2), and is not equal to the repose periods between each fire fountain as incorrectly assumed for the 1998 eruption at Etna (Harris & Neri 2002). The height of the degassing reservoir is either the one of the entire reservoir if the gas content is sufficient for small bubbles to exist at its bottom, or is the one of a bubbly layer, likely to be located in the uppermost part of the reservoir.

A tool for comparing eruptive episodes and volcanoes

The bubble content, i.e. the bubble diameter and the gas volume fraction within the magma reservoir, as well as the magma geometry, i.e. the surface area and height of the degassing layer (Fig. 2), can be found from equations 5, 6, 9 and 11 once the gas flux and the dimensionless height N_1 are estimated. While the gas flux can be easily determined for each episode from visual observations or acoustic measurements, the value of the dimensionless foam height N_1 can only be constrained so far by analogy to one Strombolian episode at Shishaldin (see next section).

The gas flux and the bubble diameter d_{res} are related to one another by the gas volume fraction (equation 5), which depends on both the number of bubbles N_{bub} and on their diameter by:

$$\alpha_{res} = \frac{N_{bub}\,\pi\,d_{res}^3}{6\,V_{ref}} \quad (12)$$

where V_{ref} is the reference volume in which the gas volume fraction is averaged. If the number of bubbles is assumed to stay constant for each episode of the series, the gas flux simply varies

with d_{res}^5, due to the combined effect of changing the bubble velocity, a function of d_{res}^2, as well as the gas volume fraction, a function of d_{res}^3 (Vergniolle 1996). In these conditions, any variation in the gas flux is directly related to a change in the bubble diameter. This approximation leads to an estimate of both the bubble diameter and the gas volume fraction for each episode of a given eruption with solely knowing the dimensionless foam height N_1 for one episode.

It may also be handy to compare eruptions regardless of the exact value of the bubble diameter, which controls both the gas flux and duration of the period with fire fountains (equation 11). This implies dividing the bubble diameter during a given episode by a characteristic bubble diameter and the time by a characteristic time. Since the transition between the period with and without fire fountains leads to a drastic change in activity, the characteristic bubble diameter is taken to be the bubble diameter at that time. The characteristic time is similarly chosen to be the duration of the period prior to the transition. This dimensionalization has also the advantage of being independent of the exact value of the gas flux.

The degassing reservoir at Etna

The gas volume determined from acoustic measurements corresponds to bubbles undergoing enough oscillations at the surface to radiate sound waves. Hence, they are likely to be produced from the massive coalescence of a foam trapped at the depth of the reservoir (Vergniolle & Ripepe 2008). These bubbles, here referred to as slugs, can therefore give a direct access to the foam volume prior to its disruption (Fig. 3). In contrast, the total gas volume estimated from the height of fire fountains includes both the passive degassing in the conduit and the driving slugs, also called here active degassing (Vergniolle & Ripepe 2008). The driving slugs, which represent 38–48% of the total gas volume regardless of surface activity, are carrying within their wake the small bubbles from passive degassing (Vergniolle & Ripepe 2008). These bubbles result from moderate syn-eruptive degassing (Mangan & Cashman 1996; Polacci et al. 2006) and mostly from bubbles coming from the reservoir (Fig. 3). Among them, some are directly brought to the surface within the slug wake (Fig. 3) and arrive quasi-simultaneously of the quasi-fire fountain (Vergniolle & Ripepe 2008). Therefore, they directly result from the gas flux in the reservoir at that time. The remaining bubbles arrive in the conduit long before the eruptive episode and they should not be considered when calculating the gas

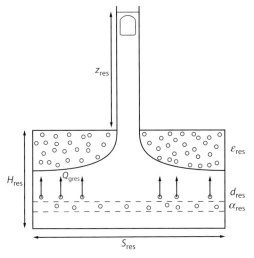

Fig. 2. Sketch of the plumbing system below the vent at Etna during an eruptive episode. The reservoir, at a depth z_{res} has a height H_{res} and a surface area S_{res}. Bubbles in the reservoir have a diameter d_{res} with a gas volume fraction α_{res} for a gas flux Q_{res}. The gas volume fraction in the foam, ε_{res}, is much higher than the gas volume fraction of the rising bubbles below the foam.

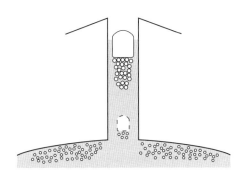

Fig. 3. Sketch for the formation of passive and active degassing. The slug represents the gas volume coming from the reservoir and detected from acoustic measurements, while the small bubbles within the slug wake represent the passive degassing, included in the surface measurements together with the gas directly from the reservoir (Vergniolle & Ripepe 2008). The small bubbles present in the volcanic conduit are pushed aside as the slug rises and may be incorporated within the slug wake, due to strong convective motions in that region (Fan & Tsuchiya 1990), as early as the base of the conduit (Vergniolle & Ripepe 2008). The gas volume fraction within the slug wake is 65% and 44% for 4 and 12 July, 2001, respectively. The depressurization induced by magma removal during fire fountains can lead to gas exsolution in the conduit. However, syn-eruptive gas exsolution is only found to be moderate (Polacci et al. 2006).

flux at that time. Because the proportion of the two bubble populations is unknown, the gas flux in the reservoir ranges between the underestimate given from acoustic measurements and the overestimate deduced from the height of the fire fountains (equations 1 and 2).

Gas flux in the reservoir from acoustic measurements

The erupted gas volume, estimated from acoustic measurements and compressed at the depth of the reservoir by the perfect gas law, is equal to $1.2 \times 10^4 \, m^3$ and to $2.0 \times 10^4 \, m^3$ on 4 and 12 July, respectively. If that gas volume has accumulated in the reservoir during the repose period before the quasi-fire fountain (a few days, Table 1), the gas flux in a magma chamber is equal to $2.7 \times 10^{-2} \, m^3 \, s^{-1}$ and $4.0 \times 10^{-2} \, m^3 \, s^{-1}$ for 4 and 12 July, respectively (Table 1). Hence, the gas flux at depth is 1.5 times larger for an episode leading to a fire fountain than for a purely Strombolian episode.

Unfortunately, technical problems on 7 July prevented us from recording sound waves and estimating the gas volume trapped at the top of the reservoir. The decrease of the gas volume during the eruption, which places 7 July on the same trend as previous Strombolian episodes (Fig. 4 in Vergniolle & Ripepe 2008), suggests that the gas volume deduced from surface activity is representative of that in the reservoir when using the same degree of unsteadiness as on 4 July, hence equal to 0.39 (Vergniolle & Ripepe 2008), as visually observed. The gas volume should be further corrected by the ratio between active and passive degassing, i.e. by a factor of 0.38 (Table 1; Vergniolle & Ripepe 2008).

The gas volume and gas flux are smaller on 7 July than on 4 July by a ratio of 2.5 and of 1.5 respectively. When combining these ratios with the gas volume and the gas flux of 4 July, $5.4 \times 10^6 \, m^3$ (Vergniolle & Ripepe 2008) and $2.7 \times 10^{-2} \, m^3 \, s^{-1}$, respectively, we obtain values of $2.2 \times 10^6 \, m^3$ and $1.9 \times 10^{-2} \, m^3 \, s^{-1}$, respectively on 7 July (Table 1).

Foam behaviour at the depth of the magma reservoir

Each eruptive episode of the 2001 eruption of Etna resembles very closely the second Strombolian phase at Shishaldin on 22–23 April 1999 (Caplan-Auerbach & McNutt 2003), although the resemblance is most marked for a purely Strombolian episode such as on 4 July. For both volcanoes, the number of explosions stays mostly constant, although explosions are more frequent at Etna, with 40 explosions per five minutes, than at Shishaldin with 10 explosions per 800 s (Vergniolle *et al.* 2004). Strombolian explosions at Shishaldin also correspond to similar bubble lengths and over-pressures (24 m and 0.15 MPa), for a duration and a repose time of 24 hours and 60 hours, respectively (Vergniolle *et al.* 2004). Since the activity observed at the surface also results from the dynamics of the foam trapped in the reservoir (Jaupart & Vergniolle 1989), the similarity in behaviour between Etna on 4 July and Shishaldin on 22–23 April suggests that the foam dynamics is the same (Fig. 2) despite very different ejected gas volumes, repose periods and viscosities (500 Pa s at Shishaldin (Vergniolle *et al.* 2004)).

At Shishaldin, the dimensionless foam height N_1 (equation 6) was first estimated during the Subplinian plume. This phase, by being driven from the reservoir by a foam at the transition between stable and instable, had allowed us to fix the value of N_1 at $1/C_{geom}$ (Vergniolle subm.), where C_{geom} is a geometrical factor defined as (Jaupart & Vergniolle 1989):

$$C_{geom} = \left(-\ln\left(\frac{S_{vent}}{S_{res}}\right) + \frac{S_{vent}}{S_{res}} - 1 \right)^{1/4} \quad (13)$$

The decompression following the plume increased the bubble diameter and the gas flux in the reservoir, leading to the value of N_1 equal to $1.2/C_{geom}$ (equation 6) for the Strombolian phase (Vergniolle subm.). A surface area between $0.1\ km^2$ and $1\ km^2$, as for Etna and Kilauea, respectively (see next section), leads to a value of the geometrical facctor C_{geom} between 0.64 and 0.59, respectively. In the following, the dimensionless foam height N_1 is taken to be equal to 0.76 for a Strombolian phase at Etna and this value represents a minimum because the explosions are more frequent there than at Shishaldin.

The high dissolved water content at Etna, 3.4 wt%, combined with a temperature of 1140 °C (Clocchiatti & Métrich 1984; Kamenetsky & Clocchiatti 1996), suggest that the magma viscosity could be as low as 1 Pa s (Giordano & Dingwell 2003; Métrich *et al.* 2004; Spilliaert *et al.* 2006). However, the magma may have a phenocryst content similar to that of scoriae, from 19% to 40% (Polacci *et al.* 2006), suggesting that a value of 10 Pa s is probably the best estimate of the *in-situ* viscosity (Table 3; Dingwell 1998).

A gas flux of $2.7 \times 10^{-2}\ m^3\ s^{-1}$, as on 4 July (Table 1), leads to the bubble diameter d_{res} of 0.54 mm (equation 6) when using a dimensionless foam height N_1 equal to 0.76 and a surface tension in magma $\sigma = 0.36\ kg\ s^{-2}$ (Table 3;

Proussevitch *et al.* 1993). On 12 July, a gas flux of $4.0 \times 10^{-2}\ m^3\ s^{-1}$ suggests that the bubble diameter must be larger by a factor 1.07 (0.59 mm) compared with its value on 4 July when assuming the same number of bubbles for both episodes (equations 5 and 12). On 7 July, the bubble diameter, which is reduced by a factor 1.07 from its value on 4 July when assuming a constant number of bubbles (equations 5 and 12) and a gas flux of $1.9 \times 10^{-2}\ m^3\ s^{-1}$, is equal to 0.50 mm.

The comparison between 4 and 12 July (Figs 9 and 10 in Vergniolle & Ripepe 2008) suggests that an eruptive episode evolves from purely Strombolian towards a fire fountain if the foam coalescence at the depth of the reservoir becomes more active. This is in perfect agreement with our estimates of the gas flux at depth, which is 1.5 times larger for an episode leading to a fire fountain than for a purely Strombolian one. Furthermore, the dimensionless height N_1 (equation 6), which quantifies the intensity of the foam coalescence, is indeed larger on 12 July, 0.89, than on 4 July (Table 1). In contrast, on 7 July, the dimensionless foam height, N_1, equal to 0.65 (equation 6), corresponds to a foam just starting to be unstable (Table 1). This is in excellent agreement with the observation of a small thermal on that day.

The surface area of the magma reservoir

While the gas volume estimated from acoustic measurements is the sole gas volume produced entirely by the foam coalescence (Vergniolle & Ripepe 2008), episodes during which the foam is close to its stability do not lead to a total withdrawal of the foam during its coalescence (Jaupart & Vergniolle 1989). Among acoustic measurements, those performed on 12 July are the closest to the case of a stable foam coalescence, as marked by the highest value of the dimensionless foam height N_1 at 0.89 (Table 1).

The entire gas volume expelled on 12 July once compressed at the reservoir depth, $1.9 \times 10^4\ m^3$, is hence interpreted as being the gas volume trapped in the foam accumulated at the top of the reservoir (Figs 1 and 2). Equations 8 and 9, combined with a gas volume of $1.9 \times 10^4\ m^3$ give a surface of the magma reservoir equal to $\cong 0.20\ km^2$. This value becomes $\cong 0.038\ km^2$ and $\cong 0.10\ km^2$ when using the gas volume on 7 and 4 July, respectively (Table 1). Therefore, the surface area of the reservoir exceeds $0.20\ km^2$.

A surface area, above $0.20\ km^2$, is very compatible with previous studies, which suggests that the activity at the southeast crater is fed by a very small and local magma reservoir (Le Cloarec & Pennisi 2001; Corsaro & Pompilio 2004; Bonaccorso 2006). This is also very compatible with the dimensions of the collapse crater at the summit, $\cong 0.072$ km^2 (Harris *et al.* 2000; Harris & Neri 2002).

The gas volume fraction in the magma reservoir

Equation 5 leads to a value of the gas volume fraction of 0.62% at the top of the reservoir for a gas flux of 2.7×10^{-2} m^3 s^{-1}, a bubble diameter of 0.54 mm and a reservoir area of $\cong 0.10$ km^2 as on 4 July (Table 1). On 7 July, it becomes 0.50% for a gas flux of 1.9×10^{-2} m^3 s^{-1} and a bubble diameter of 0.50 mm (Table 1). The gas volume fraction increased on 12 July at 0.39% for a gas flux and a bubble diameter of 4.0×10^{-2} m^3 s^{-1} and 0.59 mm, respectively (Table 1). However, the gas volume fraction on 4 and 7 July becomes 0.31% and 0.25% when using a reservoir area of $\cong 0.20$ km^2 as on 7 July (Table 1). As previously explained, these gas volume fractions are much smaller than the maximum gas volume fraction estimated from primitive fluid inclusions (Métrich *et al.* 2004), between 4.4 and 7.7% for pure CO_2 and 11% for pure H_2O.

The height of the degassing reservoir

In the absence of acoustic measurements, the gas flux in the reservoir is calculated for each episode from the maximum gas volume, obtained from the height of the fire fountains (equation 2) and compressed at the depth of the reservoir with the perfect gas law. The gas flux decreases from 0.64 m^3 s^{-1} on 8 June to 0.13 m^3 s^{-1} on 7 July, when excluding the two lowest values on 6 June and 15 June (Fig. 4). The gas volume on 9 May, which is also one of the smallest values (Fig. 4 in Vergniolle & Ripepe 2008), may rather be a precursory event for the impending eruption than a proper episode among the series of cyclic fire fountains. That interpretation, based on the abnormal repose period (Fig. 14 in Vergniolle & Ripepe 2008), only means that the vent is open. The second lowest gas flux on 15 June results from using the duration of the paroxysm (40 minutes), rather than that of the entire episode (between 1 hour and 40 minutes to 7 hours Vergniolle & Ripepe 2008). On 12 July, the gas flux suddenly triples before decreasing again while the activity changes to a flank eruption (Allard *et al.* 2006; Behncke *et al.* 2006). Note that the above gas fluxes should be corrected by the degree of unsteadiness (Table 1).

The decrease of the gas flux between 6 June and 7 July (29 days), is similar to that observed during the Mauna Ulu and Pu'u 'O'o eruption (Vergniolle & Jaupart 1990; Vergniolle 1996).

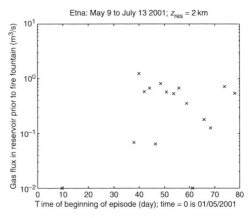

Etna: May 9 to July 13 2001; $z_{res} = 2$ km

Fig. 4. Gas flux (m^3 s^{-1}) at the depth of the reservoir, assumed to be 2 km during the 2001 eruption and deduced from surface activity (BGVN 2001; Vergniolle & Ripepe 2008). It decreases, from 0.64 m^3 s^{-1} (8 June, episode 3, day 38) to 0.13 m^3 s^{-1} (7 July, episode 14, day 68), with a peak value at 0.36 m^3 s^{-1} on 12 July (episode 15, day 74) which has the largest gas volume, 1.9×10^8 m^3. It decreases again on last episode (17 July, episode 16, day 79) while turning to a flank eruption. The low gas flux on 6 June (episode 2, day 38) is related to the abnormal repose period related to the choice of 9 May as the first episode. The second low gas flux on 15 June (episode 6, day 47) is related to using the paroxysm duration instead of the entire duration of the episode. If the two lowest gas fluxes are removed, the average gas flux is 0.30 ± 0.15 m^3 s^{-1}. Gas fluxes should be corrected by the degree of unsteadiness, between 0.39 and 0.11 (Table 1; Vergniolle & Ripepe 2008).

This duration is, hence, characteristic of the time necessary to deplete the magma reservoir of its small bubbles. When combining it with a dimensionless time τ_{ff*} equal to 1 (equation 11), the height for the degassing reservoir is equal to 97 m for a bubble diameter of 0.50 mm (Table 1). This increases to 220 m if the episode on 9 May is included in the period with cyclic fire fountains (Table 1).

The time evolution of the bubble diameter in the degassing reservoir

The bubble diameter and gas volume fraction decrease during the eruption by a factor of 1.6 and 2.1 when assuming that the number of bubbles does not change between episodes (equations 5 and 12; Fig. 5). The sudden increase of the bubble diameter on 12 July could be related to the arrival of a new magma rich in volatiles (Behncke *et al.* 2006; Allard *et al.* 2006; Spilliaert *et al.* 2006a; Métrich *et al.* 2004). Alternatively, this feature could be generated internally by a secondary peak of a bimodal bubble population. This peak should

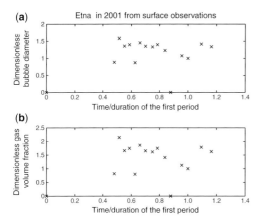

Fig. 5. (**a**) Dimensionless bubble diameter, based on the bubble diameter at minimum gas flux, as a function of a dimensionless time, based on the duration of the period with fire fountain. (**b**) Dimensionless gas volume fraction, based on the gas volume fraction at the minimum gas flux, as a function of a dimensionless time, based on the duration of the period with fire fountain.

be very sharp if solely producing the surface activity on 12 and 17 July, hence that interpretation is not favoured.

Comparison between acoustic measurements and surface activity

When assuming that the height of fire fountains is representative of the gas volume coming from the reservoir, the minimum gas flux, $5.2 \times 10^{-2}\,\mathrm{m^3\,s^{-1}}$ once corrected by a degree of unsteadiness equal to 0.39 (Table 1), represents the critical gas flux. Hence, a dimensionless foam height N_1 equal to 0.64 leads to the bubble diameter and the gas volume fraction equal to 0.40 mm and 2.9% (equations 5 and 6). When using a duration for the period with fire fountains, τ_{ff}, equal to 29 days together with a dimensionless time τ_{ff*} equal to 1 (equation 11), the height for the degassing reservoir is equal to 61 m (Fig. 2).

Bubble diameter estimated from surface activity, 0.40 mm, is very similar to that deduced from acoustic measurements, 0.50 mm (Table 1), because of the weak dependence on the gas flux (equation 6). The gas volume fraction estimated from surface activity, 2.9%, exceeds the value deduced from acoustic measurements, 1.3%, by a factor of 2.2 (Table 1). The height of the degassing layer is reduced by a factor of 1.6 while the surface area is increased by a factor of 2 (Table 1). Therefore, in the absence of acoustic measurements, the bubble characteristics in the reservoir can be found with a good order of magnitude when using

the surface activity and the degree of unsteadiness (Table 1).

Sensitivity analysis on acoustic measurements

The first cause of error in estimating the gas volume from the reservoir comes from the best fit between the measured acoustic pressure and the synthetic waveforms, but that only amounts to 20% (Vergniolle *et al.* 2004). The second cause of error is related to the non-recording of the first hour of the activity, as for 4 July. But at the beginning, activity is reduced, hence we only expect an underestimate of another 20% error for 4 July and none for the continuous recording on 12 July. Therefore, we expect a 40% accuracy on our estimate on gas volume, at the depth of the reservoir, on 4 July and 20% on 12 July. Estimates on bubble diameter, gas volume fraction and reservoir area are mainly related to the assumption made on the reservoir depth. If the reservoir lies at 4 km, the bubble diameter, reservoir area, gas volume fraction and reservoir height become 0.60 mm, 0.061 km², 0.29% and 140 m on 7 July. Although there is a factor of 1.7 between both estimates of the reservoir area and the gas volume fraction, our results are very robust to estimate the bubble diameter, known with an error of 20% for a magma chamber lying between 2 and 4 km below the vents.

A second cause of uncertainties comes from assuming that the magma is a liquid with a viscosity equal to 10 Pa s when including a crystal content between 19% and 40% (Dingwell 1998; Polacci *et al.* 2006). For a pure liquid, the viscosity could be as low as 1 Pa s (Giordano & Dingwell 2003; Métrich *et al.* 2004). This leads to a bubble diameter of 0.89 mm, a gas volume fraction of 0.0089% and a reservoir area of ≅0.18 km². Although the magma viscosity has been decreased by one order of magnitude, the bubble diameter and the reservoir area have both only increased by a factor of 1.8. The uncertainty, however, becomes significant for the gas volume fraction, with a decrease by a factor of 56, and for the height of the degassing layer, now equal to 3.1 km instead of 97 m.

A third cause of uncertainties could be related to the choice of the dimensionless foam height N_1 equal to 0.76 (equation 6) by analogy to Shishaldin. Although that approximation is extremely well constrained at Shishaldin (Vergniolle in press), we compare the results for N_1 equal to 0.65 on 7 July with those obtained for values of 0.76 and 0.89 (equation 6). The bubble diameter and the surface area of the reservoir increase by a factor of 1.2 and 1.4 to 0.61–0.70 mm and

0.046–0.054 km², respectively (equations 6 and 8). The height of the degassing reservoir increases by a factor of 1.4 and 2.0 to 150–200 m (equation 11) while the gas volume fraction decreases by a factor of 1.7 and 2.7 to 0.29–0.18% (equation 5).

Physical processes in the reservoir at Kilauea

The Mauna Ulu and Pu'u 'O'o eruptions are re-analysed with new constraints on the reservoir surface area (Yang et al. 1992) and new estimates on the average gas volume, giving the values of the gas flux, bubble diameter and gas volume fraction independently at the depth of the reservoir. However, these two eruptions are flank eruptions and the area involved in the production of the fire fountains is more complex than for a pure summit eruption. The Kilauea Iki eruption, never yet analysed within that framework, also has the advantage of being a summit eruption. That eruption can then provide a direct comparison for the surface area of the summit reservoir, obtained either by using the height of the fire fountains or deformations (Yang et al. 1992).

The Kilauea Iki eruption

The eruption, which began at the summit on 14 November 1959, produced 16 vigorous fire fountains, up to 300 m high, with an extremely long first episode compared with others (Richter et al. 1970). The last fire fountain stopped on 20 December 1959 and the eruption turned to a flank eruption, which lasted from 14 January to 19 February 1960 (Richter et al. 1970).

On 11 December 1959, huge bubbles of lava 6 to 9 m in diameter began to burst at the vent (Richter et al. 1970). The bubbling slowly evolved into a pulsating fountain that shot up to 60 m or so for a few seconds and then died down to practically nothing. Then it would rise to a few hundreds of metre for a few more seconds and die again. After 40 minutes of this sporadic behaviour, the 'dies out' ceased and the fountain jetted to heights of 200 m to 300 m. Pulsating fire fountains were often reported, such as at 7 hr 00 on 13 December with pulsations between 30 and 200 m (Richter et al. 1970). The fountain vacillated in this manner for $2\frac{1}{2}$ hours before steady activity was maintained at the vent. For about one minute before the fountain died on 17 December 1959, the lava lake drained back towards the vent. When the fountain stopped, the orifice was open and the stream of returning lava poured over the lip (Richter et al. 1970). The magma backflow was also apparent for two hours with a counterclockwise

motion around the fire fountain on 5 December, several hours before the end of the episode (Richter et al. 1970).

The simultaneous occurrence of an upwards magma flow within the annular core of the fire fountain and downward magma motion due to gravity is the mark of an annular flow (Wallis 1969). The large bubbles seen at the initiation of the fire fountain were of a similar diameter to the vent, measured at 15 by 12 m on 22 December 1959 two days after the last fire fountain (Richter et al. 1970). Therefore, a transition between a slug and an annular flow also exists at Kilauea, but it is completed in 40 minutes rather than in several hours such as at Etna. The fire fountains at Kilauea were very steady compared with Etna (Vergniolle & Ripepe 2008), with 40 minutes of transition for a total duration varying between 2 and 30 hours when excluding the first episode (Richter et al. 1970). Significant sound, although not recorded, was produced during that eruption (Richter et al. 1970).

The duration of the fire fountains slowly decreased from 30 hours to 2 hours, while the height of the fire fountains stays roughly constant (Fig. 6). The gas volume and the repose period between fire fountains decreased from 1.0×10^9 m³ to 1×10^8 m³ and from 100 hours to 9 hours, respectively (Fig. 7). The gas flux at the depth of the reservoir, assumed here at 2 km (Yang et al. 1992), roughly decreased from 20 m³ s⁻¹ to 3.4 m³ s⁻¹ (Fig. 8). This is similar to what was observed during the Mauna Ulu and Pu'u

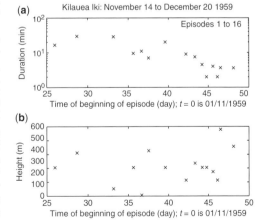

Fig. 6. Fire fountain characteristics during the Kilauea Iki eruption as a function of time (Richter et al. 1970). (**a**) Duration (min), (**b**) maximum height (m). Note that height and duration of fire fountains have been put equal to 0 if there were no observations. The duration of fire fountains decreases and has an average of 4.1 ± 2.4 hr. Its height stays roughly constant, 300 ± 130 m.

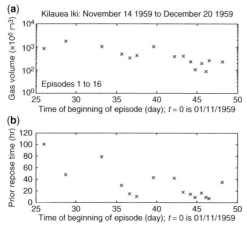

Fig. 7. The Kilauea Iki eruption. (**a**) Gas volume released at atmospheric pressure (10^6 m^3) as a function of time of the beginning of episode (time equal 0 is 1 November 1959. (**b**) Repose periods prior to fire fountains (hr) (Richter *et al.* 1970). Note that the gas volume and repose time have been put equal to 0 if there were no observations of fire fountains. Gas volumes decrease and had an average value of $1.0 \times 10^9 \pm 2.0 \times 10^9$ m^3. The repose periods prior to fire fountains also decrease, at 31 ± 27 hr. Gas volumes should be corrected by the degree of unsteadiness, between 0.83 and 0.59 (Table 2).

'O'o eruptions (Vergniolle & Jaupart 1990; Vergniolle 1996). Therefore, the minimum gas flux prior to the transition between periods with and without fire fountains is the closest to the critical gas flux.

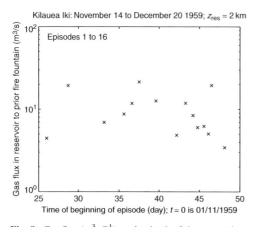

Fig. 8. Gas flux (m^3 s^{-1}), at the depth of the reservoir, assumed to be 2 km during the Kilauea Iki eruption. It roughly decreases until reaching 3.5 m^3 s^{-1}. The average gas flux is 9.8 ± 5.6 m^3 s^{-1} but these values should be corrected by the degree of unsteadiness, between 0.83 and 0.59 (Table 2).

In fact, observations of the last episode, whose height solely related to a few spattering events, never exceeded 30 m (Richter *et al.* 1970), suggesting that the transition between an unstable and a stable foam occurred prior to it. That observation places a strong constraint on the minimum gas flux in the reservoir.

The minimum gas flux varies between 2.0 m^3 s^{-1} and 2.9 m^3 s^{-1} once corrected by the degree of unsteadiness, between 0.59 and 0.83 (Table 2; Figs 7a and 8). The last episode being a very well-developed fire fountain, suggests the use of dimensionless foam height N_1 equal to 0.82 (equation 6) at Kilauea by analogy with Etna. This leads to a bubble diameter ranging between 0.20 mm and 0.18 mm (equation 6), for a magma viscosity, surface tension and gas volume fraction within the foam equal to 10 Pa s (Schmincke 2004), 0.36 kg s^{-2} (Proussevitch *et al.* 1993) and 0.6 (Jaupart & Vergniolle 1989), respectively (Table 3). The gas volume at the transition between fire fountains and effusive phases, $1.4-1.9 \times 10^8$ m^3 at the vent, is representative of the critical gas volume trapped within the foam at depth. Equations 8 and 9, combined with a gas volume of $2.5-3.6 \times 10^5$ once corrected for the reservoir pressure, give a surface area for the magma reservoir between 0.94 km^2 and 1.2 km^2 (equations 8 and 9).

This is likely to be overestimated because the passive degassing has been included into the foam volume (Fig. 3) and this factor may amount to 2.1 by analogy to Etna (Table 1). When applying these values to Kilauea, the surface area of the reservoir becomes between 0.45–0.58 km^2 (Table 2). Although the correcting factor is unknown at Kilauea, the surface area of the reservoir, possibly between 0.40 km^2 and 1.2 km^2, is in excellent agreement with the estimate of $\cong 1 \times 1$ km^2, found from the location of the centres of deflations (Yang *et al.* 1992). The gas volume fraction in the reservoir varies between 100% and 37% for a surface area between 0.45 km^2 and 1.2 km^2 (equation 5; Table 2). The degassing layer, very rich in gas, is extremely thin, between 19 m and 22 m, when using the duration of the period with cyclic fire fountains, 37 days, together with a dimensionless time τ_{ff*} equal to 1 (equation 11; Table 2). This height, if it is representative of the entire magma reservoir, suggests that the degassing reservoir at that time is only located in the uppermost portion of the magma reservoir.

The unrealistic value obtained for a reservoir area of 0.45 km^2, 100%, is reduced by two-thirds, 67%, when using a dimensionless foam height N_1 equal to 0.94 (equations 5 and 6) to account for the very steady character of fire fountains at Kilauea, the mark of a very efficient foam coalescence, compared to those at Etna. This is still much above the maximum gas volume fraction, 7.7% and 11% (equation 4; Table 2), suggesting that the surface area of the reservoir is at least 1 km^2 and that the dimensionless foam height may exceed 0.94. When using 1.2 for that parameter, the gas volume fraction becomes 12–19% while the bubble diameter, the surface area and height of the reservoir solely increase to 0.29–0.26 mm, 1.3–1.7 km^2, 38–32 m, respectively (equations 5, 6, 8 and 11). The very good agreement with results from fluid inclusions (equation 4) suggests that the surface area of the reservoir is probably at least 1 km^2 and the dimensionless foam height N_1 above 1.2. Although the exact value of the dimensionless foam height N_1 is unknown, the degassing layer is both very rich in gas and extremely thin, while the surface area of the reservoir is in excellent agreement with results from deformations (Yang *et al.* 1992).

The Mauna Ulu eruption

The Mauna Ulu eruption started on 24 May 1969 and produced 12 fire fountains until 30 December 1969 (Swanson *et al.* 1979). On 31 December 1969, the large fire fountains disappeared, although very low fountainings of short durations still occurred at the vent for several days. The absence of large fire fountains had marked the transition to the second stage and the eruption finished on 15 October 1971 (Swanson *et al.* 1979).

The gas volume expelled at the vent and the gas flux are deduced from the fire fountain maximum

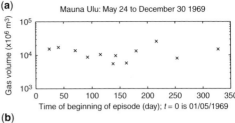

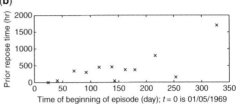

Fig. 10. The Mauna Ulu eruption. (**a**) Gas volume released at atmospheric pressure (10^6 m^3) as a function of time of the beginning of episode (time equal 0 is May 1 1969. (**b**) Repose periods prior to fire fountains (hr) (Swanson *et al.* 1979). Note that the gas volume and repose time have been put equal to 0 if there were no observations of fire fountains. The gas volume, roughly constant, has an average value of $1.2 \times 10^{10} \pm 0.57 \times 10^{10}$ m^3. The repose periods prior to fire fountains, roughly constant, have an average value of 19 ± 19 days. Gas volumes should be corrected by the degree of unsteadiness, between 0.83 and 0.59 (Table 2).

height and duration (Figs 9 and 10a) when using a vent area of 400 m^2 (Swanson *et al.* 1979) and the repose periods prior to an episode (Figs 10a and 11). The minimum gas flux varies between 2.7 m^3 s^{-1} and 3.9 m^3 s^{-1} when corrected by the degree of

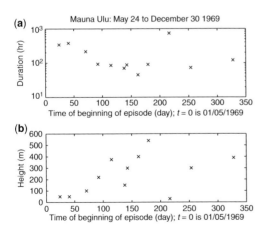

Fig. 9. Fire fountain characteristics during the Mauna Ulu eruption as a function of time (Swanson *et al.* 1979). (**a**) Duration (hr), (**b**) maximum height (m). Note that height and duration of fire fountains have been put equal to 0 if there were no observations. The duration of fire fountains decreases and has an average of 195 ± 204 hr for a height equal to 240 ± 170 m.

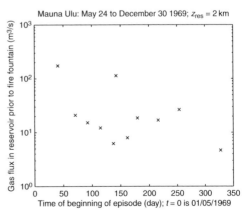

Fig. 11. Gas flux (m^3 s^{-1}), at the depth of the reservoir, assumed to be 2 km, during the Mauna Ulu eruption. It roughly decreases from 170 m^3 s^{-1} to 4.7 m^3 s^{-1}. These values should be corrected by the degree of unsteadiness, between 0.83 and 0.59 (Table 2).

unsteadiness (Table 2). Equation 6 leads to a bubble diameter between 0.18 mm and 0.17 mm when using a dimensionless foam height N_1 equal to 0.82 (equation 6). The surface area, $56-76$ km^2, largely exceeds that of the summit reservoir, 1 km^2, suggesting that the collecting area includes the area below the east-rift zone, as already suggested by the location of deformation centres (Yang et al. 1992). The gas volume fraction, between 0.98% and 1.3% (equation 5; Table 2), is indeed less than the maximum gas volume fraction (equation 4). The decrease of the gas flux during the period with fire fountains, 221 days, is characteristic of a degassing reservoir whose height ranges between 93 m and 79 m (equation 11; Table 2).

The Pu'u 'O'o eruption

The Pu'u 'O'o eruption started on 3 January 1983 and 47 episodes of fire fountains occurred until 26 June 1986 (Wolfe et al. 1987, 1988; Heliker & Mattox 2003). The fire fountains at Pu'u 'O'o, typically 400 m high, also sometimes exhibit large variations in height, $50-800$ m over timescales as little as two hours (Wolfe et al. 1987, 1988; Sparks et al. 1997). The episode 48, on 18 July 1986, with its very low fountaining at 30 m and the opening of a new fisssure, marked the transition to a solely effusive phase (Heliker & Mattox 2003). This transitional episode is similar to episode 17 of the Kilauea Iki eruption (Richter et al. 1970) and

also results from a foam just becoming stable in the magma reservoir.

Details on the height, the duration of fire fountains and their repose periods can be found in Figures 12 and 13 drawn from Wolfe et al. (1987, 1988) and from Heliker & Mattox (2003). The decrease of the average gas flux (Fig. 11) leads to a minimum gas flux of $0.47-0.66$ m^3 s^{-1} once corrected by the degree of unsteadiness (Table 2). The bubble diameter is equal to $0.29-0.26$ mm for a gas volume fraction of $0.74-0.98\%$ for a reservoir area of $5.2-7.1$ km^2 (equations 5, 6 and 8; Table 2). As for Mauna Ulu, the collecting area includes the area below the summit and below the east-rift zone, as already suggested by the location of deformations centres (Yang et al. 1992). The height of the degassing reservoir is equal to $1300-1100$ m when based on a characteristic time for gas depletion of 1237 days (equation 11; Table 2).

Comparison between Kilauea Iki, Mauna Ulu and Pu'u 'O'o eruptions

The minimum gas flux varies from $2.0-2.9$ m^3 s^{-1} to $2.7-3.9$ m^3 s^{-1} and $0.47-0.66$ m^3 s^{-1} for the Kilauea Iki, Mauna Ulu and Pu'u 'O'o eruptions, respectively (Table 2). The gas volume fraction is

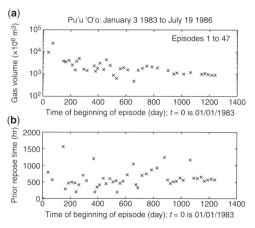

Fig. 13. The Pu'u 'O'o eruption. (**a**) Gas volume released at atmospheric pressure (10^6 m^3) as a function of time of the beginning of episode (time equal 0 is 1 May 1969. (**b**) Repose periods prior to fire fountains (hr) (Wolfe et al. 1987, 1988; Heliker & Mattox 2003). Note that the gas volume and repose time have been put equal to 0 if there were no observations of fire fountains. The gas volume decreases and has an average value of $1.6 \times 10^9 \pm 1.2 \times 10^9$ m^3 when excluding the first four episodes. The repose period prior to fire fountains, roughly constant, has an average value of at 24 ± 10 days. Gas volumes should be corrected by the degree of unsteadiness, between 0.83 and 0.59 (Table 2).

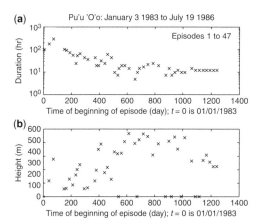

Fig. 12. Fire fountain characteristics during the Pu'u 'O'o eruption as a function of time (Wolfe et al. 1987, 1988; Heliker & Mattox 2003). (**a**) duration (hr), (**b**) maximum height (m). Note that height and duration of fire fountains have been put equal to 0 if there were no observations. The duration of fire fountains decreases and has an average of 24 ± 20 hr for a height equal to 230 ± 160 m when excluding the first four episodes.

much larger for the first eruption, 37–49%, than for the two other eruptions, 0.74–1.3% while the bubble diameter is fairly similar, 0.17–0.29 mm (Table 2). The height of the degassing reservoir increases, from 16–12 m to 93–79 m and then to 1300–1100 m (Table 2).

The maximum value exceeds the value obtained from modelling the deflations of the summit reservoir between 1970 and 1985 by a spherical body with a cross-sectional area of 1×1 km^2 (Yang et al. 1992). However, although the centre of deflations, modelled as a point source, usually spreads over a depth between 2 and 3 km, a single value at 7.5 km exists among the ten deflationary periods (Yang et al. 1992). Furthermore, the highly simplified, three-dimensional model of the complex magma reservoir shows that it lies between 2 and 6 km below the summit, with a roughly elliptical cross-section and a volume between 5 km^3 and 10 km^3 (Schmincke 2004). Therefore, a height of 1300–1100 m (Table 2) is in very good agreement with the height of the entire magma reservoir.

The relatively smooth decrease of the gas flux, seen during the Kilauea Iki, Mauna Ulu and Pu'u 'O'o eruptions (Figs 8, 11 and 14), suggests that the number of bubbles is constant for all episodes of a given eruption (equations 5 and 12). These eruptions can be compared by dividing the bubble diameter with its minimum value and the time by the duration of the fire fountains (Table 2; Fig 15). The maximum bubble diameter, by exceeding its minimum value by a factor of 2, is similar for the three eruptions (Fig. 15). The decrease in time of the bubble diameter is however different between the Kilauea Iki and Mauna Ulu or Pu'u 'O'o

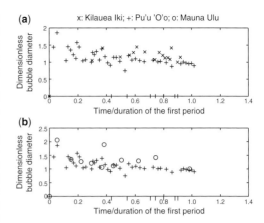

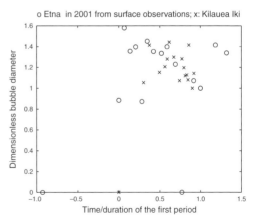

Fig. 15. Dimensionless bubbble diameter, based on the bubble diameter at minimum gas flux (0.19 mm, 0.18 mm and 0.28 mm for Kilauea Iki, Mauna Ulu and Pu'u 'O'o eruptions), as a function of a dimensionless time, based on the duration of the period with fire fountain for Kilauea (37 days, 7 months and 42 months). (**a**) Comparison between the summit eruption of Kilauea Iki and the flank eruption of Pu'u 'O'o. (**b**) Comparison between the two eruptions located on the east rift zone, Mauna Ulu and Pu'u 'O'o. The decrease in the bubble diameter in time is very different for the summit and the flank eruptions, while both flank eruptions show a similar decrease.

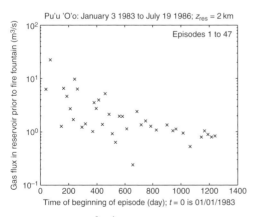

Fig. 14. Gas flux (m^3 s^{-1}), at the depth of the reservoir, assumed to be 2 km, during the Pu'u 'O'o eruption. It roughly decreases from 23 m^3 s^{-1} to 0.83 m^3 s^{-1}. These values should be corrected by the degree of unsteadiness, between 0.83 and 0.59 (Table 2).

Fig. 16. Dimensionless bubbble diameter, based on the bubble diameter at minimum gas flux (0.19 mm and 0.50 mm at Etna and Kilauea, respectively), as a function of a dimension less time, based on the duration of the period with fire fountain for the 2001 eruption of Etna and Kilauea Iki (29 days and 37 days, respectively). The decrease in the bubble diameter in time is very similar for both volcanoes, except for the last two episodes on Etna, which have a different origin.

eruptions (Fig. 15) while both eruptions located on the east rift zone are similar (Fig. 15). The decrease in time of the bubble diameter during the Kilauea Iki eruption is extremely similar to that of Etna (Fig. 16), suggesting again that both volcanoes undergo, in their reservoir, similar physical processes associated with degassing.

Comparison between Kilauea and Etna

Fire fountains at Kilauea, although sometimes displaying strong variations in their height (Richter et al. 1970; Swanson et al. 1979; Wolfe et al. 1987, 1988), have the mark of a very well-developed annular flow in contrast to the surface activity at Etna, more established at the transition between slug and annular flow (Figs 1 and 2 in Vergniolle & Ripepe 2008).

The degassing reservoir at Etna and Kilauea

We now use the Kilauea Iki eruption as being most representative of the summit reservoir of Kilauea. The minimum gas volume at Etna, between $2.2 \times 10^6 \, \text{m}^3$ and $1.4 \times 10^7 \, \text{m}^3$ (Vergniolle & Ripepe 2008) is at least one order of magnitude below the values estimated at Kilauea, $1.3 \times 10^8 \, \text{m}^3$. This results mostly from the reservoir area being one order of magnitude larger at Kilauea, $0.94–1.2 \, \text{km}^2$, than at Etna, $>0.20 \, \text{km}^2$, with a difference by a factor of 2 in bubble diameters, $0.40–0.59 \, \text{mm}$ and $0.18–0.20 \, \text{mm}$ for Etna and Kilauea, respectively (Tables 1 and 2). The small reservoir area at Etna compared with Kilauea also explains the two orders of magnitude difference in the minimum gas flux, $1.9 \times 10^{-2} \, \text{m}^3 \, \text{s}^{-1}$ and $2.0–2.9 \, \text{m}^3 \, \text{s}^{-1}$, respectively (Tables 1 and 2). The gas volume fraction is much smaller at Etna, $0.25–0.39\%$, than at Kilauea, $37–49\%$ (Tables 1 and 2). Although no explanation yet exists for the large gas volume fraction at Kilauea Iki, the reservoir area is probably one of the major parameters responsible for differences in the surface activity, between long-lasting fire fountains at Kilauea and more sporadic fire fountains at Etna.

Another main difference between both volcanoes is related to the temporal evolution of the height of the degassing reservoir over several eruptions. It increases by a factor of $5.2–12$ for each decade separating these large eruptions of Kilauea. In contrast, it decreases at Etna across a series of eruptions spanning over decades (1989–2001), from $900–1500 \, \text{m}$ to $8.9–15 \, \text{m}$, while the surface area of the magma chamber appears to stay constant (Vergniolle in prep.). The maximum height is obtained when new magma is reinjected,

such as following the initiation of a decade cycle in 1995, and the minimum reached when the content in volatiles has been quasi-exhausted, such as prior to the transition of a new cycle like in 1989 and 2001 (Vergniolle in prep.). In that framework, the Kilauea Iki eruption could mark the end of a cycle and the quasi-exhaustion of the gas in the summit reservoir. This interpretation is very compatible with the collapse of the summit caldera between 7 February and 11 March 1960 (Richter et al. 1970) and with the constantly increasing summit tilt from mid-1960 to 1974 (Schmincke 2004). The large height reached by the degassing reservoir during the Pu'u 'O'o eruption, $1300–1100 \, \text{m}$ (Table 2) could correspond to a new magma reinjection, in agreement with a very long-lasting eruption, ≥ 21 years (Heliker & Mattox 2003). Furthermore, if the maximum height reached by the degassing reservoir represents the height of the entire magma reservoir, the similarity in their values, $3100–3700 \, \text{m}$ and $900–1500 \, \text{m}$ at Kilauea (Table 2) and Etna in 1995 (Vergniolle in prep.), respectively, suggests that the magma reservoir could be one order of magnitude larger at Kilauea than at Etna, as a consequence of the different reservoir areas. This is indeed the case with a volume of the magma reservoir equal to $0.5 \, \text{km}^3$ at Etna (Le Cloarec & Pennisi 2001) and between $5 \, \text{km}^3$ and $10 \, \text{km}^3$ at Kilauea (Schmincke 2004).

The inter-eruptive gas flux at Kilauea and Etna

Another way to compare Kilauea and Etna is to estimate the gas volume expelled during eruptions leading to fire fountains and to calculate a gas flux based on the repose period prior to the eruption. This leads to a minimum inter-eruptive gas flux because a significant amount of gas may sometimes be expelled owing to the long-lasting effusive phase such as during the Mauna Ulu or Pu'u 'O'o eruptions.

The 2001 eruption at the southeast crater of Etna expelled a total gas volume between $0.76 \times 10^8 \, \text{m}^3$ and $4.2 \times 10^8 \, \text{m}^3$ once corrected by the degree of unsteadiness (Table 1; Vergniolle & Ripepe 2008). This gas volume is very close to the gas volume coming from the reservoir, due to moderate syn-eruptive gas exsolution (Mangan & Cashman 1996; Polacci et al. 2006). If that gas had been accumulated in the reservoir since the end of previous eruption, the inter-eruptive gas flux ranges between $2.5 \times 10^{-3} \, \text{m}^3 \, \text{s}^{-1}$ and $9.5 \times 10^{-3} \, \text{m}^3 \, \text{s}^{-1}$ (Table 3).

The repose period prior to the Kilauea Iki eruption is either 4 years and 9 months or 7 years and 4 months depending on whether previous eruption is taken to be a summit eruption or not

(Table 2; Richter *et al.* 1970). A total gas volume of $1.0 \times 10^{10} - 1.4 \times 10^{10}$ m^3 corresponds to an inter-eruptive gas flux between $0.13-0.18$ m^3 s^{-1} and $0.084-0.12$ m^3 s^{-1}, respectively (Table 2). During the 9.3 years separating the Kilauea Iki and Mauna Ulu eruptions, a gas volume of $0.87-1.2 \times 10^{11}$ m^3 has been produced by an inter-eruptive gas flux of $1.1-1.6$ m^3 s^{-1} (Table 2).

A total gas volume of $6.2 \times 10^{10} - 8.8 \times 10^{10}$ m^3 was expelled during the Pu'u 'O'o eruption, without accounting for the eight unobserved episodes. However, their total gas volume can be estimated by multiplying the average gas volume expelled per episode, 2.7×10^9 m^3, with the number of unobserved episodes, giving a total of $7.5 \times 10^{10} - 11 \times 10^{10}$ m^3 (Table 2). A repose period of 10.3 years leads to an inter-eruptive gas flux of $0.77-1.1$ m^3 s^{-1} and $0.95-1.4$ m^3 s^{-1} when adding the unobserved episodes (Table 2). Although the inter-eruptive gas flux should be below the critical gas flux, i.e. assumed to be the minimum gas flux, $0.47-0.66$ m^3 s^{-1} (Table 2), both are of the same order of magnitude, as expected from a very long-lasting eruption, ≥ 24 years.

The total gas volume expelled at Kilauea, $1.0 \times 10^{10} - 1.4 \times 10^{10}$ m^3, is a few orders of magnitude above that of Etna, $1.2 \times 10^8 - 4.2 \times 10^8$ m^3, leading to the same difference in the inter-eruptive gas flux, with $0.13-0.18$ and $2.5 \times 10^{-3} - 9.5 \times 10^{-3}$ m^3 s^{-1} at Kilauea and Etna, respectively (Table 3).

Conclusion

The formation of Strombolian explosions, thermals and fire fountains, observed during the 2001 eruption of Etna, has been sucessfully interpreted by the behaviour of a foam trapped at the top of the magma reservoir. The gas flux at the depth of the reservoir is larger for an episode leading to a fire fountain, 4.0×10^{-2} m^3 s^{-1}, than for one solely with Strombolian explosions, 2.7×10^{-2} m^3 s^{-1}. This suggests that foam coalescence is more efficient when a fire fountain is produced. The main assumption to determine the bubble diameter, the gas volume fraction and the reservoir surface has been to fix the value of the dimensionless number based on the foam thickness N_1 to a value of 0.76 for a Strombolian episode by analogy with the surface activity at Shishaldin, where it is very well constrained (Vergniolle in prep.). It was also assumed that the difference in gas flux was solely related to changes in the bubble diameter and not to changes in the number of bubbles.

Applying that model, together with a process of gas transport in the conduit, has given us an estimate on bubble diameter, 0.50 mm, and gas volume fraction, 0.25%, at the top of the magma reservoir. The height of the degassing reservoir, deduced from the duration of the period with fire fountains, was found be very small, between 97 m and 220 m, suggesting that the degassing reservoir solely corresponds to the uppermost portion of the magma reservoir. The surface area of the magma reservoir, above 0.20 km^2, was estimated from the episode during which the foam coalescence was the most active.

The same framework was used to analyse the Kilauea Iki, Mauna Ulu and Pu'u 'O'o eruptions, leading to estimates on bubble diameter, gas volume fraction and height of the degassing reservoir. The summit eruption of Kilauea Iki gave an unbiased estimate of the surface area of the summit reservoir compared with the other eruptions located on the east-rift zone. The excellent agreement between the prediction of the foam model, $0.94-1.2$ km^2, and that resulting from deformations, 1 km^2 (Yang *et al.* 1992), reinforces the validity of the foam model both qualitatively and quantitatively. The bubble diameter was similar for the three eruptions, $0.17-0.29$ mm, while the gas volume fraction was very large for the summit eruption, $37-49\%$. The height of the degassing reservoir increases by a factor $5.2-12$ for each decade separating eruptions, in contrast with Etna (Vergniolle in prep.). The maximum height reached by the Pu'u 'O'o eruption, $1300-1100$ m, suggests that the entire magma reservoir may have been used to produce that eruption, while during the Kilauea Iki eruption the bubbles were concentrated in a extremely thin layer, $16-12$ m, at the uppermost part of the reservoir.

The inter-eruptive gas flux at Kilauea, between $0.13-0.18$ m^3 s^{-1} and $0.77-1.1$ m^3 s^{-1}, is also larger than at Etna, $2.5 \times 10^{-2} - 9.5 \times 10^{-3}$ m^3 s^{-1}. However, the decrease in time of the bubble diameter, extremely similar for Kilauea Iki and Etna, suggests that indeed the same physical processes related to the degassing of a reservoir are at work for both volcanoes.

Thorough reviews by L. Wilson and H. Pinkerton have greatly improved the manuscript. I thank S. Lane and J. Gilbert for careful editing. I also thank the CNRS, INSU, IPGP and MAE for their support via PNRN and ACI, BQR and grant numbers 122/2000 and 02000146. This is an INSU contribution number (408) and an IPGP contribution number (2311).

References

Aki, K. & Ferrazzini, V. 2001. Comparison of Mount Etna, Kilauea, and Piton de la Fournaise by a quantitative modeling of their eruption histories. *Journal of Geophysical Research*, **106**, 4091–4102.

ALLARD, P., BURTON, M. & FILIPPO, F. 2005. Spectroscopic evidence for a lava fountain driven by previously accumulated magmatic gas. *Nature*, **433**, 407–410.

ALLARD, P., BEHNCKE, B., D'AMICO, S., NERI, M. & GAMBINO, S. 2006. Mount Etna 1993–2005: Anatomy of an evolving eruptive cycle. *Earth Science Reviews*, **78**, 85–114.

ALPARONE, S., ANDRONICO, D., LODATO, L. & SGROI, T. 2003. Relationship between tremor and volcanic activity during the southeast crater eruption on Mount Etna in early 2000. *Journal of Geophysical Research*, **108**, 2241, ESE 6-1-6-13.

BEHNCKE, B. & NERI, M. 2003. The July-August eruption of Mt Etna (Sicily). *Bulletin of Volcanology*, **65**, 461–476.

BEHNCKE, B., NERI, M., PECORA, E. & ZANON, V. 2006. The exceptional activity and growth of the Southeast crater, Mt Etna (Italy) between 1996 and 2001. *Bulletin of Volcanology*, **69**, 149–173.

BERTAGNINI, A., CALVARI, S., COLTELLI, M., LANDI, P. & POMPILIO, M. 1990. The 1989 eruptive sequence. *In*: BARBERI, F., BERTAGNINI, A. & LANDI, P. (eds) *Mt Etna: The 1989 Eruption*. Gruppo Nationale Vulcanologia, Italy, 10–22.

BONACCORSO, A. 2006. Explosive activity at Mt Etna summit craters and source modeling by using high-precision continuous tilt. *Journal of Volcanology and Geothermal Research*, **158**, 221–234.

BRUCE, P. M. & HUPPERT, H. E. 1989. Thermal controls of basaltic fissure eruptions. *Nature*, **342**, 665–667.

BULL. GLOBAL VOLCANISM NETWORK (BGVN) 2001. Sistema Posidon, Instituto Nazionale di Geofisica e Vulcanologia-Sezione di Catania, Italy, Etna. *Smithsonian Institution*, Etna activity reports, **26**.

CALVARI, S. & RESEARCH STAFF OF THE INSTITUTO NAZIONALE DI GEOFISICA E VULCANOLOGIA-SEZIONE DI CATANIA, ITALY 2001. *EOS, Transactions of AGU*, **82**, 653–656.

CAPLAN-AUERBACH, J. & MCNUTT, S. R. 2003. New insights into the 1999 eruption of Shishaldin volcano based on acoustic data. *Bulletin of Volcanology*, **65**, 405–417.

CARACAUSI, A., ITALIANO, F., PAONITA, A. & RIZZO, A. 2003. Evidence of deep magma degassing and ascent by geochemistry of peripheral gas emissions at Mount Etna (Italy): assessment of magmatic reservoir pressure. *Journal of Geophysical Research*, **108**, ECV 2-1-ECV 2–15.

CARBONE, D., ZUCCARELLO, L., SACCOROTTI, G. & GRECO, F. 2006. Analysis of simultaneous gravity and tremor anomalies observed during the 2002–2003 Etna eruption. *Earth and Planetary Science Letters*, **245**, 616–629.

CHIARABBA, C., AMATO, A., BOSCHI, E. & BARBERI, F. 2000. Recent seismicity and tomographic modeling of the Mount Etna plumbing system. *Journal of Geophysical Research*, **105**, 10923–10938.

CLOCCHIATTI, R. & MÉTRICH, N. 1984. La crystallisation des pyroclastes des éruptions etnéennes de 1669 (Mt Rossi) et 1982 (Mt Silvestri). *Bulletin of Volcanology*, **47**, 908–928, 1984.

CLOCCHIATTI, R., CONDOMINES, M., GUENOT, N. & TANGUY, J.-C. 2004. Magma changes at Mount Etna: the 2001 and 2002–2003 eruptions. *Earth and Planetary Science Letters*, **226**, 397–414.

COLTELLI, M., DEL CARLO, P. & VEZZOLI, L. 1998. Discovery of a Plinian basaltic eruption of Roman age at Etna volcano, Italy. *Geology*, **26**, 1095–1098.

CORSARO, R. A. & POMPILIO, M. 2004. Magma dynamics in the shallow plumbing system of Mt Etna as recorded by compositional variations in volcanics of recent summit activity (1995–1999). *Journal of Volcanology and Geothermal Research*, **137**, 55–71.

DAWSON, P. B., CHOUET, B. A., OKUBO, P. G., VILLASENOR, A. & BENZ, H. M. 1999. Three-dimensional velocity structure of Kilauea caldera, Hawai'i. *Geophysical Research Letters*, **26**, 2,805–2,808.

DINGWELL, D. B. 1998. Recent experimental progress in the physical description of silicic magma relevant to explosive volcanism. *In*: GILBERT, J. S. & SPARKS, R. S. J. (eds) *The Physics of Explosive Volcanic Eruptions*, Geological Society, London, Special Publications, **145**, 9–26.

DIXON, J. E., STOLPER, E. M. & HOLLOWAY, J. R. 1995. An experimental study of water and carbon dioxide solubilities in mid-ocean ridge basaltic liquids. Part 1: calibration and solubility models. *Journal of Petrology*, **36**, 1607–1631.

GERLACH, T. M., MCGEE, K. A., ELIAS, T., SUTTON, A. J. & DOUKAS, M. P. 2002. Carbon dioxide emission rate of Kilauea volcano: implications for primary magma and the summit reservoir. *Journal of Geophysical Research*, **107**, 2189, ESV 3, 1–15.

GIORDANO, D. & DINGWELL, D. B. 2003. Viscosity of hydrous Etna basalt: implications for Plinian-style basaltic eruptions. *Bulletin of Volcanology*, **65**, 8–14.

GRESTA, S., RIPEPE, M., MARCHETTI, E., D'AMICO, S., COLTELLI, M., HARRIS, A. J. L. & PRIVITERA, E. 2004. Seismo-acoustic measurements during July-August 2001 eruption of Mt Etna volcano, Italy. *Journal of Volcanology and Geothermal Research*, **137**, 219–230.

HARRIS, A. J. L, MURRAY, J. B., ARIES, S. E., *ET AL.* 2000. Effusion rate at Etna and Krafla and their implications for eruptive mechanisms. *Journal of Volcanology and Geothermal Research*, **102**, 237–270.

HARRIS, A. J. L. & NERI, M. 2002. Volumetric observations during paroxysmal eruptions at Mount Etna: pressurised drainage of a shallow chamber or pulsed supply? *Journal of Volcanology and Geothermal Research*, **116**, 79–95.

HELIKER, C. & MATTOX, T. N. 2003. The first two decades of the Pu'u 'O'o-Kupaianaha eruption: chronology and selected bibliography. *US Geological Survey Professional Papers*, **1676**, 1–28.

JAUPART, C. & VERGNIOLLE, S. 1988. Laboratory models of Hawaiian and Strombolian eruptions. *Nature*, **331**, 58–60.

JAUPART, C. & VERGNIOLLE, S. 1989. The generation and collapse of a foam layer at the roof of a basaltic magma chamber. *Journal of Fluid Mechanics*, **203**, 347–380.

KAMENETSKY, V. & CLOCCHIATTI, R. 1996. Primitive magmatism of Mount Etna: insights from mineralogy and melt inclusions. *Earth and Planetary Science Letters*, **142**, 553–572.

LADELFA, S., PATANE, G., CLIOCCHIATTI, R., JORON, J.-L. & TANGUY, J.-C. 2001. Activity of Mount Etna preceding the February 1999 fissure eruption: inferred mechanism from seismological and geochemical data. *Journal of Volcanology and Geothermal Research*, **105**, 121–139.

LAIGLE, M. & HIRN, A. 1999. Explosion-seismic tomography of a magmatic body beneath Etna: volatile discharge and tectonic control of volcanism. *Geophysical Research Letters*, **26**, 2665–2668.

LAIGLE, M., HIRN, A., SAPIN, M. & LEPINE, J.-C. 2000. Mount Etna dense array local earthquake P and S tomography and implications for volcanic plumbing. *Journal of Geophysical Research*, **105**, 21633–21646.

LECLOAREC, M. F. & PENNISI, M. 2001. Radionucleides and sulfur content in Mount Etna plume in 1983–1995: new constraints on the magma feeding system. *Journal of Volcanology and Geothermal Research*, **108**, 141–155.

MANGAN, M. T. & CASHMAN, K. V. 1996. The structure of basaltic scoria and reticulite and inferences for vesiculation, foam formation, and fragmentation in lava fountains. *Journal of Volcanology and Geothermal Research*, **73**, 1–18.

MÉTRICH, N., COCCHIATTI, R., MOSBASH, M. & CHAUSSIDON, M. 1993. The 1989–1990 activity of Etna: magma mingling and ascent of H_2O-Cl-S-rich basaltic magma. Evidence from melt inclusions. *Journal of Volcanology and Geothermal Research*, **59**, 131–140.

MÉTRICH, N., ALLARD, P., SPILLIAERT, N., ANDRONICO, D. & BURTON, M. 2004. 2001 flank eruption of the alkali- and volatile-rich primitive basalt responsible for Mount Etna's evolution in the last three decades. *Earth and Planetary Science Letters*, **228**, 1–17.

MURRU, M., MONTUORI, C., WYSS, M. & PRIVITERA, E. 1999. The locations of magma chambers at Mt Etna, Italy, mapped by b-values. *Geophysical Research Letters*, **26**, 2553–2556.

MYSEN, B. O. 1977. The solubility of H_2O and CO_2 under predicted magma genesis conditions and some petrological and geophysical implications. *Reviews of Geophysics and Space Physics*, **15**, 351–361.

PARFITT, E. A. & WILSON, L. 1995. Explosive volcanic eruptions: IX The transition between Hawai'ian-style lava fountaining and Strombolian explosive activity. *Geophysical Journal International*, **121**, 226–232.

PARFITT, E. A., WILSON, L. & NEAL, C. A. 1995. Factors influencing the height of Hawai'ian lava fountains: Implications for the use of fountain height as an indicator of magma content. *Bulletin of Volcanology*, **57**, 440–450.

PARFITT, E. A. 2004. A discussion of the mechanisms of explosive basaltic eruptions. *Journal of Volcanology and Geothermal Research*, **134**, 77–107.

PATANÈ, D., CHIARABBA, C., COCINA, O., DE GORI, P., MORETTI, M. & BOSCHI, E. 2002. Tomographic images and 3D earthquake locations of the seismic

swarm preceding the 2001 eruption Mt Etna eruption: evidence for dyke intrusion. *Geophysical Research Letters*, **29/10**, 135-1–135-4, doi:10.1029/2001GL014391, 135-1–135-4.

PATANÈ, D., BARBERI, G., COSINA, O., DE GORI, P. & CHIARABBA, C. 2006. Time-resolved seismic tomographic detects magma intrusion at Mount Etna. *Science*, **313**, 821–823.

POLACCI, M., CORSARO, R. A. & ANDRONICO, D. 2006. Coupled textural and compositional characterization of basaltic scoria: insights into the transition from Strombolian to fire fountain activity at Mount Etna. *Geology*, **34**, 201–204.

PRIVITERA, E., SGROI, T. & GRESTA, S. 2003. Statistical analysis of intermittent volcanic tremor associated with the September 1989 summit explosive eruptions at Mount Etna, Sicily. *Journal of Volcanology and Geothermal Research*, **120**, 235–247.

PROUSSEVITCH, A. A., SAHAGIAN, D. L. & KUTOLIN, V. A. 1993. Stability of foams in silicate melts. *Journal of Volcanology and Geothermal Research*, **59**, 161–178.

RICHTER, D. H., EATON, J. P., MURATA, K. J., AULT, W. U. & KRIVOY, H. L. 1970. Chronological narrative of the 1959–60 eruption of Kilauea volcano, Hawaii. *US Geological Survey Professional Papers*, **537**, 1–70.

RIPEPE, M. & GORDEEV, E. 1999. Gas bubble dynamics model for shallow volcanic tremor at Stromboli. *Journal of Geophysical Research*, **104**, 10, 639–10,654.

RIPEPE, M., COLTELLI, C., PRIVITERA, E., GRESTA, S., MORETTI, M. & PICCININI, D. 2001a. Seismic and infrasonic evidences for an impulsive source of the shallow volcanic tremor at Mount Etna, Italy. *Geophysical Research Letters*, **28**, 1071–1074.

RIPEPE, M., CILIBERTO, S. & DELLA SCHIAVA, M. 2001b. Time constraints for modeling source dynamics of volcanic explosions at Stromboli. *Journal of Geophysical Research*, **106**, 8713–8727.

SACCOROTTI, G., ZUCCARELLO, L., DEL PEZZO, E., IBANEZ, J. & GRESTA, S. 2004. Quantitative analysis of the tremor wavefield at Etna volcano, Italy. *Journal of Volcanology and Geothermal Research*, **136**, 223–245.

SABLE, J. E., HOUGHTON, B. F., DEL CARLO, P. & COLTELLI, M. 2006. Changing conditions of magma ascent and fragmentation during the Etna 122 BC basaltic eruption: evidence from clast microtextures. *Journal of Volcanology and Geothermal Research*, **158**, 333–354.

SCARTH, A. 1994. *Volcanoes*. UCL Press, London.

SCHMINCKE, H.-U. 2004. *Volcanism*. Springer-Verlag, Berlin, Germany.

SPARKS, R. S. J., BURSIK, M. I., CAREY, S. N. ET AL. 1997. *Volcanic Plumes*, Wiley, Chichester.

SPILLIAERT, N., METRICH, N. & ALLARD, P. 2006a. S-Cl-F degassing pattern of water-rich alkali basalt: modelling and relationship with eruption styles on Mount Etna volcano. *Earth and Planetary Science Letters*, **248**, 772–786.

SPILLIAERT, N., ALLARD, P., METRICH, N. & SOBOLEV, A. 2006b. Melt inclusion record of the conditions of ascent, degassing and extrusion of

volatile-rich alkali basalt during the powerful 2002
flank eruption of Mount Etna (Italy). *Journal of Geophysical Research*, **111**, B04203, doi:10.1029/
2005JB003934.

STOLPER, E. & HOLLOWAY, J. R. 1988. Experimental
determination of the solubility of carbon dioxide in
molten basalt at low pressure. *Earth and Planetary
Science Letters*, **87**, 397–408.

SWANSON, D. A., DUFFIELD, D. A., JACKSON, D. B. &
PETERSON, D. W. 1979. Chronological narrative
of the 1969–1971 Mauna-Ulu eruption of Kilauea
volcano, Hawaii. *US Geological Survey Professional
Papers*, **1056**.

THURBER, C. H. 1987. Seismic structure and tectonics of
Kilauea volcano. *US Geological Survey Professional
Papers*, **1350**.

VERGNIOLLE, S. & JAUPART, C. 1990. Dynamics of
degassing at Kilauea volcano, Hawaii. *Journal of
Geophysical Research*, **95**, 2793–2809.

VERGNIOLLE, S. & BRANDEIS, G. 1994. Origin of the
sound generated by Strombolian explosions. *Geophysical Research Letters*, **21**, 1959–1962.

VERGNIOLLE, S. & BRANDEIS, G. 1996. Strombolian
explosions, 1, A large bubble breaking at the surface
of the lava column as a source of sound. *Journal of
Geophysical Research*, **101**, 20433–20448.

VERGNIOLLE, S. 1996. Bubble size distribution in magma
chambers and dynamics of basaltic eruptions. *Earth
and Planetary Science Letters*, **140**, 269–279.

VERGNIOLLE, S., BOICHU, M. & CAPLAN-AUERBACH,
J. 2004. Acoustic measurements of the 1999 basaltic
eruption of Shishaldin volcano, Alaska: 1) Origin of
Strombolian activity. *Journal of Volcanology and
Geothermal Research*, **137**, 109–134.

VERGNIOLLE, S. & CAPLAN-AUERBACH, J. 2004.
Acoustic measurements of the 1999 basaltic eruption
of Shishaldin volcano, Alaska: 2) Precursor to the Subplinian activity. *Journal of Volcanology and Geothermal Research*, **137**, 135–151.

VERGNIOLLE, S. & CAPLAN-AUERBACH, J. 2006. Basaltic thermals and Subplinian plumes: Constraints from
acoustic measurements at Shishaldin volcano,
Alaska. *Bulletin of Volcanology*, **68**, 611–630.

VERGNIOLLE, S. & RIPEPE, M. 2008. From Strombolian
explosions to fire fountains at Etna Volcano (Italy):
what do we learn from acoustic measurements?. *In*:
LANE, S. J. & GILBERT, J. S. (eds) *Fluid Motions
in Volcanic Conduits: A Source of Seismic and
Acoustic Signals*. Geological Society, London,
Special Publications, **307**, 103–124.

WALLACE, P. 2005. Volatiles in subduction zone
magmas: concentrations and fluxes based on melt
inclusion & volcanic gas data. *Journal of Volcanology
and Geothermal Research*, **140**, 217–240.

WALLIS, G. B. 1969. *One Dimensional Two-Phase Flows*.
McGraw-Hill, New York.

WILLIAMS, H. & MCBIRNEY, A. R. 1979. *Volcanology*.
Freeman Cooper, San Francisco.

WILSON, L. 1980. Relationships between pressure,
volatile content and ejecta velocity. *Journal of Volcanology and Geothermal Research*, **8**, 297–313.

WOLFE, E. W., GARCIA, M. O., JACKSON, D. B.,
KOYANAGI, R. Y., NEAL, C. A. & OKAMURA, A.
T. 1987. The Pu'u 'O'o eruption of Kilauea volcano,
episodes 1–20, January 3, 1983 to June 8, 1984,
in Volcanism in Hawaii. *US Geological Survey
Professional Papers*, **1350**, 471–508.

WOLFE, E. W., NEAL, C. A., BANKS, N. G. & DUGGAN,
T. J. 1988. Geologic observations and chronology of
eruptive events, in the Pu'u 'O'o eruption of Kilauea
volcano, episodes 1–20, January 3, 1983 to June 8,
1984. *In*: WOLFE, E. (ed.) *US Geological Survey
Professional Papers*, **1463**, 471–508.

YANG, X., DAVIS, P. M., DELANEY, P. T. & OKAMURA,
A. T. 1992. Geodetic analysis of dike intrusion and
motion of the magma reservoir beneath the summit
of Kilauea volcano, Hawai'i: 1970–1985. *Journal of
Geophysical Research*, **97**, 3305–3324.

Modelling the rapid near-surface expansion of gas slugs in low-viscosity magmas

M. R. JAMES, S. J. LANE & S. B. CORDER

Lancaster Environment Centre, Lancaster University, Lancaster, LA1 4YQ
(e-mail: m.james@lancaster.ac.uk)

Abstract: The ascent of large gas bubbles (slugs) in vertical cylindrical conduits and low-viscosity magmas is simulated using 1D mathematical and 3D computational fluid dynamic (CFD) models. Following laboratory evidence, the 1D model defines a constant rise velocity for the slug base and allows gas expansion to accelerate the slug nose through the overlying fluid during ascent. The evolution of rapidly expanding gas slugs observed in laboratory experiments is reproduced well and, at volcano scales, predicts at-surface overpressures of several atmospheres without requiring any initial overpressure at depth. The near-surface dynamics increase slug nose velocities through the overlying magma by a factor of $c.$ 2.5 and the gas expansion results in pre-burst magma surface velocities of $c.$ 35 m s^{-1}. To examine pressure distributions and the forces exerted on a conduit, 3D CFD simulations were carried out. At volcano scales, the vertical single forces during final slug ascent to the surface are $c.$ 10^6 N, two orders of magnitude smaller than those associated with very-long-period seismic events at Stromboli. This supports a previous interpretation of these events in which they are generated by gas slugs flowing through changes in conduit geometry, rather than being the direct result of slug eruption processes.

At volcanoes characterized by low-viscosity magma, degassing can occur through the periodic release of large gas slugs, resulting in Strombolian-style eruptions (Chouet *et al.* 1974; Blackburn *et al.* 1976). These types of magmatic system are often open and persistently active, allowing long-term studies of degassing rates and conduit flow processes. Continuous geophysical monitoring including video and thermal video, gas, infrasonic and seismic data collection (e.g. Aster *et al.* 2004; Auger *et al.* 2006; Marchetti *et al.* 2006) is now carried out at several such volcanoes, providing a wealth of data from which significant advances in our understanding of degassing, conduit geometry and near-surface magma rheology can be made. In order to combine these data and to include other geophysical, geochemical (Oppenheimer *et al.* 2006; Burton *et al.* 2007) and petrological or textural approaches (e.g. Polacci *et al.* 2006; Armienti *et al.* 2007; Lautze & Houghton 2007), physical models of the formation, ascent and burst of the bubbles are required. Such models would thus represent a framework to enable an integrated understanding of the dynamics of these persistent volcanic systems.

The aim of the work here is to present two models describing the final ascent of gas slugs (up to, but not including, their burst) with emphasis on understanding the form and magnitude of any geophysical effects produced, and to assess model sensitivity to changes in system parameters. Given that a full understanding of flow processes in

complex conduit geometries (e.g. multiple inclined dykes) will require large and detailed fluid dynamic models (which, in turn, will be non-trivial to validate experimentally), we start here by considering a straightforward vertical pipe geometry. This facilitates comparisons with laboratory data and allows models to draw on a significant body of engineering literature. We concentrate on the relatively unexplored effects of the very rapid near-surface gas expansion and associated conduit-scale liquid accelerations, leaving the incorporation of initial bubble coalescence and slug formation to future work. Slugs are represented as large single bubbles, as would be anticipated if they formed by foam collapse or at a depth sufficient that gas expansion would have caused coalescence of any initial component smaller bubbles. At Stromboli, this is supported by gas chemistry data which suggest that slugs form at depths of up to 0.8 to 2.7 km (Burton *et al.* 2007). Nevertheless, we anticipate that, to first order, our results would also be applicable to ascending foam rafts as long as their density was significantly less than that of the surrounding liquid.

Previous mathematical models of slug ascent for volcanic scenarios have been given (Vergniolle 1998; Seyfried & Freundt 2000) but not widely investigated and these models could not simulate accurately the laboratory experiments given here. More sophisticated 3D lattice-Boltzmann, diffuse interface and Bhatnagar-Gross-Krook numerical models are under development (D'Auria *et al.*

From: LANE, S. J. & GILBERT, J. S. (eds) *Fluid Motions in Volcanic Conduits: A Source of Seismic and Acoustic Signals.* Geological Society, London, Special Publications, **307**, 147–167.
DOI: 10.1144/SP307.9   0305-8719/08/$15.00 © The Geological Society of London 2008.

2004; O'Brien & Bean 2004, 2006; D'Auria 2006) but have yet to be quantitatively validated. Here we use two approaches; developing a relatively straightforward 1D model, to provide a quick solution to the first-order physics and using a 3D CFD model (constructed with a commercial fluid dynamics package) for a more full description of the flow. The applicability, relative usefulness of the different models and, more importantly, the uncertainties involved with modelling these processes are demonstrated.

Modelling slug flow is important in industrial settings where multiphase flows are used to transport fluids and catalyse chemical reactions. Much of the industrial research is oriented at horizontal or low-angle pipes, or flowing liquids, with these scenarios being applicable to oil pipelines (e.g. Gregory & Scott 1969; Wallis 1969; Dukler & Hubbard 1975; Fabre & Liné 1992). Furthermore, associated models often describe time-averaged parameters of continuous flow rather than the progress of a single gas slug. Nevertheless, considerable research exists on the ascent of individual gas slugs up liquid-filled vertical pipes (Davies & Taylor 1950; White & Beardmore 1962; Brown 1965; Wallis 1969; Polonsky *et al.* 1999). Computational fluid dynamic (CFD) models of individual gas slugs have been developed (Mao & Dukler 1990, 1991; Clarke & Issa 1997; Taha & Cui 2006*a, b*) but detailed modelling of multiphase flows, including slug flow, is complex and an ongoing area of research.

When compared with engineering-oriented simulations, volcanic scenarios have the added complexity of the large length scales applicable, which result in highly non-steady conditions due to extremely rapid near-surface accelerations. Within the engineering literature, limited slug expansion due to gas decompression during ascent (in standard laboratory-scale apparatus) has often been noted. With one recent exception (Sousa *et al.* 2006), the resulting effects have usually been corrected for (e.g. in order to calculate steady-state slug nose velocities: Laird & Chisholm 1956; Nicklin *et al.* 1962; White & Beardmore 1962), rather than being the main focus of the work. The authors are unaware of any previous engineering research on the dynamics of rapidly expanding slugs as studied here, and the only experimental data have been from our own preliminary studies (Lane *et al.* 2005; James *et al.* 2006*a*) simulating the expansions applicable to near-surface regions of volcanoes (e.g. in the last few tens of metres of ascent). As an example of the relevant magnitudes, in a generic laboratory experiment under standard conditions, slugs may ascend at *c.* 0.2 m s^{-1} through a pressure gradient of *c.* 10^4 Pa m^{-1} (i.e. air slugs in water, with a viscosity of 0.001 Pa s, density 1000 kg m^{-3} and

surface tension of 0.07 Pa, within a 3.4-cm diameter tube; see Wallis (1969) for the calculation of slug velocities), giving a decompression rate of 2 kPa s^{-1}. For a timescale representative of the potential for gas expansion, one could consider the duration required for a theoretical volume doubling during decompression to surface pressure (i.e. from 200 to 100 kPa), in this case *c.* 50 s. In comparison, for a volcanological scenario, ascent rates may be *c.* 2 m s^{-1}, through pressure gradients of *c.* 20 kPa m^{-1} (e.g. steam slugs in magma with a viscosity of 500 Pa s, density *c.* 2600 kg m^{-3} and surface tension of 0.4 Pa, within a 4-m diameter conduit (Seyfried & Freundt 2000) giving decompression rates of 50 kPa s^{-1}, and hence effective volume doubling timescales of *c.* 2 s.

In order to assess the accuracy of model results under relevant expansion conditions, laboratory experiments were designed to allow slug ascent under low ambient surface pressures to be observed. For the theoretical laboratory example presented above (and neglecting any decompression-related phase changes), a 'volume doubling to surface pressure' timescale of 2 s could be achieved by using a surface pressure of *c.* 4 kPa. However, note that this analysis is presented as an illustrative example only, rather than a formal scaling relation. Actual fluid accelerations will be driven by the magnitude of volumetric expansion (as opposed to volume ratios) so, given a timescale similar to the volcanological scenario, the smaller laboratory length scales will result in significantly smaller accelerations. Furthermore, the unsteady nature of the flow in these regions means that traditional dimensionless scaling arguments (which were derived for steady-state conditions) cannot be guaranteed to scale expansion-dominated laboratory results accurately to volcanic dimensions.

Methods and models

In this work, both laboratory- and volcano-scale scenarios are modelled using the fluid and conduit parameters given in Table 1 (unless otherwise stated), with a Newtonian liquid phase being used throughout. Appropriate parameters for a generic basaltic magma are used for the volcano-scale scenario, as used in previous experimental work (Seyfried & Freundt 2000).

In volcanological fluid flow problems, high temperature basaltic systems can be regarded as some of the most tractable, due to the liquid phase being reasonably represented as a Newtonian fluid in many cases. Consequently, the use of Newtonian liquids has been standard in order to facilitate related types of experiment and analysis (Vergniolle & Jaupart 1990; Vergniolle 1998;

Table 1. *Fluid and conduit parameter values*

	Laboratory experiments	Volcanic scenario
	Ultragrade19 vacuum oil (BOC Edwards)	Basalt magma
Liquid phase		
Density, ρ (kg m^{-3})	862	2600
Viscosity, μ (Pa s)	0.124	500
Surface tension, σ (N m^{-1})	0.032	0.4
Gas phase*	Air	H_2O
Molecular mass (g mol^{-1})	28.9	18.0
Specific heats ratio[†], γ	1.4	1.1[‡]
Temperature, T (K)	293	1370
Conduit radius, r_c (m)	0.013	1.5
Initial slug depth, h_0 (m)	*c.* 1.78	100
Dimensionless parameters (for steady slug ascent conditions)[§]		
Morton number, Mo	0.08	10^9
Eötvös number, Eo	179	10^6
Froude number, Fr	0.321	0.315
Inverse viscosity, N_f	91	85
Slug ascent velocity, U_s (m s^{-1}) (calculated from equation 8)	0.162	1.71

*The models used assume negligible gas densities (with respect to those of the liquids) and consider only gas volume and pressure. The values provided here are used only to calculate equivalent gas masses for comparative purposes.
[†]The best model fits to experimental data were obtained assuming isothermal conditions. Consequently, unless specifically noted, $\gamma = 1.0$ was used rather than the values here, which would be used under adiabatic conditions.
[‡]For hot gases (Lighthill 1978).
[§]See equation 8 and adjacent text for definitions and descriptions of the dimensionless numbers.

Seyfried & Freundt 2000; James *et al.* 2004, 2006*b*) to these presented here. At temperatures below *c.* 1130 °C, basalt rheology becomes increasingly non-Newtonian with further cooling (Shaw 1969). However, at conduit temperatures, and for the large flow motions investigated here, departures from a Newtonian model are likely to represent only second order uncertainties with respect to those from other poorly constrained parameters (for instance, the general magnitude of the apparent viscosity). In the following section, the laboratory experiments are described first, then details of the 1D and 3D CFD models are given.

Laboratory experiments

The laboratory flow tube used was a *c.* 2-m-long vertical borosilicate glass tube of internal diameter *c.* 25 mm, sealed at the base (with the exception of a syringe for gas injection), and connected to a vacuum pump at the top (Fig. 1a). A vacuum chamber provided a buffer volume (*c.* 0.1 m^3) to maintain a near-constant pressure at the liquid surface during slug expansion, and was linked to the flow tube by a flexible connection including a trap to prevent ingress of any liquid droplets into the chamber. The tube contained a depth of *c.* 1.7 m of vacuum oil (Table 1) for the liquid

phase. With a vapour pressure of 10^{-6} Pa at 20 °C, the use of the vacuum oil allowed the apparatus to be run at low pressures without either boiling or vapour adding significant mass to injected air bubbles as they expanded.

Liquid pressure at the base of the tube was recorded by an ASG pressure sensor (BOC Edwards). Flow-induced vertical motions of the apparatus were measured by suspending the apparatus on a spring with the base resting on a force transducer (S-100, Strain Measurement Devices Ltd.), which was calibrated for displacement. All sensors were logged at 5 kHz by a 12-bit National Instrument DAQ board in a PC. To carry out an experiment, the surface pressure was set to the required value and the pump was isolated from the system prior to a pocket of gas (*c.* 6 ml, at near hydrostatic pressure) being injected with the syringe into the base of the tube. This procedure initiated a few transient oscillations in the gas and liquid column but these were rapidly damped out by the liquid viscosity as the slug ascended. The ascent and expansion of the gas slugs was recorded using two digital video cameras (Canon MV750i and XL1 in Frame mode) to cover the tube region. A Basler A602f Firewire camera (up to *c.* 400 frames per second) was also used to image the rapid near-surface slug expansion and burst. The

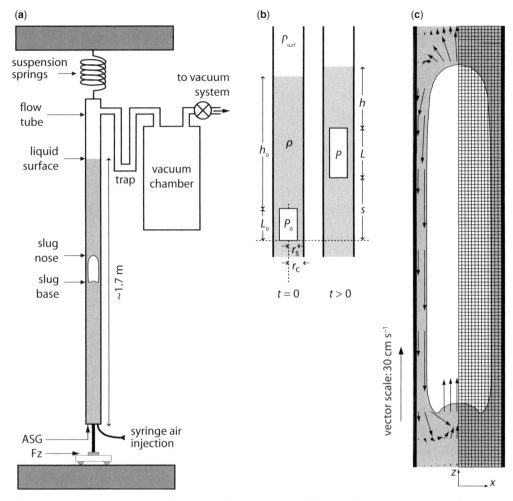

Fig. 1. (a) Experimental apparatus, (b) canonical 1D model and (c) CFD model. During experiments (a), fluid pressures were measured at the base of the apparatus by a pressure sensor (ASG) and apparatus vertical motion was measured by a force sensor (Fz). In (b), the parameters used in the 1D model (equations 1 to 7) are shown at ($t = 0$), and after ($t > 0$), the model start. The FLOW-3D® model is illustrated in (c), where a 13-cm-high $y = 0$ slice through the results of a quarter-tube simulation (mirrored in the $x=0$ plane) are overlain by liquid velocity vectors on the left-hand side and by a representation of the computational mesh on the right-hand side. The results depicted are 1.2 s into the simulation of a laboratory experiment (with a surface pressure of 319 Pa), at which time the slug nose is *c.* 45 cm below the liquid surface (not shown).

camera images were synchronized to the sensor data by including a binary counter linked to the DAQ board's scan clock within the recorded field of view. Spatial scaling (to convert pixels to metres) was achieved by using 10-cm-interval marks along the length of the tube.

From an extensive suite of experiments carried out, two (with surface pressures of 1 and 319 Pa) are dominantly used here to assess the models; a detailed discussion of the experimental data is left for other work. The specific experiments selected provide an extreme example of expansion (at

1 Pa) and a volcanologically relevant scenario (at 319 Pa). For the *c.* 1.7 m head of liquid, the static pressure at the base of the liquid column was *c.* 14.4 kPa greater than the pressure at its surface. Thus, with a surface pressure of 319 Pa, the total amount of gas expansion possible in the experiment (a factor of 46) is equivalent to that of gas rising from a depth of *c.* 175 m under volcanological conditions.

In Figure 2, example images are given from two experiments, one carried out at atmospheric pressure (Fig. 2a), illustrating a slug which has

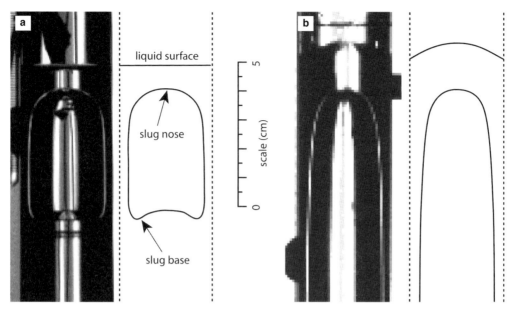

Fig. 2. Images of gas slugs approaching the liquid surface in laboratory experiments. In (**a**), a short slug (*c.* 14 mg of gas) is expanding slowly in an experiment carried out at atmospheric pressure (*c.* 10^5 Pa). The image is backlit through the flow tube and the slug outline is predominantly shown by refracted light from the gas–liquid interface. The sketch to the right indicates the position of the liquid surface, slug outline and exterior of the tube (dashed lines). Although the shape of the slug is horizontally distorted by refraction through the circular tube, the standard morphology under mixed viscous and inertial control (a domed nose and base) can be easily observed. In (**b**), a similar image shows the nose of a much longer slug (*c.* 1 mg of gas) expanding to a surface pressure of 319 Pa (the apparent decrease in image resolution is caused by the wider angle camera lens used in order to capture more of the long slug body). At this point in time, the slug nose was moving upward at *c.* 2.4 m s^{-1}. Note the longer curved length of the nose and the upward doming of the liquid surface, testifying to the dynamic nature of the slug expansion.

undergone only a very small expansion and one carried out with a surface pressure of 1 Pa (Fig. 2b) demonstrating the effects of very high rates of slug expansion.

1D model

Vergniolle (1998) described a dynamic model for slug ascent within a vertical cylindrical conduit, in which the slug nose rises at a constant velocity with respect to the overlying liquid. The bubble length, and hence its volume and pressure, was calculated by equating the rate of change of momentum of the liquid above the slug to viscous, gravitational and pressure forces. However, liquid volume in the falling liquid film around the slug was not considered and application of this model to the experimental data proved problematic for several reasons.

In our low pressure experiments, despite the rapid expansion of the slugs, the video data indicated that the slug base velocity was constant (to within measurement error) throughout the ascent (see the 'Comparison of laboratory and model results' section). Additionally, although shape changes to the slug nose region were observed as slugs approached the surface (Fig. 2), the thickness of the falling film was deemed to be effectively constant (if anything, the shape changes at the slug nose could have been represented as a film thickening). This observation implies that, as the slug expands, the volume of liquid maintained within the falling film must increase and, in order to do that without decreasing the slug base velocity, liquid flux past the slug nose must also increase. Thus, a constant nose velocity with respect to the overlying liquid should not be assumed for rapidly expanding slugs and a more convenient reference frame to use would be that of a constant-velocity slug base.

Furthermore, in previous work, the inertial force resulting from the acceleration of the liquid above the slug has been calculated from the total rate of change of momentum (Vergniolle (1998) equation 10, and also implied by Seyfried & Freundt (2000) equation 11). In this formulation, liquid mass initially above the slug, which then flows past the slug nose (i.e. loss of liquid mass from above the slug), is effectively considered to 'leave' its momentum above the slug nose. A realistic mechanism for this to occur cannot be envisaged and one implication is that it permits regions of parameter space in which slugs at pressures less than the liquid surface pressure are still increasingly expanding; a clearly non-physical result. Although any simplified model of the momentum flux around and above the slug nose is necessarily going to require significant assumptions, using an alternative scheme in which the momentum flows with the mass (and hence is lost from the liquid above the slug) avoids this non-physical scenario.

Here, in line with these observations, we describe a new 1D model, similar to that of Vergniolle (1998) in that it is based on the forces exerted on the liquid above the slug, but also accounting for liquid volume surrounding the slug and using a constant slug base, rather than nose, velocity. The slug is represented as a gas cylinder of constant radius r_s and initial length L_0, within a vertical cylindrical conduit of internal diameter r_c (Fig. 1b). Maintaining this simplified geometrical representation of slug morphology used by previous workers (Vergniolle 1998; Seyfried & Freundt 2000) allows the first order dynamics associated with the depressurization expansion of long slugs to be captured with relative ease because, for the slug sizes of interest, variations from a cylindrical shape (e.g. the curved or seiching base and domed nose) are volumetrically relatively small.

With the liquid surface an initial distance h_0 above the top of the slug, conservation of liquid volume at any given time yields

$$h_0 \pi r_c^2 + L_0 \pi (r_c^2 - r_s^2) = (h + s) \pi r_c^2$$
$$+ L \pi (r_c^2 - r_s^2) \qquad (1)$$

where s is the distance between the slug base and its initial position, and h is the height of fluid above the top of the gas cylinder (Fig. 1b). Simplification of equation 1 allows h to be expressed as

$$h = h_0 - U_s t - (L - L_0)(1 - A') \qquad (2)$$

where $A' = (r_s/r_c)^2$ and, for a slug with a constant base velocity, U_s, at time t, $s = U_s t$.

The acceleration of the liquid above the slug can then be defined in terms of the pressure, gravitational and viscous forces acting on the liquid cylinder, given by $F_p = \pi r_s^2 (P - P_{surf})$, $F_g = -\pi r_s^2 \rho h g$ and F_v respectively, where P is the slug gas pressure, P_{surf} is ambient surface pressure, ρ is the liquid density and g is the acceleration due to gravity. For laminar flow in a pipe, the viscous drag force can be derived from the pressure drop

under Poiseuille flow (Batchelor 1967), to give

$$F_v = -8 \pi \mu l_p U_f \qquad (3)$$

where l_p is the pipe length and U_f is the mean flow velocity along the pipe. An estimate of the viscous drag experienced by the liquid above the slug can thus be made by assuming that it flows at a volume flux equal to that of the gas expansion, giving

$$F_v \approx -8 \pi \mu h \dot{L} A'. \qquad (4)$$

Equating the product of mass and acceleration (of the centre of mass of the liquid column directly over the gas slug) to the sum of forces gives

$$\pi r_s^2 \rho h \frac{d^2 (U_s t + L + \frac{1}{2} h)}{dt^2} = F_p + F_g + F_v. \qquad (5)$$

Substituting for the forces and expanding the differential (using equation 2 to define h in terms of L) gives

$$\frac{1}{2} \pi r_s^2 \rho h (1 + A') \ddot{L} = \pi r_s^2 (P - P_{surf})$$
$$- \pi r_s^2 \rho h g - 8 \pi \mu h \dot{L} A' \qquad (6)$$

which, if the slug is represented as a perfect gas ($PV^\gamma = $ constant, where V is gas volume and γ the ratio of specific heats, in this case leading to $P = P_0 (L_0/L)^\gamma$, for a constant radius cylinder and an initial slug pressure, P_0), can be reduced to

$$\frac{1}{2} \rho (1 + A') \ddot{L} = P_0 L_0^\gamma L^{-\gamma} h^{-1} - \rho g$$
$$- P_{surf} h^{-1} - 8 \mu \dot{L} r_c^{-2}. \qquad (7)$$

Note that h is a function of L (equation 2) and of the (constant) slug parameters U_s and r_s, the calculation of appropriate values for which is discussed below. For all the simulations carried out, the initial bubble pressure was given by the static head only, i.e. $P_0 = \rho g h_0 + P_{surf}$. Equation 7 is solved numerically in Matlab®, using an explicit Runge-Kutta (4, 5) formula (the Dormand-Prince pair; Dormand & Prince 1980) and, from the solutions, the positions of the slug and liquid surfaces, as well as the slug gas pressure, P, are obtained. Starting conditions are defined by L_0 and by an initial estimate of the rate of slug expansion, $\dot{L}_0$. For this, the initial expansion rate is assumed to be much less than the slug velocity ($\dot{L}_0 \ll U_s$) and dominated by the ascent-related reduction in static pressure (i.e. $\dot{P} \approx \rho g \dot{h}$). Differentiating $PV^\gamma =$ constant then allows an initial rate of expansion to be estimated as $\dot{L}_0 \approx \rho g U_s L_0/(\gamma P_0)$ and using this non-zero value minimizes initial oscillations of the slug at the start of simulations.

Slug ascent velocity

For the flow regimes of interest, both inertial and viscous forces are important and defining an appropriate value for U_s is non-trivial. Existing work (White & Beardmore 1962; Wallis 1969) provides dimensionless relationships for slug nose velocity which are valid for conditions of no gas expansion (and hence also represent slug base velocities under the same conditions). Given that in our laboratory experiments, slug base velocities were not detected to change with expansion (and changed negligibly with surface pressure), we can thus utilize these existing determinations of nose velocity as a slug base velocity in our expanding-slugs model.

A general approach is provided by Wallis (1969) in which ascent velocity for a buoyant phase rising through stagnant liquid in a vertical pipe is given in terms of the dimensionless Froude number, Fr, a ratio of inertial to buoyancy forces. In the case of a gas slug with negligible density with respect to the surrounding liquid

$$U_s = Fr\sqrt{gD} \tag{8}$$

where D is the internal pipe diameter. Other dimensionless numbers are defined to characterize parameter space into regions of different control (Wallis 1969); the Eötvös number relating gravitational and surface forces, $Eo = \rho g D^2/\sigma$, and the Morton number, relating viscous and surface forces, $Mo = g\mu^4/(\rho\sigma^3)$, with the ratio $(Eo^3/Mo)^{1/4}$ also being used as a dimensionless inverse viscosity, N_f (Fabre & Liné 1992). If surface tension effects are negligible ($Eo > 100$, applicable for both the experiments and for volcanic scenarios) and $N_f > 300$, slugs are under inertial control and $Fr = 0.345$. In contrast, for $Eo > 100$ and $N_f < 2$, slugs are under viscous control, and $Fr = 0.01 N_f$ (Wallis 1969). Over the intervening region of mixed viscous and inertial effects

$$Fr = 0.345(1 - e^{-N_f/34.5}). \tag{9}$$

For the experiments described here (Table 1), $N_f = 91$, $Eo = 179$, putting slugs in this region of mixed control with negligible surface tension effects. Applying equations 8 and 9 to the experimental conditions gives a slug ascent velocity of

0.162 m s^{-1}. This is only slightly greater than a graphical estimate made from the Eo-Fr-Mo plot of White and Beardmore (1962) which gives $Fr = 0.305$, corresponding to $U_s = 0.154$ m s^{-1} (identical within measurement error to the value measured from the video data). Hence, although the measured values can be used for modelling the laboratory experiments, equation 8 is used for determining slug base velocity for volcanic scenarios.

Slug radius and falling film thickness

The other fluid-related parameter requiring a calculated value for the 1D model is the slug radius, r_s. For long slugs, the falling film surrounding the slug approaches a constant thickness, δ_∞, with increasing distance from the slug nose (Brown 1965). At this thickness, the liquid velocities within the film are steady and the thin film is fully supported (i.e. is no longer accelerating) by the viscous shear stress on the tube wall. Batchelor (1967) showed that, in this case, and for films sufficiently thin that their curvature round the slug can be neglected ($\delta_\infty \ll r_c$), conservation of mass can be used to approximate the falling film thickness as

$$\delta_\infty = \sqrt[3]{\frac{3\mu r_c U_s}{2\rho g}}. \tag{10}$$

This was used by Vergniolle (1998), along with $U_l = 0.345(gD)^{1/2}$ for inertially controlled slugs. With the U_s values calculated above, equation 10 provides a range of values for the laboratory experiments of 3.5 to 3.6 mm.

An alternative was used by Seyfried and Freundt (2000), who employed an expression for the equilibrium film thickness derived by Brown (1965) in which

$$\delta_\infty = \frac{-1 + \sqrt{1 + 2Nr_c}}{N} \tag{11}$$

where

$$N = \sqrt[3]{14.5\frac{\rho^2 g}{\mu^2}} \tag{12}$$

and typographical errors in Seyfried and Freundt (2000, their equations 5 and 6) are corrected here. Brown (1965) gives a surface-tension-derived limit for the application of this of $Eo > 5(1 - \delta_\infty/r_c)^{-2}$ and, from comparison with experimental data, a viscous limit of $N > 30/r_c$, below which it was assumed that the potential flow was being significantly altered by viscous effects. For the experiments, this approach gives $\delta_\infty = 3.2$ mm. This is valid under the surface tension criterion (the right-hand side of equation 10 evaluates to 8.8 and $Eo = 179$), but N is calculated to be 1890 and the viscous limit of $30/r_c$ evaluates to 2300, so strictly this exceeds, although is close to, the limit of applicability.

For the experiments, the effective thin film thickness can be calculated by applying constant liquid volume criteria (i.e. equation 1) to measurements of the slug length and the liquid surface position during the ascent of a slug. This approach provided a best fit value of $r_s = 9.5$ mm, i.e. $\delta_\infty = 3.5$ mm, for experiments carried out at atmospheric pressure in which slug nose shape changes are minimized. The models thus provide a good estimate of this, with equation 10, evaluated using the measured slug ascent velocity, being the closest. Consequently, we retain equation 10 for calculation of slug radii in volcanological scenarios.

3D CFD model

3D computational fluid dynamics simulations were carried out using FLOW-3D® (release 9.2, http://www.flow3d.com), a general purpose finite difference (finite volume) CFD program. FLOW-3D® is based on solving the Navier-Stokes equations over a specified Cartesian grid and specializes in free-surface flows.

Given the strong contrast in density and viscosity between the liquid and the gas phase, the gas phase can be modelled as a 'void' region, governed by $PV^\gamma = $ constant. For such void regions, internal fluid flow, and consequently also shear stresses at boundaries, is not simulated. Thus, in contrast to analytical approaches, there is no constant stress condition across the interface between the gas (void) and the liquid regions. Such free surface interfaces are tracked using the volume-of-fluid approach (Hirt & Nichols 1981) in which the fluid volume fraction for each cell is stored. Free surfaces are then indicated by cells partially filled with fluid (or between full and empty cells) and their slopes and curvatures can be computed from the fluid volume fractions in neighbouring cells. The interface positions are controlled by the gas and liquid static pressures, liquid dynamics and surface tension. Although surface tension plays little role in slug ascent within the experiments (for important surface tension effects, $Eo < 100$ (Wallis 1969) but, for the experiments, $Eo = 179$) and could be neglected entirely at the volcano scale ($Eo = 10^6$, Table 1), it was included within the model to allow a comparison of slug ascent velocities over a wide region of

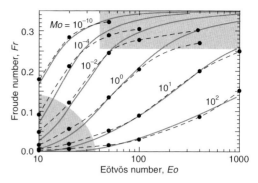

Fig. 3. Dimensionless slug ascent velocities. The grey lines represent curves of constant Morton number (labelled with the appropriate Mo values) in Eo–Fr space, redrawn from White & Beardmore (1962). The black symbols joined by dashed curves (purely to guide the reader), represent the results of 3D CFD simulations carried out with a liquid of density 1000 kg m^{-3}, surface tension of 0.1 N m^{-1}, with combinations of viscosity values of 1.79×10^{-3}, 56.5×10^{-3}, 0.179, 0.565, 1.79 and 5.65 Pa s and tube diameters of 0.0101, 0.0143, 0.0226, 0.0319, 0.0639 and 0.101 m. For each simulation, Fr was calculated from the resulting base ascent velocity of the slug. The grey-shaded areas are regions of parameter space for which the slug velocities are either consistently under- or overestimated, with the upper, high-Fr area representing the region of strong inertial control.

parameter space for model validation (Fig. 3, and see next section).

The liquid was represented as an incompressible Newtonian fluid and turbulence was accounted for with either a Prandtl mixing length model (a standard simplified parameterization in which turbulence is estimated from the local shear rate and a mixing length describing the distance over which velocity changes are equal to the mean turbulent fluctuation) or a renormalized group (RNG) model. Gas slugs have been successfully modelled previously with an RNG turbulence model (Taha & Cui 2006a) but in FLOW-3D®, the RNG option is only usable with explicit calculation of viscous stresses, which can considerably slow calculations. Using the Prandtl model generally decreased run times by allowing implicit (successive under-relaxation) viscous solution. A no-slip condition was applied at the liquid–solid boundary and heat exchange between the phases was neglected. The pressure–velocity equations are iterated by an implicit successive over-relaxation solver, which includes a self-corrective procedure (Hirt & Harlow 1967) to minimize accumulation of iteration errors. The time step, dt, convergence and stability criteria (which include maintaining dt smaller than

0.45 × the Courant limit) are automatically and dynamically adjusted throughout each simulation to optimize solution.

Although FLOW-3D® can work within a cylindrical coordinate system (which could have been utilized with the straightforward geometries used here), a Cartesian mesh was selected to enable more complex conduit geometries to be investigated comparably in the future. In order to reduce the total mesh size, and hence minimize computational time, axial symmetry was utilized, with quarter-tube simulations being defined with the Z axis along the conduit axis, and symmetry boundary conditions prescribed at the fluid or void-filled mesh boundaries ($X=0$ and $Y=0$). All force results presented have been subsequently rescaled back to full-tube conditions. For most of the scenarios of interest, a cubic mesh with 8×8 cells in $X–Y$ produced results very similar to those of more densely meshed simulations, so this relatively coarse meshing was used to minimize run times. However, for some simulations in which surface tension was important or conduit shear forces needed to be accurately resolved, a denser mesh of 14×14 cells in $X–Y$ (and a corresponding increase in Z) was used.

In the simulations carried out, the starting conditions consisted of a vertical, cylindrical tube, closed at the base and open at the top, containing liquid to a specified height and with an initial bubble represented as a smaller, concentric cylindrical void region at the base of the tube, as indicated in Figure 1b for the 1D model. The closed base faithfully models the laboratory experiments but is clearly less directly relevant within a volcanic scenario. However, its presence does not alter the flow after stable slug ascent has been established and it provides a convenient mechanism for calculating vertical pressure forces. In a volcanic scenario, these forces would be coupled to the surrounding rock at any non-vertical region of the conduit (James *et al.* 2006*b*), so the model can be viewed as providing the magnitude, but not the location, of the vertical component of pressure forces exerted on the conduit.

Simulations were run on a desktop PC (a 300 MHz Pentium 4, with 2 Gb of ram) and generally took a few hours to complete, with larger, denser simulations taking up to several days. Forces and pressures at multiple points within the system were output at defined time intervals (normally every 0.1 s) throughout the simulations.

3D CFD slug velocities

The 3D CFD simulations produced slug shapes (e.g. domed nose and falling film thicknesses,

Fig. 1c) that were entirely consistent with those observed in the laboratory (Fig. 2). In particular, the topography of the slug base provides a sensitive indicator of *Fr*, and this was well simulated by the 3D CFD model. In order to more quantitatively investigate the accuracy of the simulations, slug ascent velocities over different regions of parameter space were compared with previously published values. To reduce simulation time, only relatively small lengths of tube were simulated (sufficient only to establish a stable slug velocity) with most quarter-tube simulations being between 173 and 400 cells in Z, depending on the anticipated ascent velocity. For $Mo < 10^{-2}$ and low values of *Eo* (indicating an increasing importance of surface forces with respect to viscous forces), surface tension was resolved better, and hence velocity accuracies were increased, by using 14×14 cells in $X–Y$, with appropriate decreases in cell size in Z and corresponding increases in run time. Slug expansion was unwanted, so full atmospheric pressure was prescribed (10^5 Pa) at the liquid surface. Nevertheless, slug velocities were ascertained from the slug base to avoid any remaining expansion-related velocity increases that the slug nose velocity would have been susceptible to.

In Figure 3, the results are compared with the benchmark *Fr-Eo-Mo* dimensionless graphical correlations determined by White & Beardmore (1962) from multiple laboratory datasets (grey curves). It can be seen that for most Froude numbers <0.25, the ascent velocities calculated reproduce the previously published laboratory data well. However, velocities are underestimated for larger Froude numbers, an issue that was confirmed by consultation with the manufacturer of FLOW-3D® (but note that constant slug base velocities emerged from the CFD model, in line with the experimental observation). Consequently, slug ascent under conditions of strong inertial control is not reproduced as accurately. At the lowest Eötvös and Froude numbers (strong viscous control), velocities are somewhat overestimated, but this region of parameter space is far from the laboratory or volcanological scenarios considered in this work.

It is worth noting that Taha & Cui (2006*a*) demonstrated a CFD model capable of fully reproducing the White & Beardmore (1962) results. However, such simulations (carried out with FLUENT®) used a relatively small translating computational domain, moving so that it was stationary with respect to the slug. This efficient approach is not applicable here where the interest in slug expansion and the associated liquid motions requires the entire fluid-filled region to be modelled simultaneously.

Comparison of laboratory and model results

In Figure 4, the ascent of a gas slug is illustrated by laboratory data (symbols) and the two different models (lines) for a gas slug expanding into surface pressures of 1 Pa (a, b) and 319 Pa (c, d). The laboratory data were obtained from the digital video and illustrate the position of the liquid surface, the slug nose, and the slug base (the lowest point on the slug), as indicated in Figure 1a. During slug ascent, the data illustrate a constant base velocity and the rapid slug expansion as it approaches the surface.

For the 1D model, the base velocity of the gas slug was defined by the value obtained from the laboratory video data and, for the majority of the ascent, the model can reproduce the slug nose and liquid surface positions well. The model was fitted (by eye) by varying γ (as a proxy for variable degrees of heat transfer) and the mass of gas injected (i.e. changing L_0). In Figure 4, the results are shown using gas masses which best fit the data when $\gamma = 1.0$, representative of isothermal gas expansion. Although the masses of gas involved are very small (c. 1 mg) and the surface to volume ratios of the slugs are very high (potentially allowing heat to be absorbed rapidly), expansion is unlikely to be completely without temperature change. However, there is a trade-off between the values of γ and L_0 such that the result of increasing γ can be somewhat offset by also increasing L_0. The choice of $\gamma = 1.0$ was made in this case because of its real physical meaning and the fact that the gas

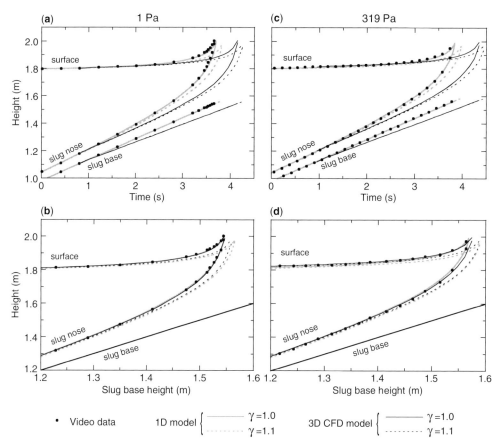

Fig. 4. Plots of gas slug ascent video data (symbols) with 3D CFD (black curves) and 1D (grey curves) model results overlain, for experiments carried out with surface pressures of 1 Pa (**a, b**) and 319 Pa (**c, d**). The solid curves give the best fit results for isothermal conditions (i.e. $\gamma = 1.0$), obtained by varying the slug gas mass. For the parameters appropriate to the laboratory model, the 3D model underestimates the ascent velocity by c. 14% (a, c) but plotting the same data against slug base height rather than time (b, d), demonstrates that the slug expansions are well reproduced. The dashed curves show results for simulations carried out using $\gamma = 1.1$ in order to demonstrate the sensitivity of the results to variations in the degree of heat exchange.

Table 2. *Parameter values for the experiments and models in Figure 4*

	Laboratory experiment	1D model	3D CFD model
Slug radius, r_s (cm)	*0.95*	0.95	*0.987*
Slug base velocity, U_s (m s^{-1})	*0.151*	0.151	*0.131*
	0.154	0.154	*0.131*
Effective gas mass* (mg)	*1.1*[†]	1.33	1.33
	1.1[†]	1.41	1.45

Figures in upright type denote predetermined constant values, determined either from the experimental apparatus or as inputs to the models. Figures in italics represent measurements or model outputs. Where two figures are given the upper one is for a surface pressure of 1 Pa and the lower one for a surface pressure of 319 Pa.
*Both models assume negligible gas densities and consider only gas volume and pressure. For comparative purposes, volumes have been converted to masses using the parameters given in Table 1.
[†]Minimum values calculated from injection volumes and assuming hydrostatic pressure only.

masses implied by the model were larger than those directly estimated for the experiments, from the volumes injected with the syringe (Table 2). Note that these estimated injected masses will be minimum values (a one-way valve on the injection mechanism means that the injection pressures, and hence masses, will have been slightly larger than calculated, assuming static pressure alone), so some cooling is indeed likely but attempts to fit the data assuming adiabatic expansion (i.e. using $\gamma = 1.4$, Table 1) produced significantly poorer results. To illustrate the sensitivity of the model to the degree of heat transfer, the same scenario was also run with $\gamma = 1.1$ to represent a nearly isothermal system, with the results shown by the dashed grey curves in Figure 4.

The 3D CFD model was fitted to the data (again, by eye) by assuming isothermal conditions (for direct comparison with the 1D results) and by varying the initial gas volume (i.e. L_0). The most obvious result is that the slug ascent velocity (as measured at the slug base and which is independent to the relatively small changes in L_0) is underestimated by c. 14% (Fig. 4a, c) consistent with the results shown in Figure 3. Although this produces a very poor fit to the data when plotted against time, plotting the data against slug base position (Fig. 4b, d) shows that the expansion of the slug is reproduced with some accuracy. The similarities between the 3D CFD results and those from the 1D model, including the estimates of gas mass involved (Table 2), provide mutual support between these different approaches.

Variations with gas mass

The 1D model has a simplicity and speed that makes it suitable for easy investigation of parameter space and, in Figure 5, ascent profiles for a range of gas slug sizes (Fig. 5a) and surface pressures (Fig. 5b) are compared with laboratory experiments. For each run, the model was fitted by

varying L_0 as previously, and can reproduce the observed expansions well. In Figure 5c, similar ascent profiles determined for volcanic-scale slugs of different gas mass are illustrated.

A further advantage of the 1D model (over the CFD approach) is that the solver allows the definition of event criteria, at which a solution is specifically calculated, allowing run results to be accurately compared at identical points of the modelled process. Output from the 3D CFD model occurs at a defined time interval, so comparisons between model runs are dependent on the relative timing of the process with respect to the output intervals and can thus appear artificially noisy if single points (e.g. maxima or minima) of rapidly changing parameters are of interest.

Here, for a comparative point in the slug ascent process, we define a 'slug burst' criterion based on the thickness of the remaining liquid over the slug. For the volcanological scenario, a thickness of 0.1 m of magma is used. This is a reasonable value based on clast sizes ejected in Strombolian eruptions (e.g. Chouet *et al.* 1974) and of the order of magnitude (i.e. >0.01 m and <1 m) appropriate for draining basalt films. However, it is well within the region of interaction between the velocity field around the slug nose and the liquid surface ($h \approx 2r_c$), where the 1D model assumptions are likely to be poor representations of the strongly 2D flow. Nevertheless, the use of a consistent point of evaluation will allow relative comparison of runs and absolute accuracies can be assessed by observation of the sensitivity of the results to parameters.

In Figure 6, slug overpressures, slug length, surface height changes and slug nose velocities evaluated for this 'burst' point are given for slugs of different gas masses, ascending in different radii conduits. Although at depth, overpressure is defined as $P - \rho g h - P_{surf}$, at the burst point, in order to reflect surface measurable effects, overpressure is given by $P - P_{surf}$. All grey shaded regions

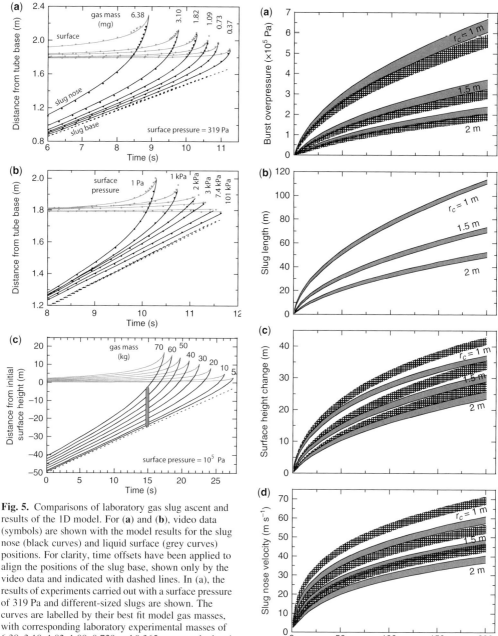

Fig. 5. Comparisons of laboratory gas slug ascent and results of the 1D model. For (**a**) and (**b**), video data (symbols) are shown with the model results for the slug nose (black curves) and liquid surface (grey curves) positions. For clarity, time offsets have been applied to align the positions of the slug base, shown only by the video data and indicated with dashed lines. In (a), the results of experiments carried out with a surface pressure of 319 Pa and different-sized slugs are shown. The curves are labelled by their best fit model gas masses, with corresponding laboratory experimental masses of 6.38, 3.10, 1.82, 1.09, 0.729 and 0.365 mg, as calculated from nominal injection volumes of 35, 17, 10, 6, 4 and 2 ml respectively. Note that video data are not available for heights >2.18 m above the tube base. In (b), similar results are given for a suite of experiments carried out with different surface pressures using an injected volume of 6 ml (*c.* 1 mg) for all pressures except 101 kPa, in which 10 ml (*c.* 14 mg) was used. In (c), slug ascent curves are given for different equivalent gas masses under volcanological conditions (Table 1). As an illustration, the position and length of a 70 kg slug is given schematically by the grey shaded region at 15 s.

Fig. 6. Changes in slug characteristics at the time of burst with varying gas mass (as given by the 1D model, for volcanological conditions). The grey shaded areas show the results calculated for the labelled conduit radius and degrees of heat transfer between full (isothermal conditions, $\gamma = 1.0$, upper black bounding curves) and none (adiabatic conditions, $\gamma = 1.1$, lower black bounding curves). For these simulations, r_s and U_s were calculated from equations 10 and 8 respectively.

represent simulations carried out with the viscosity and magma density of the standard volcano-scale parameters (Table 1, 500 Pa s and 2600 kg m^{-3} respectively), with the upper bound being isothermal conditions (i.e. $\gamma = 1.0$) and the lower bound, adiabatic conditions. In Figure 6a and 6c, the effects of reducing the viscosity to 300 Pa s (a value suggested to be appropriate for Stromboli (Vergniolle et al. 1996)) are given by the hatched regions. Note that this decrease in viscosity has negligible effect on slug length (so has not been plotted in Fig. 6b) and, counter-intuitively, decreases the calculated overpressure at burst.

Given that the burst criterion of 0.1 m is poorly constrained and previous work (Vergniolle & Brandeis 1996) has suggested that even smaller values could be appropriate, based on the smaller (centimetric) clast sizes imaged by Chouet et al. (1974) during Strombolian eruptions, simulations were also carried out using a 'burst criterion' of $h = 0.01$ m. The results only showed significant differences from the previous runs for the calculated slug nose velocities (given by the hatched regions in Fig. 6d and reflecting the large accelerations occurring just prior to burst), but not for the other parameters plotted. Thus, although the poor constraints on an appropriate h value to define slug burst are recognized, most characteristics appear relatively independent of this variable.

Variations with conduit radius and starting conditions

The different curves in Figure 6 indicate that conduit radius has a strong influence on burst conditions. In Figure 7, variations of ascent time (for the slug nose ascending through 50 m of magma), overpressure at burst, liquid surface height change and slug nose velocity are shown for slugs of gas masses 50, 100 and 150 kg in conduits of radius 0.5 to 2 m. Both ascent time and overpressure (Fig. 7a) are shown to increase significantly for conduit radii <1 m, whereas nose velocity and

surface height changes (Fig. 7b) demonstrate much more linear increases.

In Figure 8, the relative independence of burst overpressures to the initial bubble conditions (depth and any starting overpressure) are shown. For the volcanological conditions given in Table 1, an initial depth <30 m is required in order to significantly reduce burst pressures, indicating that it is the final c. 30 m of ascent in which most of the dynamic pressurization must occur. In Figure 8b, this is further demonstrated by the fact that for sufficient start depths (here, modelled at 100 m, and for isothermal systems, i.e. $\gamma = 1.0$) burst pressures do not reflect the starting pressure. Initial overpressures will cause oscillations of the slug (Vergniolle 1998), but will be effectively dissipated by the time the slug reaches the surface. Figure 8 also shows the adiabatic case ($\gamma = 1.1$, Table 1) to illustrate the sensitivity of the results to variations in heat transfer.

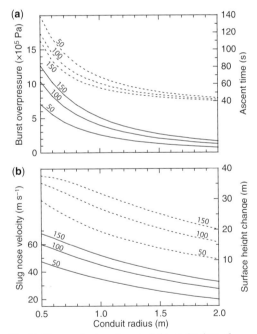

Fig. 6. (*Continued*) Slugs were deemed to burst when the fluid depth above the slug, h, was 0.1 m (with the slug nose velocities given being the average velocity between $h = 0.2$ m and 0.1 m). Further simulations were carried out by individually changing (i) magma viscosity to 300 Pa s and (ii) by using a burst condition of $h = 0.01$ m. Where the results showed significant deviation from the original solutions, they are shown by hatched regions. Decreasing viscosity produced significant changes only in overpressure (**a**), and surface height (**c**), and changing the burst criterion produced significant change only to the final slug nose velocity (**d**), now calculated between $h = 0.1$ and 0.01 m.

Fig. 7. Changes in slug characteristics at the time of burst with varying conduit radius (as given by the 1D model for isothermal volcanological conditions). Characteristic parameters have been evaluated for slugs of three different gas masses, which label the curves (in kg). In (**a**), the overpressure at burst (solid curves) and ascent time (dashed curves) are shown to be strong functions of conduit radius at small radius values. In (**b**), slug nose velocity (solid curves) and the surface height change (dashed curves) are shown to vary more linearly with conduit radius.

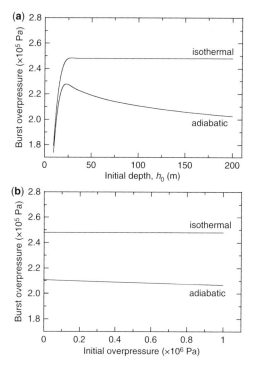

Fig. 8. Burst overpressures as functions of the initial depth of the slug (**a**) and the initial overpressure (**b**), calculated by the 1D model for volcanological conditions. Curves are given for both isothermal ($\gamma = 1.0$) and adiabatic ($\gamma = 1.1$) conditions to demonstrate the relative effect of heat transfer; all other parameters as given in Table 1.

Conduit force components

Although the 1D model reproduces the observed slug expansions well (as indicated by the comparisons with experimental data, Figs 4–5) and can be used for relatively straightforward investigations of parameter space, it cannot be used to characterize pressures and forces on the conduit because the dynamics of the liquid flow around the slug are not calculated. In contrast, the 3D models allow calculation of full moment and single force components at any position. Note that here, the conduit is assumed to be effectively rigid, so solid deformations are not calculated. This is in contrast to analytical approaches which have been employed to look at deeper seismic events such as tremor, in which, with only a single incompressible fluid phase, oscillations are stimulated through deformation of the conduit walls (Julian 1994; Balmforth *et al.* 2005). The use of a rigid conduit is thought to be a reasonable assumption for near-surface two-phase scenarios, given that the volume and relative compressibility of the gas determine that liquid flow

due to pressure-related gas volume changes would significantly outweigh those associated with conduit volume changes.

In Figure 9a, the pressure and force transients generated by the ascent and burst of a slug within the laboratory experiments are illustrated. The pressure trace (recorded at the base of the tube) shows the effect of increasing slug length (as increasing pressure drop), slug burst (as high-frequency transient) and drainage of the liquid film surrounding the burst slug (as a slow pressure recovery). The net force on the apparatus, as illustrated by apparatus displacement, shows no discernible effect until a sudden upward transient at slug burst, followed by a damped oscillation of the apparatus. The details of these data are expanded in Figure 9b, with model results added. To enable direct graphical comparison, the model-time has been scaled to account for the underestimation of the ascent velocity by the 3D CFD model (i.e. model times have been multiplied by the ratio of modelled to observed slug velocities, 0.131/0.154). Furthermore, only the last *c.* 0.6 m of slug ascent was modelled (rather than the full *c.* 1.7 m of the experiment) so the model time was offset to coincide the slug burst points.

The plot shows that the pressure drop below the slug is reproduced accurately in both form and magnitude by the model. After this rapid pressure decrease, oscillations are shown in the model-pressure that appear to be similar to those in the experimental data. In the experiments, the pressure oscillations are induced by the vertical resonant oscillation of the entire apparatus on the spring and force transducer but, as conduit motion is not included in the 3D CFD model, this cannot be the case for the simulation. No physical process has been attributable to the model oscillations and it is thus assumed that they are numerical artefacts triggered by the instantaneous nature of the bubble depressurization on burst (when the overlying liquid membrane first ruptures), produced because internal flow of the gas phase was not simulated.

Although the model does not include apparatus displacement, the net vertical force calculated over the domain (Fig. 9b, bottom panel) is similar in form to the observed displacement, with a single main upward transient. The rapid nature of the apparatus transient (which appears to be faster than the resonant frequency of the apparatus) prevents accurate determination of the force involved, but the data suggest that the magnitude is *c.* 1 N, which is not deemed to be significantly greater than the modelled 0.3 to 0.4 N.

In Figure 10, the vertical forces acting on a conduit are shown for the standard volcanological scenario (Table 1). Figure 10a gives a time–height representation of the development of the vertical

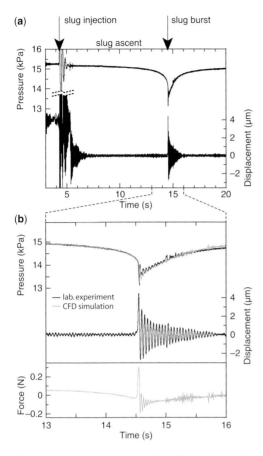

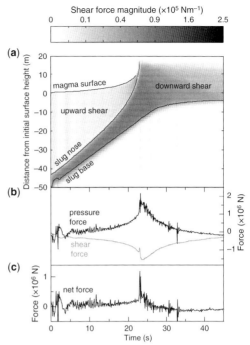

Fig. 9. Pressure changes and vertical displacements of the apparatus during laboratory experiments. (**a**) Fluid pressure at the base of the apparatus (upper curve) and vertical apparatus displacement (lower curve) recorded during the ascent and burst of a gas slug (mass c. 1 mg) into a 319 Pa atmosphere. The data show the initial pressure transients associated with injection of the gas slug and the large displacement offset (at 5 to 6 s) produced when the injection syringe was replaced on the apparatus. After these initial transients cease, the fluid pressure below the slug decreases as the slug expands, reflecting the increasing volume of liquid within the falling film surrounding the slug, effectively being supported by viscous shear forces on the tube walls. No net displacement of the apparatus is detected until a rapid upward transient, starting within 0.1 s of slug burst, excites a damped oscillation of the apparatus. In (**b**), data from around the time of the slug burst are enlarged to show further detail and for comparison with 3D CFD model results (grey curves). To facilitate direct comparison, model-time has been rescaled and offset to coincide the times of slug burst (see text for more details). The 3D CFD model does not calculate apparatus displacements so, in the lowest panel, the calculated net vertical force on the computational domain is plotted as a proxy. The single upward force transient agrees well with the vertical apparatus displacement recorded.

Fig. 10. Conduit forces modelled for the volcanic scenario. A gas slug equivalent to 40 kg gaseous H_2O is initiated with the slug base at 50 m from the magma surface. In (**a**), the magnitude of the vertical shear (force per vertical metre of the conduit) acting on the conduit wall is plotted over the time period of the gas slug ascent (from 0 to c. 22 s), the slug burst, and subsequent onset of film drainage. The solid black curves show the vertical position of the liquid surface, the slug nose and the slug base, with the nose and surface data close to the time of burst omitted for clarity. Above the slug, the liquid motion exerts an upward shear on the conduit wall, and surrounding and just below the slug, the conduit experiences a downward shear. In (**b**), the sum of these shear forces on the conduit are plotted with time and the change in the pressure force exerted at the (notionally closed) base of the conduit. In (**c**), the sum of these components, and thus the net vertical force over the entire conduit is plotted, showing the same upward transient as observed in the laboratory models. In this example, the transient magnitude is c. 10^6 N.

shear forces acting on the conduit during the slug ascent. Above the slug nose, liquid being accelerated upward by the slug expansion induces upward shear on the conduit, which significantly increases in magnitude within the region of rapid slug expansion near the surface (c. 22 s). In the region surrounding the slug, the falling liquid film produces downward shear on the conduit. The implication of this is that, in the vertical direction, the conduit experiences the passage of the slug as an ascending region of dilation (at the slug nose), followed by a region of compression at the slug base.

In Figure 10b, the sum of the vertical shear force components is provided at each time step (grey curve), along with changes in the pressure force experienced at the base of the conduit (black curve). Initial transients excited as the slug starts to ascend can be seen in both curves and numerical noise is notably higher amplitude in the pressure force trace (calculated from significantly fewer cells than the net shear force). As the slug expands and more liquid is viscously supported on the conduit wall, the pressure below the slug decreases (inducing the upward pressure force illustrated).

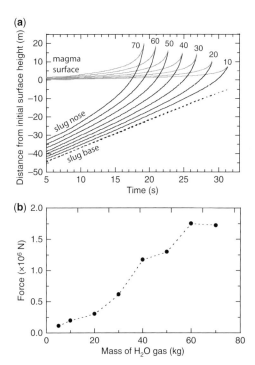

Fig. 11. (a) Slug ascent profiles calculated by 3D CFD, labelled by equivalent gas mass, for the volcanic scenario (Table 1). The position of the liquid surface (grey curves), the slug nose (black curves) and the slug base (dashed lines) are shown. Simulations were initiated with slugs at the base of a 50-m depth of magma (as shown in Fig. 10), and results are shown for times after initial transients have been damped out. The individual simulations have been shifted (in time) to align the ascent of the slug base. Slug base velocities during the different simulations are sufficiently similar that the paths overlay on the plot but there is a small but consistent increase with gas mass, from 1.49 to 1.53 m s^{-1} for slugs of 10 and 70 kg respectively. In **(b)**, the magnitude of the upward force transient (experienced by the conduit) calculated at the burst of these slugs is shown against equivalent gas mass.

During the bulk of the ascent, this is almost fully compensated for by the shear, as illustrated by the zero net force in Figure 10c. However, just prior to burst, an upward force transient is observed, just as in the simulations of the laboratory experiment (Fig. 9b), but, at the volcano-scale, the force magnitude is *c.* 10^6 N. Note that this is strongly related to the intense upward shear on the conduit imparted by the liquid at and ahead of the slug, rather than a straightforward pressure-loss effect occurring at burst. An implication of this is that reduction or damping of these forces due to magma compressibility and any deformation of the lower conduit is likely to be negligible, or at least a second order effect.

Carrying out simulations for varying gas masses allows assessment of the force magnitude change. In Figure 11, ascent curves produced from the 3D CFD model for slugs of different gas mass are illustrated, and the magnitude of the force transients associated with the slug burst are shown (Fig. 11b). Due to the rapid nature of the transient, the precise values (but not the general magnitude or the trend) of the forces determined vary with the output frequency of the model. With forces output at 0.1 s intervals, forces are smaller but less noisy than if obtained at 0.01 s. In Figure 11b, a compromise of 0.01-s-data, subsequently averaged by a seven-point sliding window, was used.

Discussion

The 1D and 3D CFD models presented have proved complementary and capable of reproducing the fundamental aspects of slug ascent and expansion under laboratory conditions. Given that the 1D model contains no end-effects relating to the interaction of the liquid velocity field with the liquid surface, near-surface expansions are simulated surprisingly well (Fig. 4). Extending the model to include end effects would involve significant complications (and there are no appropriate parameterizations to describe changes in the slug nose shape), so are deemed unnecessary at this stage. The 3D CFD model has proved invaluable in allowing accurate evaluation of the conduit force and pressure changes involved, although issues remain concerning the velocity underestimation at high Froude numbers. Simulations involving a non-Newtonian fluid are a future extension of work in this area and would build on similar slug research within the engineering field (Carew *et al.* 1995; Welsh *et al.* 1999; Xie *et al.* 2003; Sousa *et al.* 2004, 2005).

Ascent and expansion

One of the potential observables at volcanoes is the position and motion of the fluid surface during slug ascent, which can be calculated from both models (Figs 4, 5 and 11). As a slug ascends and expands, the liquid surface rises at a rate which reflects the volumetric expansion of the gas. Due to the dynamic pressurization of the slug only becoming significant in the upper few tens of metres, at depths greater than these expansion is dominated by the reducing static load. Consequently, initial liquid surface motions reflect the slug position, ascent velocity (a function of conduit geometry and liquid parameters) and the gas mass. If slugs start to rise at depths sufficient to establish relatively steady ascent, then the final burst conditions are also strong functions of gas mass, as shown by Figure 6.

Thus, detailed records of magma surface motion could be used to ascertain gas masses, ascent velocities and rheological parameters by constraining ascent models. Note that the greatest surface velocities only occur in the last few seconds, so measurements would have to be relatively frequent. Such liquid surface velocity data have been obtained by Doppler radar at Erebus (Gerst et al. 2007, 2008) for slugs busting at the surface of a lava lake. Although the flared nature of the conduit (i.e. rapid near-surface opening into the lake) permits unobstructed observation of the burst, the straight cylindrical conduit geometry modelled here means that our results cannot be appropriately applied directly to the Erebus data. In cases where the magma surface cannot be observed, model estimates of the magnitude of surface movement (Figs 6c and 7b) can provide constraints for its position, with applications for the interpretation of seismic and infrasonic data.

As slugs reach the surface, the expansion-related accelerations result in slug nose velocities that (in an external reference frame) are approximately an order of magnitude greater than the slug base velocities (Figs 6d and 7b). In the reference frame of the liquid above the slug, nose velocity also increases and, depending on the overall duration of slug ascent, this could significantly reduce estimated ascent times (which are likely to naturally reflect nose rather than base velocities). For example, ascending through the last 10 m of magma in our volcano-scale scenario, the 1D model indicates an average slug nose velocity of 4.4 m s^{-1} (compared to the slug base velocity of 1.7 m s^{-1}). A similarly elevated apparent slug velocity of 3.7 m s^{-1} was interpreted by Ripepe et al.(2001) from analysis of image, seismic and infrasonic data from Stromboli. Such velocities were difficult to reconcile with calculated 'standard' slug velocities of 1.8 m s^{-1} (Ripepe et al. 2001), but are in line with the nose velocity increases calculated here. This supports the importance of this aspect of our 1D model and indicates that expansion should be accounted for when combining

such near-surface geophysical data (Harris & Ripepe 2007).

At the time just prior to burst, magma in the liquid film above the slug nose will be also be travelling upward, but at a lower velocity than the slug nose due to the increased downward liquid flux into the falling liquid film around the slug. For nose velocities much greater than slug base velocity, the liquid surface velocity can be approximated by the product of the nose velocity and the ratio of the slug and conduit cross-sectional areas (c. 0.5 for the volcano-scale scenario used). This suggests that magma in the liquid film can have velocities of several tens of metres per second, a considerable fraction of measured ejecta velocities which are typically up to c. 100 m s^{-1}, but can be significantly less (Harris & Ripepe 2007). Thus, final (but preburst) slug expansion velocities need to be accounted for when calculating slug parameters from measurements of ejecta velocity (Chouet et al. 1974; Blackburn et al. 1976; Ripepe et al. 1993; Hort et al. 2003; Dubosclard et al. 2004).

In the work described here, the emphasis has been on rapid slug expansion and the associated dynamics, but similar volumetric analyses could be equally applied to the much more gradual changes during gas pistoning events (e.g. such as those documented at Hawaii; Swanson et al. 1979; Tilling 1987; Johnson et al. 2005). In descriptions of gas pistoning, it is often unclear as to whether the observed magma levels reflect the ascent of many small bubbles (possibly to be trapped by a geometric or rheological boundary, in a 'gas accumulation' phase) or the relatively slow ascent of a large slug-bubble. Further insight could be gained by comparing surface position data with appropriate slug ascent models.

For accurate time-based comparisons between simulations and field data, reliable slug rise velocities are critical. In the 1D model, velocities are determined using tried-and-tested engineering parameterizations (Wallis 1969), but this restricts accurate comparisons to idealized scenarios of vertical cylindrical conduits. In contrast, although the 3D CFD model provides the flexibility to vary conduit geometry, its performance in reproducing slug velocities for high Froude numbers (Fr > 0.25) was disappointing. However, for the generic volcanic scenario used (which includes $\mu = 500$ Pa s, giving $Fr = 0.32$), Froude numbers are less than 0.25 if liquid viscosity is >950 Pa s. This is well within the range of viscosities which may be relevant at many appropriate volcanoes, for example, viscosity estimates at Erebus are between 1 and 10 kPa s (Dibble et al. 1984) and, at Stromboli, geochemical investigations indicate 14 kPa s (Metrich et al. 2001) and 100 kPa has been suggested if the presence of small bubbles in the magma is accounted for (Ripepe & Gordeev 1999). These greater viscosities would imply ascent velocities one or two orders of magnitude less than those that are commonly accepted and consistent with seismics, infrasonic data and general observations of the dynamic nature of the system (*c.* 2 m s^{-1}). Consequently, this viscosity range and the resulting inconsistencies underscore the requirement for an integrated approach to improve our current understanding.

Burst overpressures

Slug burst overpressures have been previously correlated with initial slug depth (Ripepe *et al.* 2001), the duration of slug ascent (Lautze & Houghton 2007) and gas flux (Marchetti *et al.* 2006), with initial slug overpressure (Vergniolle 1998) often also regarded as having a role. In our models, all slugs are initiated at depths sufficient to allow stable slug flow to establish itself and consequently, initial depth has no significant effect on burst overpressure (Fig. 8a). Furthermore, initial overpressures were explicitly excluded by initializing slugs at the local static pressure (the sum of the surface pressure and the liquid static head). Nevertheless, the volcano-scale simulations indicate that at-surface excess slug pressures of *c.* 10^5 Pa are readily achievable (e.g. with slugs of *c.* 50 kg in a 1.5-m-radius conduit, Fig. 6a) purely as a result of dynamic pressurization during ascent. This is in good agreement with order of magnitude estimates made at Stromboli of between 10^4 Pa for the relatively passive burst of small (*c.* 0.5-m-sized) bubbles (Ripepe *et al.* 2001; Marchetti *et al.* 2006) to 10^5 Pa for Strombolian eruptions (Vergniolle & Brandeis 1996; Ripepe *et al.* 2001; Ripepe & Marchetti 2002; Vergniolle & Caplan-Auerbach 2006), mainly from infrasonic data.

The variation of overpressure with conduit radius shown in Figure 7 could also help explain infrasonic measurements made by Ripepe & Marchetti (2002) at Stromboli, in which smaller slugs appeared to be more overpressured than larger ones. Their data indicated that explosions occurring at the NE crater were eight times more overpressured, but half the volume of, those from the SW crater. Thus, converting the overpressure values given (4 × 10^5 and 0.5 × 10^5 Pa) into absolute pressures (5 × 10^5 and 1.5 × 10^5 Pa) suggests that the ratio of gas masses involved was *c.* 1.7. Figure 7a indicates that for a gas mass ratio of two (50 and 100 kg), an overpressure ratio of nearly five can be achieved by the smaller slug ascending a 0.5-m radius conduit and the larger one a 2-m radius conduit. We consider only ratios here, rather than absolute values because, unless slugs are bursting at the very top of the conduit

(which is not usually the case), then these infrasonic-derived overpressures will be minimum values, and post-burst gas expansion within the conduit would also need to be accounted for. The ability to determine gas mass directly from infrasonic data would allow degassing rates from these events to be monitored for comparison with more traditional gas flux measurements such as FTIR spectroscopy (e.g. Burton *et al.* 2003, 2007; Oppenheimer *et al.* 2006) and to allow continuous measurement through periods when other techniques cannot be used, for example in cloud or at night.

It is intuitive to anticipate that a dynamic, low viscosity system would produce greater overpressures than a more viscously controlled one (e.g. Lautze & Houghton 2007). However, this is not the case for the 1D model (Fig. 7a), which reflects the importance of viscous forces in retaining pressure within ascending slugs. Note that changing viscosity from 500 to 300 Pa s also had no discernible effect on final slug length (Fig. 7b) so, from a volumetric point of view, the variation in calculated burst overpressure with viscosity is not reflecting slug length changes, but variations in the thickness of the falling film around the slug.

Conduit forces

Although much slower to run than the 1D model, the 3D CFD model was required in order to quantify the forces and pressures exerted on the apparatus in the laboratory experiments. For volcano-scale simulations, the forces involved (*c.* 10^6 N) are several orders of magnitude lower than those interpreted from VLP seismic data from Stromboli (Chouet *et al.* 2003). Interestingly, this is in agreement with previous estimates based on first order scaling arguments (James *et al.* 2004; Lane *et al.* 2005) and supports the hypothesis that VLP signals are produced as a result of flow regime instability as slugs pass though a change in the conduit geometry at some depth (James *et al.* 2006*b*) rather than reflecting near-surface slug expansion or the burst process itself. Nevertheless, as instrumentation at volcanoes increases and improves, and seismic data analysis allows for moving and distributed source functions, seismics generated by slug expansion and burst may also play a part in estimating the gas masses involved.

Conclusions

The 1D model reproduced bulk expansions well and can be used to investigate relationships between ascent parameters and surface geophysical data acquired during slug bursts. This has highlighted the relationships between overpressure, gas mass, conduit radius and liquid surface motions. Allowing the velocity of the slug nose through the overlying liquid to vary during the ascent is a key difference to previous models, and is supported by laboratory and field measurements.

For the volcanic scenario used, the models have shown that a gas slug of c. 100 kg will be overpressured at the surface by c. 2×10^5 Pa. This is the result of dynamic expansion-related compression which dominantly occurs in the last c. 30 m of ascent and is effectively decoupled from the bubble pressure at greater depths. Close to the surface and just prior to burst, gas expansion will have accelerated the liquid above the slug nose to a velocity of c. 30 m s^{-1}, with the slug nose rising at an additional c. 4.4 m s^{-1} within it (greater than two-and-a-half times the stable slug base velocity).

Ultimately, computational fluid dynamic models will provide powerful tools, capable of incorporating most of the parameters measurable at volcanoes. In particular, they are the route to interpreting LP and VLP seismic source functions determined in terms of fluid flow processes. Although the scenarios modelled here were specifically chosen to represent the most straightforward cases, FLOW-3D® has the capabilities to simulate complex conduit geometries and liquid rheologies, to include thermal effects (including convection, radiative cooling and rheological variation) and could also be used to track geochemical mixing. Consequently, such CFD tools are well placed to form the backbone of integrated approaches but, as illustrated by the low slug velocities determined here, care needs to be taken to validate models appropriately.

MRJ was funded by the Royal Society and SBC by Lancaster University. J. Phillips and L. D'Auria are thanked for thorough reviews which significantly improved the manuscript. B. Chouet is also thanked for comments on an early version of the text. The CFD models were improved by various valuable contributions from A. Chandorkar, D. Souders (Flow Science Inc.) and K. Doyle (CFD Solutions Ltd.).

References

ARMIENTI, P., FRANCALANCI, L. & LANDI, P. 2007. Textural effects of steady state behaviour of the Stromboli feeding system. *Journal of Volcanology and Geothermal Research*, **160**, 86–98.

ASTER, R., MCINTOSH, W., KYLE, P. *ET AL.* 2004. Real-time data received from Mount Erebus volcano, Antarctica. *EOS Transactions of AGU*, **85**, 97–101.

AUGER, E., D'AURIA, L., MARTINI, M., CHOUET, B. & DAWSON, P. 2006. Real-time monitoring and massive inversion of source parameters of very long period seismic signals: An application to Stromboli Volcano, Italy. *Geophysical Research Letters*, **33**, L04301.

BALMFORTH, N. J., CRASTER, R. V. & RUST, A. C. 2005. Instability in flow through elastic conduits and volcanic tremor. *Journal of Fluid Mechanics*, **527**, 353–377.

BATCHELOR, G. K. 1967. *An Introduction to Fluid Dynamics*. Cambridge University Press, Cambridge.

BLACKBURN, E. A., WILSON, L. & SPARKS, R. S. J. 1976. Mechanics and dynamics of Strombolian activity. *Journal of the Geological Society London*, **132**, 429–440.

BROWN, R. A. S. 1965. The mechanics of large gas bubbles in tubes I. Bubble velocities in stagnant liquids. *The Canadian Journal of Chemical Engineering*, **43**, 217–230.

BURTON, M., ALLARD, P., MURÈ, F. & OPPENHEIMER, C. 2003. FTIR remote sensing of fractional magma degassing at Mount Etna, Sicily. *In*: OPPENHEIMER, C., PYLE, D. M. & BARCLAY, J. (eds) *Volcanic Degassing*. Geological Society, Special Publications, **213**, 281–293.

BURTON, M., ALLARD, P., MURE, F. & LA SPINA, A. 2007. Magmatic gas composition reveals the source depth of slug-driven Strombolian explosive activity. *Science*, **317**, 227–230.

CAREW, P. S., THOMAS, N. H. & JOHNSON, A. B. 1995. A physically based correlation for the effects of power law rheology and inclination on slug bubble rise velocity. *International Journal of Multiphase Flow*, **21**, 1091.

CLARKE, A. & ISSA, R. I. 1997. A numerical model of slug flow in vertical tubes. *Computers & Fluids*, **26**, 395–415.

CHOUET, B., HAMISEVICZ, N. & MCGETCHIN, T. R. 1974. Photoballistics of volcanic jet activity at Stromboli, Italy. *Journal of Geophysical Research*, **79**, 4961–4976.

CHOUET, B., DAWSON, P., OHMINATO, T. *ET AL.* 2003. Source mechanisms of explosions at Stromboli Volcano, Italy, determined from moment-tensor inversions of very-long-period data. *Journal of Geophysical Research-Solid Earth*, **108**.

D'AURIA, L. 2006. Numerical modelling of gas slugs rising in basaltic volcanic conduits: inferences on Very-Long-Period event generation. *International Workshop: The Physics of Fluid Oscillations in Volcanic Systems*, Lancaster, U.K. http://www.es.lancs.ac.uk/seismicflow/.

D'AURIA, L., CHOUET, B. & MARTINI, M. 2004. Lattice Boltzmann modeling of a gas slug rise in a volcanic conduit: inferences on the Stromboli dynamics. *IAVCEI 2004 General Assembly*, Pucon, Chile. Abstr. # s08b_pf_119.

DAVIES, R. M. & TAYLOR, G. 1950. The mechanics of large bubbles rising through extended liquids and through liquids in tubes. *Proceedings of The Royal Society of London Series A-Mathematical and Physical Sciences*, **200**, 375–390.

DIBBLE, R. R., KIENLE, J., KYLE, P. R. & SHIBUYA, K. 1984. Geophysical studies of Erebus Volcano, Antarctica, from 1974 December to 1982 January. *New Zealand Journal of Geology and Geophysics*, **27**, 425–455.

DORMAND, J. R. & PRINCE, P. J. 1980. A family of embedded Runge-Kutta formulae. *Journal of Computational and Applied Mathematics*, **6**, 19–26.

DUBOSCLARD, G., DONNADIEU, F., ALLARD, P. *ET AL.* 2004. Doppler radar sounding of volcanic eruption dynamics at Mount Etna. *Bulletin of Volcanology*, **66**, 443–456.

DUKLER, A. E. & HUBBARD, M. G. 1975. Model for gas-liquid slug flow in horizontal and near horizontal tubes. *Industrial & Engineering Chemistry Fundamentals*, **14**, 337–347.

FABRE, J. & LINÉ, A. 1992. Modelling of two-phase slug flow. *Annual Review of Fluid Mechanics*, **24**, 21–46.

GERST, A., HORT, M., JOHNSON, J. B. & KYLE, P. R. 2007. The first second of a Strombolian eruption: Doppler radar and infrasound observations at Erebus volcano, Antarctica. *Geophysical Research Abstracts*, **9**, 07280.

GERST, A., HORT, M., KYLE, P. R. & VOGE, M. 2008. 4D velocity of Strombolian eruptions and man-made explosions derived from multiple Doppler radar instruments. *Journal of Volcanology and Geothermal Research*, doi: 10.1016/j.jvolgeores.2008.05.022.

GREGORY, G. A. & SCOTT, D. S. 1969. Correlation of liquid slug velocity and frequency in horizontal cocurrent gas-liquid slug flow. *AIChE Journal*, **15**, 933–935.

HARRIS, A. & RIPEPE, M. 2007. Synergy of multiple geophysical approaches to unravel explosive eruption conduit and source dynamics – a case study from Stromboli. *Chemie Der Erde-Geochemistry*, **67**, 1–35.

HIRT, C. W. & HARLOW, F. H. 1967. A general corrective procedure for the numerical solution of initial-value problems. *Journal of Computational Physics*, **2**, 114–119.

HIRT, C. W. & NICHOLS, B. D. 1981. Volume of fluid (VOF) method for the dynamics of free boundaries. *Journal of Computational Physics*, **39**, 201–225.

HORT, M., SEYFRIED, R. & VOGE, M. 2003. Radar Doppler velocimetry of volcanic eruptions: theoretical considerations and quantitative documentation of changes in eruptive behaviour at Stromboli volcano, Italy. *Geophysical Journal International*, **154**, 515–532.

JAMES, M. R., LANE, S. J., CHOUET, B. & GILBERT, J. S. 2004. Pressure changes associated with the ascent and bursting of gas slugs in liquid-filled vertical and inclined conduits. *Journal of Volcanology and Geothermal Research*, **129**, 61–82.

JAMES, M. R., CORDER, S. B. & LANE, S. J. 2006*a*. Laboratory investigations of possible gas-slug related seismic source processes in low viscosity magmas. *International Workshop: The Physics of Fluid Oscillations in Volcanic Systems*, Lancaster, U.K. http://www.es.lancs.ac.uk/seismicflow/.

JAMES, M. R., LANE, S. J. & CHOUET, B. A. 2006*b*. Gas slug ascent through changes in conduit diameter: Laboratory insights into a volcano-seismic source process in low-viscosity magmas. *Journal of Geophysical Research-Solid Earth*, **111**, B05201, doi:05210.01029/02005JB003718.

JOHNSON, J. B., HARRIS, A. J. L. & HOBLITT, R. P. 2005. Thermal observations of gas pistoning at Kilauea Volcano. *Journal of Geophysical Research-Solid Earth*, **110**, B11201.

JULIAN, B. R. 1994. Volcanic tremor – nonlinear excitation by fluid-flow. *Journal of Geophysical Research-Solid Earth*, **99**, 11859–11877.

LAIRD, A. D. K. & CHISHOLM, D. 1956. Pressure and forces along cylindrical bubbles in a vertical tube. *Industrial and Engineering Chemistry*, **48**, 1361–1364.

LANE, S. J., JAMES, M. R. & CORDER, S. B. 2005. Experimental investigation of volcano-seismic forces generated during gas-slug expansion in Strombolian eruptions. *EOS Transactions of AGU*, **85**, Fall Meet. Suppl., Abstr. V31D–0652.

LAUTZE, N. C. & HOUGHTON, B. F. 2007. Linking variable explosion style and magma textures during 2002 at Stromboli volcano, Italy. *Bulletin of Volcanology*, **69**, 445–460.

LIGHTHILL, M. J. 1978. *Waves in Fluids*. Cambridge University Press, Cambridge.

MAO, Z. S. & DUKLER, A. E. 1990. The motion of Taylor bubbles in vertical tubes—I. A numerical simulation for the shape and the rise velocity of Taylor bubbles in stagnant and flowing liquids. *Journal of Computational Physics*, **91**, 132–160.

MAO, Z. S. & DUKLER, A. E. 1991. The motion of Taylor bubbles in vertical tubes: II. Experimental data and simulations for laminar and turbulent flow. *Chemical Engineering Science*, **46**, 2055–2064.

MARCHETTI, E., RIPEPE, M., ULIVIERI, G. & DELLE DONNE, D. 2006. Degassing dynamics at Stromboli volcano: Insights from infrasonic activity. *EOS Transactions of AGU*, **87**, Fall Meet. Suppl. Abstract. V43B–1796.

METRICH, N., BERTAGNINI, A., LANDI, P. & ROSI, M. 2001. Crystallization driven by decompression and water loss at Stromboli volcano (Aeolian Islands, Italy). *Journal of Petrology*, **42**, 1471–1490.

NICKLIN, D. J., WILKES, J. O. & DAVIDSON, J. F. 1962. Two phase flow in vertical tubes. *Transactions of the Institute of Chemical Engineers*, **40**, 61–68.

O'BRIEN, G. S. & BEAN, C. J. 2004. A discrete numerical method for modeling volcanic earthquake source mechanisms. *Journal of Geophysical Research*, **109**, B09301.

O'BRIEN, G. & BEAN, C. 2006. Seismicity generated by gas slug ascent: a numerical investigation. *International Workshop: The Physics of Fluid Oscillations in Volcanic Systems*, Lancaster, U.K. http://www.es.lancs.ac.uk/seismicflow/.

OPPENHEIMER, C., BANI, P., CALKINS, J. A., BURTON, M. R. & SAWYER, G. M. 2006. Rapid FTIR sensing of volcanic gases released by Strombolian explosions at Yasur volcano, Vanuatu. *Applied Physics B-Lasers and Optics*, **85**, 453–460.

POLACCI, M., CORSARO, R. A. & ANDRONICO, D. 2006. Coupled textural and compositional characterization of basaltic scoria: insights into the transition from Strombolian to fire fountain activity at Mount Etna, Italy. *Geology*, **34**, 201–204.

POLONSKY, S., SHEMER, L. & BARNEA, D. 1999. The relation between the Taylor bubble motion and the velocity field ahead of it. *International Journal of Multiphase Flow*, **25**, 957.

RIPEPE, M. & GORDEEV, E. 1999. Gas bubble dynamics model for shallow volcanic tremor at Stromboli. *Journal of Geophysical Research-Solid Earth*, **104**, 10639–10654.

RIPEPE, M. & MARCHETTI, E. 2002. Array tracking of infrasonic sources at Stromboli volcano. *Geophysical Research Letters*, **29**.

RIPEPE, M., ROSSI, M. & SACCOROTTI, G. 1993. Image-processing of explosive activity at Stromboli. *Journal of Volcanology and Geothermal Research*, **54**, 335–351.

RIPEPE, M., CILIBERTO, S. & DELLA SCHIAVA, M. 2001. Time constraints for modeling source dynamics of volcanic explosions at Stromboli. *Journal of Geophysical Research-Solid Earth*, **106**, 8713–8727.

SEYFRIED, R. & FREUNDT, A. 2000. Experiments on conduit flow and eruption behavior of basaltic volcanic eruptions. *Journal of Geophysical Research-Solid Earth*, **105**, 23727–23740.

SHAW, H. R. 1969. Rheology of basalt in the melting range. *Journal of Petrology*, **10**, 510–535.

SOUSA, R. G., NOGUEIRA, S., PINTO, A., RIETHMULLER, M. L. & CAMPOS, J. 2004. Flow in the negative wake of a Taylor bubble rising in viscoelastic carboxymethylcellulose solutions: particle image velocimetry measurements. *Journal of Fluid Mechanics*, **511**, 217–236.

SOUSA, R. G., RIETHMULLER, M. L., PINTO, A. & CAMPOS, J. 2005. Flow around individual Taylor bubbles rising in stagnant CMC solutions: PIV measurements. *Chemical Engineering Science*, **60**, 1859–1873.

SOUSA, R. G., PINTO, A. & CAMPOS, J. 2006. Effect of gas expansion on the velocity of a Taylor bubble: PIV measurements. *International Journal of Multiphase Flow*, **32**, 1182–1190.

SWANSON, D. A., DUFFIELD, W. A., JACKSON, D. B. & PETERSON, D. W. 1979. Chronological narrative of the 1969–1971 Mauna Ulu eruption of Kilauea volcano, Hawaii. *US Geological Survey Professional Papers*, **1056**.

TAHA, T. & CUI, Z. F. 2006a. CFD modelling of slug flow in vertical tubes. *Chemical Engineering Science*, **61**, 676.

TAHA, T. & CUI, Z. F. 2006b. CFD modelling of slug flow inside square capillaries. *Chemical Engineering Science*, **61**, 665–675.

TILLING, R. I. 1987. Fluctuations in surface height of active lava lakes during 1972–1974 Mauna Ulu eruption, Kilauea Volcano, Hawaii. *Journal of Geophysical Research*, **92**, 13721–13730.

VERGNIOLLE, S. 1998. Modelling two-phase flow in a volcano. *13th Australasian Fluid Mechanics Conference*, Aristoc. Offset, Monash University, Melbourne, Australia.

VERGNIOLLE, S. & JAUPART, C. 1990. Dynamics of degassing at Kilauea Volcano, Hawaii. *Journal of Geophysical Research-Solid Earth and Planets*, **95**, 2793–2809.

VERGNIOLLE, S. & BRANDEIS, G. 1996. Strombolian explosions. 1. A large bubble breaking at the surface of a lava column as a source of sound. *Journal of Geophysical Research-Solid Earth*, **101**, 20433–20447.

VERGNIOLLE, S. & CAPLAN-AUERBACH, J. 2006. Basaltic thermals and subplinian plumes: constraints from acoustic measurements at Shishaldin volcano, Alaska. *Bulletin of Volcanology*, **68**, 611–630.

VERGNIOLLE, S., BRANDEIS, G. & MARESCHAL, J. C. 1996. Strombolian explosions. 2. Eruption dynamics determined from acoustic measurements. *Journal of Geophysical Research-Solid Earth*, **101**, 20449–20466.

WALLIS, G. B. 1969. *One-Dimensional Two-Phase Flow*. McGraw-Hill, New York.

WELSH, S. A., GHIAAISAAN, S. M. & ABDEL-KHALIK, S. I. 1999. Countercurrent gas-pseudoplastic liquid two-phase flow. *Industrial & Engineering Chemistry Research*, **38**, 1083–1093.

WHITE, E. R. & BEARDMORE, R. H. 1962. The velocity of rise of single cylindrical air bubbles through liquids contained in vertical tubes. *Chemical Engineering Science*, **17**, 351–361.

XIE, T., GHIAASIAAN, S. M., KARRILA, S. & MCDONOUGH, T. 2003. Flow regimes and gas holdup in paper pulp-water-gas three-phase slurry flow. *Chemical Engineering Science*, **58**, 1417–1430.

Cyclic activity at Soufrière Hills Volcano, Montserrat: degassing-induced pressurization and stick-slip extrusion

N. G. LENSKY[1,2], R. S. J. SPARKS[3], O. NAVON[2] & V. LYAKHOVSKY[1]

[1]The Geological Survey of Israel, 30 Malkhe Israel St., Jerusalem 95501, Israel
(e-mail: nadavl@gsi.gov.il)

[2]Inst. of Earth Science, The Hebrew University, Jerusalem 91904, Israel

[3]Department of Earth Sciences, University of Bristol, Bristol, BS8 1RJ, UK

Abstract: The growth of lava domes is often associated with cyclic variations of ground deformation, seismicity and mass flux of gas and magma. We present a model of cyclic volcanic activity which is controlled by degassing of supersaturated magma, magma flow into the conduit, gas escape from the permeable magma, deformation of the conduit walls and the friction between the walls and the plug at the top of the conduit. When the difference between magma pressure and ambient pressure exceeds the static friction, motion begins, bubbles expand and overpressure relaxes. Bubble expansion builds permeability, allows gas escape and faster depressurization. Depressurization and crystallization of the magma build supersaturation and gas diffusion from melt to bubbles. Gas flux into bubbles and magma flux from the chamber act to increase pressure. The rate of extrusion is controlled by the gas pressure, driving the motion, and by the rate- and state-dependent friction along shear zones between the plug and the host rock. When the magma overpressure drops to the dynamic strength of the slip surfaces, the plug sticks and blocks the vent. As bubble volume is now constant, exsolution of gas from the supersaturated melt leads to pressurization and begins a new cycle.

The growth of lava domes is commonly characterized by repetitive cyclic patterns of ground deformation, degassing, seismicity, dome extrusion and explosive eruptions, with timescales typically of hours to days. Such cyclic patterns have been observed at the Soufrière Hills Volcano (SHV), Montserrat (Voight et al. 1999), Mount Pinatubo, Phillipines (Denlinger & Hoblitt 1999) and Merapi, Indonesia (Voight et al. 2000). At SHV, the cycles have been observed from tilt, seismic and gas emission data (Green & Neuberg 2006; Voight et al. 1998; Voight et al. 1999; Watson et al. 2000). They are also closely associated with episodic lava extrusion and repetitive vulcanian explosions (Connor et al. 2003; Druitt et al. 2002; Formenti et al. 2003).

Soufrière Hills Volcano (SHV) cycles

At the SHV, two tiltmeters were stationed in 1997 within 400 m of the growing andesitic dome and several months of observation yielded a detailed time series of data (Denlinger & Hoblitt 1999; Voight et al. 1998, 1999). The main observations were of ground tilt, seismicity and visual phenomena. The geophysical data were also supplemented with scans of gas emission (SO_2) using the COSPEC instrument (Edmonds et al. 2003; Watson et al. 2000) and recording of vulcanian

explosions (Druitt et al. 2002). Since the data have already been described in several publications, here we synthesize the key observations (Fig. 1).

Each cycle begins with an inflation of ground tilt often accompanied by a swarm of shallow earthquakes, which have been classified as hybrid events on SHV (Miller et al. 1998). Hybrid earthquakes are characterized by emergent signals, an absence of detectable S wave arrivals and energy predominantly in the 1–5 Hz range (Neuberg et al. 1998; Neuberg & O'Gorman 2002). Their sources have been located at shallow depths (<2 km) and there appears to be a continuum of hybrid events from mixed high and low frequency components to pure, almost monochromatic long period events. Hybrid earthquakes begin when the tilt goes through its inflection point, build-up to a maximum as the peak in the tilt cycle is approached and decline as the peak is passed with hybrid earthquakes being few or absent in the deflation phase of a tilt cycle (Green & Neuberg 2006). Tilt cycles can occur without hybrid swarms. Green & Neuberg (2006) also showed that the seismicity consists of several families of distinctive earthquakes, each family having similar spectral characteristics.

The peak in tilt and the following deflationary phase are associated with a variety of volcanic phenomena. The most commonly observed phenomenon is that the deflation is accompanied

From: LANE, S. J. & GILBERT, J. S. (eds) Fluid Motions in Volcanic Conduits: A Source of Seismic and Acoustic Signals. Geological Society, London, Special Publications, **307**, 169–188.
DOI: 10.1144/SP307.10   0305-8719/08/$15.00 © The Geological Society of London 2008.

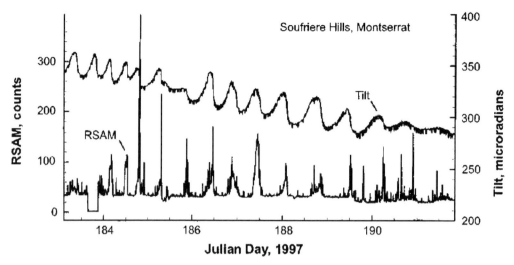

Fig. 1. Tilt and real-time seismic amplitude measurements (RSAM) showing cyclic behaviour in dome growth (after Denlinger and Hoblitt, 1999).

by onset of elevated rockfall activity of the dome and sometimes generation of pyroclastic flows. The rockfall activity in the deflation periods and the infrequent occurrence of rockfall seismicity during the inflation period were taken by Voight *et al.* (1999) to indicate much faster lava extrusion rates during deflation than during inflation. They further proposed stick-slip behaviour on the basis of these observations and this was followed by some heuristic models by Denlinger & Hoblitt (1999) and Wylie *et al.* (1999) and seismic observations consistent with stick-slip behaviour (Green *et al.* 2006). In some cases, the maximum tilt and onset of deflation were marked by intense ash and gas emissions. Ash venting commonly occurs for up to an hour after the tilt peak. Watson *et al.* (2000) found that in June 1997 SO_2 fluxes fluctuated over a cycle with the peak SO_2 emission occurring about an hour after the peak in ground tilt. Edmonds *et al.* (2003) noted that during December 1999 and January 2000, SO_2 emissions appeared to correlate with relative seismic amplitude measurements (RSAM).

The most dramatic phenomena, however, were intense vulcanian explosions which occurred at the peak in ground tilt in early August 1997 (Voight *et al.* 1998). Unfortunately, the destruction of the tiltmeter by one of these explosions prevented further observations, but it is inferred that 74 repetitive vulcanian explosions occurred in the period between 21 September to 22 October 1997, with an average cycle of ten hours (Druitt *et al.* 2002). Detailed analysis of the explosions (Druitt *et al.* 2002; Formenti *et al.* 2003) indicated that each explosion evacuated several hundred metres of the

upper conduit in a few tens of seconds. Statistical distribution of repose periods between vulcanian explosions (Connor *et al.* 2003) can be well described by a log logistic model, which requires the influence of two counteracting processes. Connor *et al.* (2003) proposed that these processes were the build-up of gas pressure in supersaturated magma and the escape of gas from magma by permeable flow. The former increases gas pressure and the latter decreases gas pressure. Jacquet *et al.* (2006) found that the same data treated as a time series showed memory effects. They interpreted the 50–60 hour memory in the time series as the repeated decompression of magma batches during their ascent in the conduit by successive vulcanian explosions. Clarke *et al.* (2007) studied the petrological characteristics of the ejecta from vulcanian explosions and found that they originated at pressures of up to 40 MPa, consistent with evacuation of the conduit to depths of up to 2 km in the largest vulcanian explosions. They observed that dense clasts gave low pressure estimates while the density of vesicular clasts increased with estimated pressure. Their results were interpreted in terms of a dense degassed cap of lava overlying a gas-rich and pressurized column of magma prior to an explosion.

The time period of the cycles

The time period of the tilt cycles varied between four and 36 hours (Voight *et al.* 1999), as measured by the intervals between peaks. The period and amplitude could remain stable for several days, but eventually changed or broke down. Data from June to August 1997 showed that the period and amplitude of the tilt could change systematically over periods of several weeks. After an abrupt onset of high amplitude and short period tilt cycles, the period gradually increased and amplitude decreased (Voight *et al.* 1999). Concurrently, the seismicity would decline and eventually almost cease. Such an episode would then be interrupted by another abrupt decrease of period and increase in amplitude, accompanied by a marked increase in seismicity. Watts *et al.* (2002) showed that these abrupt changes were associated with the extrusion of a new lobe of lava. Costa *et al.* (2007) interpreted these longer-term cycles in terms of dynamic flow in a dyke with the timescale of several weeks being controlled by the elastic deformation of the dyke walls. This inference is supported by analysis of the trends in the tilt data over the six to seven-week cycles, which are consistent with a model of a pressurized dyke with a NW–SE orientation from a pressure source at 880–1230 m depth (Hautmann *et al.* 2008).

In detail, tilt cycles are characterized by roughly similar periods of inflation and deflation. In most cases, the inflation deformation was largely recovered during the deflation, indicating almost fully recoverable strain. Voight *et al.* (1999) estimated a depth of more than 400 m below the dome as the top of the shallow pressure source. For the majority of tilt cycles, the main phenomenon of the deflation period was the onset of numerous rockfalls. This led Voight *et al.* (1999) to propose that the deflation was associated with active dome extrusion. Alternatively, Green *et al.* (2006) proposed that the tilt cycles were better explained by shear stresses acting along a conduit over a distance of *c.* 1000 m, rather than a single pressure source. Their model avoids having to invoke unreasonably high pressures and a large geologically implausible volume for the pressure source.

Models

It has been suggested that pressure build-up in the magma within the conduit is caused by: (i) variations of viscosity along the conduit (Connor *et al.* 2003; Jaquet *et al.* 2006; Melnik & Sparks 1999; Sparks 1997; Voight *et al.* 1999), (ii) influx of magma from the chamber to a conduit that is partly blocked from above by the resistance of magma (Denlinger & Hoblitt 1999; Wylie *et al.* 1999) and by (iii) volatile diffusion from the supersaturated melt into the bubbles (Sparks & Young 2002; Watson *et al.* 2000). The volatile diffusion is augmented by growth of microlite crystals from degassed melt which can achieve pressure increases of 1 MPa per 1% of crystallization in a volume-constrained, sealed system (Sparks 1997; Stix *et al.* 1997). All these cases invoke the presence

of a low permeability plug of highly viscous and partially crystallized magma at the upper part of the conduit, which confines the volume of the conduit. Relaxation of the overpressurized magma has been attributed to magma extrusion and sliding of the plug along the conduit walls when the yield strength between the plug and the wall rocks is exceeded (Connor *et al.* 2003; Jaquet *et al.* 2006; Sparks 1997).

In this paper, we present a model that accounts for the cyclic activity of a volcano using the data from the SHV as an example. We estimate the role of different processes controlling the rates of pressure build-up and relaxation in a supersaturated magma capped with a plug that sticks when pressure falls and slips when pressure is high.

Observations indicating gas pressurization and stick-slip motion in lava domes

Gas emission

Strong gas emission during the deflation periods implicates gas escape and depressurization (Watson *et al.* 2000). The association of vulcanian explosions with the inflation maxima is consistent with gas pressurization as the main cause of inflation. In general, the repetitive vulcanian explosions are thought to be caused by the same basic cyclic process as the tilt cycles (Druitt *et al.* 2002; Jaquet *et al.* 2006).

Observations from SHV and other lava domes

At SHV observations of lava dome behaviour, morphology and structure are consistent with a stick-slip style of extrusion during exogenous growth (Sparks *et al.* 2000). Common morphologies of the lava are spines and shear lobes bounded by cylindrical, relatively smooth surfaces (Cashman *et al.* 2007; Iverson *et al.* 2006; Nakada & Motomura 1999; Sparks *et al.* 2000; Watts *et al.* 2002), which can be interpreted as faulted surfaces rooted in the upper conduit (Tuffen & Dingwell 2005). They indicate that the upper conduit wall is a zone of localized deformation and that the extrusion process can be seen as faulting process acting on a rigid plug of degassed magma which occupies the upper conduit. Blocks ejected during the 16 September 1996 subplinian explosive eruption at SHV gives some insights into the conditions in the upper conduit. Robertson *et al.* (1998) documented the density and petrology of ejecta. Blocks from the lava dome and possibly uppermost conduit have low water content and high density (2200–2400 kg/m^3) and are consistent with an uppermost region of degassed lava. Rather dense

pumiceous blocks (1200–2000 kg/m^3) with low
water content (Harford & Sparks 2001) may represent deeper zones of bubbly magma. Pressure
estimations based on the shooting distances of ballistic blocks indicate pressures up to 25 MPa for
these ejecta (Robertson *et al.* 1998) consistent
with the interpretation that they represent cleaning
out of the uppermost conduit. Rare, but important,
rock types in the ejecta are highly sheared vesicular
blocks with characteristic cataclastic textures of
ground hornblende and plagioclase phenocrysts
(Fig. 2). These samples have not been described
previously, but are important in the context of this
study as they are thought to represent fault rock
formed at the conduit wall by stick-slip behaviour.

Direct evidence for plug ascent along cylindrical
shear surfaces was recently provided by Bluth &
Rose (2004), who observed gas and ash venting
from a ring-shaped set of fractures at the summit
of Santiaguito volcano. They interpret the correlation between ash bursts and measured extrusion
rates as an indication for incremental plug flow.
During an eruption, the magma ascends a few

centimetres, with most of the shear localized
along the conduit walls, about 50 m in diameter.
Time interval between venting events is of the
order of tens of minutes.

Plug ascent along localized shear surfaces was
also observed in the recent dome extrusion at
Mount St. Helens (Cashman *et al.* 2007; Iverson
et al. 2006) and at Unzen volcano (Nakada &
Motomura 1999). Seismicity associated with these
extrusions was related to stick-slip motion of the
plug. The period of the seismic drumbeats is of
the order of minutes (Iverson *et al.* 2006), much
less than the period of tilt cycles discussed in the
present paper (see discussion).

Tuffen *et al.* (2003) and Tuffen & Dingwell
(2005) described fault textures in rhyolites from
Iceland and interpreted them in terms of multiple
seismogenic cycles. They suggested that each
cycle includes stick-slip motion, which opens pathways for gas escape.

The final matter is to consider decompression of
the ascending magma during a cycle. In the case of
SHV, the time-averaged extrusion rates are well

Fig. 2. A highly sheared clast of vesicular ejecta from the 16 September 1996 explosive eruption and thought
to be a shear zone from the conduit margin. The sample has cataclastic textures with broken-down phenocrysts of
hornblende and plagioclase. The left-hand sample shows the plane of the foliation with the dark region being a single
hornblende phenocryst, which has been sheared into hundreds of pieces. The right sample is shown normal to the
foliation with elongate vesicles marking the position of cataclasite formed from phenocrysts. Note that the deformation
is very variable and a few intact dark hornblende crystals are also preserved.

constrained at $4-7 \text{ m}^3/\text{s}$ in the May–August 1997 period when the tilt cycles were documented (Sparks *et al.* 1998). At a characteristic average extrusion rate of $5 \text{ m}^3/\text{s}$, conduit area of 700 m^2 (Melnik & Sparks 2002) and average tilt cycle of 12 hours, the vertical displacement of the extruded magma is estimated at 300 m. Taking magma density at $2300 \text{ kg}/\text{m}^3$, then the change in magma-static pressure for a parcel of ascending magma is estimated at *c.* 7 MPa. Displacement may be larger if the cross-sectional area of the conduit narrows with depth (e.g. Couch *et al.* 2003*b*; Jaquet *et al.* 2006) suggesting that decompression may actually be larger. Following Voight *et al.* (1999), it is surmised that ascent and decompression occurred mainly during deflation and that flow rates were approximately twice the average. We conclude that decompression during pulsatory ascent associated with a single cycle may lead to supersa-turation of several MPa, provided that ascent is sufficiently fast, so that the system cannot maintain equilibrium. Crystallization of the melt also con-tributes to the build-up of supersaturation.

The model of cyclic activity in domes

Overview

We consider a cylindrical conduit filled with magma and topped by a rigid plug (Fig. 3). The two major pressurization processes considered are volatile diffusion from melt to bubbles and magma flow from the feeding source region into the blocked conduit. The drive for volatile diffusion is the supersaturation that develops due to decom-pression of volatile-bearing magma and due to crys-tallization which leads to enrichment of volatiles in the residual melt. As the magma pressurizes, the surrounding conduit walls deform, allowing magma expansion and reduction in the rate of press-urization. With efficient pressurization, magma pressure beneath the plug-base surpasses the strength of the plug and initiates motion. As the magma flows, bubbles expand and pressure decreases. In addition, bubbles merge and coalesce and new pathways are opened for gas escape and further depressurization. When magma pressure falls below the dynamic friction of the plug, motion stops and a new cycle begins. For low depressurization rate, a steady-state plug motion is expected.

Pressures and forces driving the plug

The largest pressure component of the magma in the conduit is associated with gravity and is termed the magma static pressure, P_{ms}. This pressure at the base of the plug (at Z_{pb}) is:

$$P_{ms} = \int_0^{Z_{pb}} \rho(z)g\,dz. \tag{1}$$

Using the average density, the equation simpli-fies to $P_{ms} = \rho g Z_{pb}$.

The ambient magma pressure just below the depth of the plug base, P_a, is a sum of the weight of the plug, P_{ms}, and additional pressure driving the plug motion, τ_{dr}:

$$P_a = P_{ms} + \tau_{dr}. \tag{2}$$

In slowly ascending bubbly magma, the gas pressure, P_g, closely approaches the ambient pressure (Lensky *et al.* 2004). Hence, the driving pressure may be well approximated by the differ-ence between the gas pressure and the magma static pressure:

$$\tau_{dr} \approx P_g - P_{ms}. \tag{3}$$

Volatile diffusion and magma flow from the deep feeding source lead to increase of the driving pressure. The pressurization stage comes to its end when the stress conditions meet the criterion for the onset of slip and plug motion, $\tau_{dr} \geq P_{mo}$. Following the Byerlee law (Byerlee 1967), slip starts when the ratio between the driving stress, P_{mo}, and normal stress, σ_N, overcomes the internal friction, f_{mo}. Typical values of f_{mo} range between 0.6 and 0.8 for granite samples (e.g. Marone 1998).

The criterion for plug arrest, P_{ar}, is the decrease of τ_{dr}/σ_N below the dynamic friction, f_{ar}, which in seismological modelling of the tectonic earthquake source is taken as 0.2–0.3 (BenZion & Rice 1993). In this study, we adopted the values of $f_{mo} = 0.6$ and $f_{ar} = 0.3$. For a first order approximation, the horizontal normal stress (σ_N) component in an elastic plug is proportional to the vertical stress, $\rho \cdot g \cdot z$, multiplied by the ratio $\nu/(1-\nu)$ (Jaeger & Cook 1969) where ν is the Poisson's ratio.

$$\sigma_N = \frac{\nu}{1-\nu}\rho g z. \tag{4}$$

The Poisson ratio varies from 0 to 0.5 with typical value of *c.* 0.3. Hence, we assume the horizontal stress (normal stress) is *c.* $0.5 \cdot \rho \cdot g \cdot z$, and the average value along the plug is $0.25 \cdot \rho \cdot g \cdot Z_{pb}$. Finally, using the static and dynamic friction values, and a plug that is 500 m long, the onset of motion occurs at driving pressure, $P_{mo} = 13.0$ MPa, and the

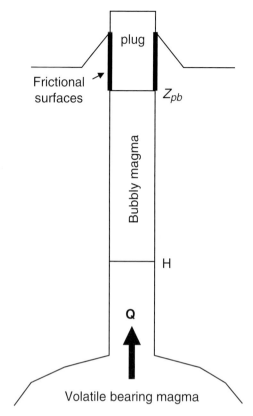

Fig. 3. Schematic diagram of the discussed system. The magma at the source contains volatiles, it flows upward in the conduit and from depth H the magma vesiculates. The plug partially blocks the flow, according to the pressurization processes and friction along plug walls — see text.

plug is arrested at $P_{ar} = 12.1$ MPa. These values are used for all model realizations presented in this study.

Pressurization

Rigid conduit (including diffusion and magma flow). We consider a batch of magma in the conduit below the plug. The magma is considered as a suspension of equally sized and evenly distributed gas bubbles of radius R, each surrounded by a spherical shell of melt of outer radius S (Proussevitch *et al.* 1993). The shells partially overlap so that the volume fraction of gas in the magma, α, equals the volume of fraction of the bubble in a single cell (bubble + melt shell). No mass is transferred between cells, and the pressure at the cell boundaries is the ambient pressure of the suspension (Lyakhovsky *et al.* 1996). The gas pressure within the bubble, P_g, is in equilibrium

with the dissolved gas at the bubble wall, $C_R = C(r = R)$, approximated by Henry's solubility law:

$$C_R = K_H P_g^{1/n} \tag{5}$$

where K_H is Henry's (for silicic magmas $n = 2$ and $K_H = 4.11 \cdot 10^{-6}\,\mathrm{Pa}^{-0.5}$ (Burnham 1975)).

When the plug sticks and a new cycle begins, ambient and gas pressure are low, but the magma is supersaturated. In this model, the conduit walls are considered rigid and the plug is blocked, hence, magma is confined to a limited volume and bubbles cannot expand. Thus, volatile exsolution into the bubbles leads to pressurization of the bubbles and the magma. The time for pressure build-up due to exsolution is controlled by the kinetics of water diffusion, as described by the equation of mass balance of the gas in the bubble:

$$\frac{d}{dt}\left(\frac{4}{3}\pi R^3 \rho_g\right) - 4\pi R^2 \rho_m D\left(\frac{\partial C}{\partial r}\right)_{r=R}, \tag{6}$$

where ρ_g and ρ_m are the densities of gas and melt and D is the diffusivity of water in the melt. Under the relevant conditions ($T = 820–880\,°C$, $P = 10^5–10^7$ Pa), the gas is approximated as an ideal gas:

$$\rho_g = P_g \frac{M}{GT}, \tag{7}$$

where P_g is the gas pressure, M is the molecular weight of water, G the gas constant and T is the temperature of the gas and melt (see values in Table 1). In the case of lava domes, with their high viscous melt and low ascent rate, the diffusive flux of volatiles into bubbles can be regarded as quasi-static, (Lensky *et al.* 2004; Lyakhovsky *et al.* 1996; Navon *et al.* 1998):

$$\left(\frac{\partial C}{\partial r}\right)_{r=R} = \frac{C_S - C_R}{R}, \tag{8}$$

where C_R and C_S are the volatile concentration at the bubble-melt interface and at the shell boundary. Initially, the concentration of volatiles away from the bubble is equal to the initial concentration, $C_S = C_i$. Later, as water diffuses into the bubbles, C_S declines. During the numerical procedure C_S is adjusted so that the total volatile mass in the finite cell (bubble + melt shell) is conserved and $(\partial C/\partial r)_{r=S} = 0$, while the concentration gradient is approximated by equation 8 (e.g. Lyakhovsky *et al.* 1996). Substituting equations 7–8 into

Table 1. *Parameters used in the equations and calculations*

Extrusion rate	(m^3/s)	$4-7\ (5)^{a,b}$
Conduit area	(m^2)	700^c
Tilt cycle (average)	(hr)	$4-36\ (12)^d$
Vertical magma movement during a cycle	(m)	300^e
Melt density	$\rho_m\ (kg/m^3)$	2300
Radius of bubble	$R\ (\mu m)$	$130-220^f$
Henry's constant for solubility law	$K_H\ (Pa^{-0.5})$	4.11×10^{-6g}
Temperature (gas and melt)	$T\ (^\circ C)$	850^h
Pressures: a — ambient, g — gas, ms — magmastatic pressure at plug base, sr — steady-state pressure at relaxation, ar — plug arrest, mo — plug motion	$P_a, P_g, P_{ms}, P_{sr},$ P_{ar}, P_{mo}	
Gas constant	$G\ (J/K/mol)$	8.1345
Molecular weight of water	$M\ (kg/mol)$	0.018
Diffusivity	$D\ (\mu m^2/s)$	$4.1-3.7^i$
Gas volume fraction	α	$0.1-0.3$
Bubble number density	(m^{-3})	10^{10j}
Internal friction	f_{mo}	$0.6-0.8^k$
Friction (steady state)	f_0	$0.2-0.3^l$
Coefficient in the rate- and state-dependent friction	$a-b$	0.1^m
Initial slip velocity	$v_0\ (m/s)$	0.05
Height of bubbly region	$H\ (m)$	1000^n
Maximum expansion of the edifice	$\Delta V_{em}\ (m^3)$	2×10^{5o}
Crystal fraction: e — equilibrium, d — dynamic, stick — at stick stage	$\beta_e, \beta_d, \beta_{stick}$	p
Permeability	$K\ (m^2)$	$5 \times 10^{-12} \cdot \alpha^{3.5q}$
Gas viscosity	$\eta\ (Pa\ s)$	1.5×10^{-5q}
Biot modulus	$M_b\ (Pa)$	10^{11}
Length scale for gas escape	$L\ (m)$	300
Coefficient B for variable permeability	$B\ (1/s)$	$0.5 \times \alpha^{3.5q}$

[a]Sparks *et al.* 1998; [b]values in parenthesis are the default values used in the calculations; [c]Melnik & Sparks 2002; [d]Voight *et al.* 1999; [e]5 m^{3}s^{-1} × 12 × 3600 s/700 m^2; [f]Voight *et al.* 2006; Murphy *et al.* 2000; [g]Burnham 1975; [h]Murphy *et al.* 2000; [i]calculated according to Zhang & Behrens 2000; [j]$3\alpha/(4\pi R^3)$; [k]Byerlee 1967; [l]Ben-Zion & Rice 1993; [m]Blanpied *et al.* 1991; [n]estimated; [o]extruded volume during a cycle; [p]Couch *et al.* 2003b; [q]Costa 2006.

equation 6 yields the equation for the rate of pressure build-up:

$$\frac{dP_g}{dt} = \frac{3\rho_m D}{R^2}\frac{GT}{M}(C_S - C_R). \qquad (9)$$

Initially, $C_R < C_S$, and with time C_R increases while C_S decreases until equilibration is achieved when $C_R = C_S$ and diffusive flux ceases. The timescale can be estimated if C_S is assumed to remain constant. In this case, gas pressure exponentially approaches the equilibrium gas pressure, P_0, $(P_g = P_0 - (P_0 - P_{ar})e^{-t/\tau})$ with a timescale

$$\tau = (3/2)(R^2/D)(\rho_m/\rho_{g,0})K_H\sqrt{P_0}.$$

The gas pressure, P_0, and density, $\rho_{g,0}$, at equilibrium are calculated following Lensky *et al.* (2006). Onset of plug motion occurs when P_g exceeds P_{mo} after time $t_{mo} = \tau \ln (P_0 - P_{ar})/(P_0 - P_{mo})$.

The above approximation underestimates the diffusive flux at the initial stage of pressurization, since C_S actually decreases with time, and hence, the onset of plug motion is achieved earlier than this analytical approximation. For typical values for SHV diffusion coefficient $D \sim 1\ \mu m^2/s$ and bubbles radius of 100 μm, $C_0\rho_m/\rho_{g,0} \sim 1$, leading to τ-values of the order of hours, which is the upper limit for the duration of diffusive controlled pressurization.

Since the magma in the upper conduit is evolved and partially degassed, we estimate bubble size based on observations on clasts ejected during vulcanian eruptions. The bubble size chosen (130–220 μm) are between the size of observed vesicles in SHV dome and pumice samples (Clarke *et al.* 2007; Formenti & Druitt 2003; Murphy *et al.* 2000) and the size of bubbles in the magma chamber, for which a radius of c. 45 μm was estimated by Voight *et al.* (2006) based on deformation data. Vesicularity varies between 10–30%, leading to bubble number density of c. 10^{10} m^{-3}, consistent with these observations. The temperature is kept constant at 850 $^\circ C$ (Murphy *et al.* 2000) and the water content of the melt varies in the range

1.5–1.6 wt%. Since the melt (but not the bulk rock) is rhyolitic, we calculate diffusion coefficients using the equation of Zhang & Behrens (2000). The diffusivity in these conditions is around 4×10^{-12} m^2/s. Hence the pressurization timescale (R^2/D) is of the order of a few hours; smaller bubbles of 50 μm (number density of 10^{12} m^{-3}) will reduce the timescale to an hour.

Pressure build-up is very sensitive to the initial volatile content (C_i), which reflects the potential energy for diffusive pressurization. Figure 4a presents the pressure build-up due to diffusion of volatiles in a rigid conduit (i.e. – constant volume) with water content at the range of $C_i = 1.5$–1.6 wt%. Higher C_i means that the last equilibration was at greater depth and that decompression rate was higher.

Another process that influences the pressure in the conduit is the flow of magma into, or out of, the conduit. Magma is supplied from the feeding source region, but when pressure in the conduit increases, backflow into the chamber is also possible. This process is controlled by the difference between the magmatic overpressure in the source, P_{op}, and the ambient pressure. Assuming that the flux of the magma, Q, is proportional to the pressure difference with a flow rate coefficient K_{mf}, and that $P_a \sim P_g$,

$$Q = K_{mf}(P_{op} - P_g). \qquad (10)$$

For incompressible melt, the rate of gas volume change is equal to the flux of magma. Combining equation 10 with equation of state of the ideal gas equation 7, the rate of pressurization governed by the magma flow is:

$$\frac{dP_g}{dt} = K_{mf}\left(P_{op} - P_g\right)\frac{P_g}{H\alpha}, \qquad (11)$$

where $H\alpha$ stands for the total volume of gas bubbles per unit area of the conduit cross-section. While K_{mf} is an important parameter that controls the dynamic of a volcano, it is not well confined by observations. High K_{mf} values mean that the magma can easily flow into the conduit when $P_{op} > P_g$ and enhances pressurization, but it also means easy backflow when diffusion-controlled pressurization is efficient and there is a smaller chance to move the plug. At low K_{mf}, flow into and out of the conduit is limited and degassing remains the major contributor to pressure build-up in the conduit. Following Melnik & Sparks (2005), we assume that P_{op} does not change significantly over the period of a cycle (hours). Integrating for the total mass of the magma extruding the conduit per cycle, and approximating $P_{op} - P_g \sim P_{mo} - P_{ar}$ of the

order of 1 MPa, enables estimation of the value of K_{mf}. For SHV, we take 300 m ascent during a cycle of about 10 hrs, which gives $K_{mf} \sim 300$ m/$(3.6 \times 10^4$ s$)/10^6$ Pa $\sim 10^{-8}$ m/Pa/s. For K_{mf} value estimated using equation 10, the characteristic time of the flow-controlled pressurization (equation 11) is the ratio between the gas volume in the conduit $H\alpha$ and overpressure during magma flow, $\tau = H\alpha/K_{mf}/P_g$.

Figure 4b demonstrates the flow-controlled pressurization for different values of magmatic overpressure, P_{op}, neglecting the diffusive-controlled pressurization. For these simulations, $P_{op} = P_{ar} + a(P_{mo} - P_{ar})$ is chosen to fall between P_{mo} and P_{ar} with $a = 0.8$, or equal to P_{mo} ($a = 1$), or above P_{mo} ($a = 1.2$). Higher P_{op} means higher final pressure and higher chance for onset of plug motion. In addition, the figure demonstrates the role of the vesicularity, α, on the rate of pressurization due to magma flow. The higher vesicularity is, the slower the pressurization. When α approaches zero, the pressurization time reduces significantly. This may explain the very short pressurization time that Iverson et al. (2006) calculated in their model, which neglects vesicularity of magma.

Deformable conduit (diffusion, magma flow, and crystallization)

When the plug sticks and pressure increases, the conduit walls deform. During the slip phase, the volcanic edifice relaxes due to magma extrusion. In this end member case, we assume that the reversible volume change associated with the tilt of the edifice during the stick stage is equivalent to the volume of magma extruded during the slip stage. If the expansion of the edifice (ΔV_e) is proportional to pressure, then:

$$\frac{\Delta V_e}{\Delta V_{em}} = \frac{P_g - P_{ar}}{P_{mo} - P_{ar}}, \qquad (12)$$

where ΔV_{em} is the maximum volume change of the volcanic edifice. The ratio between ΔV_e and the volume of the bubbly region in the conduit (V_{co}), at P_{ar} prior to its expansion, is the volumetric strain of the magma:

$$\varepsilon = \frac{\Delta V_{em}}{V_{co}} \frac{P_g - P_{ar}}{P_{mo} - P_{ar}}, \qquad (13)$$

and the expansion strain rate is:

$$\dot{\varepsilon} = 3\frac{\dot{S}}{S} = 3\frac{\dot{R}}{R}\alpha \approx \frac{\Delta V_{em}}{V_{co}}\frac{1}{P_{mo} - P_{ar}}\frac{dP_g}{dt} \qquad (14)$$

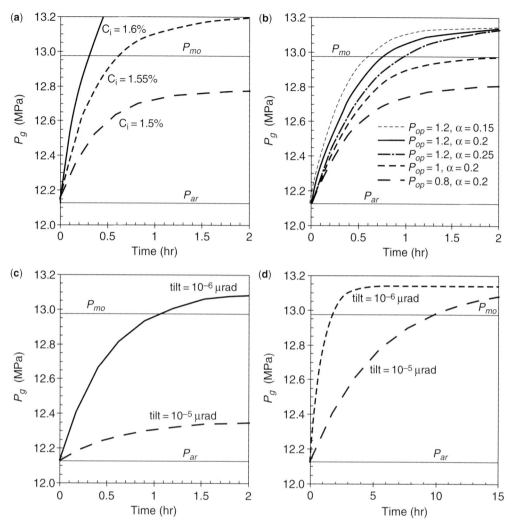

Fig. 4. Pressurization of magma below a blocked conduit by diffusion (**a, c**) and by magma flow (**b, d**) in a non-deformable conduit (a, b) and in a deformable conduit (c, d). (a) Pressurization due to diffusive flux in a rigid blocked conduit. The pressure is very sensitive to variation in the initial volatile content in the melt shell; higher volatile content results in a higher pressure and earlier onset of plug motion ($P_g > P_{mo}$). (b) Pressurization due to magma flux into the rigid conduit. Higher magma overpressure (P_{op}) yields higher final pressure, thus higher potential for onset of plug motion. (c) The effect of deformation on pressurization is clear. Higher deformation extends the pressurization time and reduces the final pressure and the chance to move the plug (the solution is based on A with $C_i = 1.6\%$). (b) Similar to (c), but for pressurization driven by magma flow instead of diffusion. Here again the influence of deformation is clear, time to onset of plug motion extends to 10 hrs (the solution is based on (b) with $P_{OP} = 1.2$, $\alpha = 0.2$).

Combining equations 14 and 9, we obtain:

$$\frac{dP_g}{dt} = \frac{3\rho_m D}{R^2} \frac{GT}{M}(C_S - C_R) + K_{mf} \times (P_{op} - P_g)$$

$$\times \frac{P_g}{H\alpha} - \frac{\Delta V_{em}}{V_{co}\alpha} \frac{P_g}{P_{mo} - P_{ar}} \frac{dP_g}{dt}. \quad (15)$$

Figure 4c presents the pressurization of magma due to volatile diffusion into the bubbles (no magma inflow) in a deformable conduit. The solution is very sensitive to the amount of tilt. Low tilt (i.e. low deformation) means that the gas flux into bubbles results in pressurization rather than expansion. In this case, a tilt of 1 microradian (μrad) is low enough so that diffusion-driven pressurization alone can initiate plug motion. However, more deformable walls (tilt of 10 μrad) leads to exhausting of volatiles before P_{mo} is reached (unless supersaturation is maintained by magma crystallization).

Figure 4d presents the pressure build-up due to magma inflow from the source region to a deformable conduit, neglecting the effect of diffusive influx of volatiles. The solution is very sensitive to the amount of deformation (tilt). Higher tilt leads to slower pressurization, since in addition to compressing the bubbles, the inflowing magma now fills the extra volume created by the deformation of the conduit walls. Since we assume that the overpressure at the source is not affected by the tilt at the top of the conduit, then all solutions will potentially reach that overpressure. If P_{mo} is lower than that overpressure, then the plug will move after a few hours. In this solution, the time-scale is longer than the previous timescale (compare Fig. 4d with 4a–c).

Petrographic studies of the dome samples and ejecta reveal extensive groundmass crystallization in the upper conduit, particularly at pressures of 30 MPa or less (Clarke *et al.* 2007; Couch *et al.* 2003*a*; Sparks 1997; Sparks *et al.* 2000). Crystallization increases water concentration in the residual melt and enhances supersaturation and degassing. In order to quantify the role of crystallization, we estimate the amount of extra water that can be delivered during crystallization and compare the timescales of crystallization and diffusion: if the former is much faster, then the diffusion timescale remains as is; if it is slower, then the crystallization timescale will govern the pressurization.

As the crystallizing assemblage carries negligible water, the mass of water released due to crystallization (per volume of magma) is $\Delta M_g = C_0 \rho_m \Delta\beta$, where $\Delta\beta$ is the fraction of magma that crystallized. The water concentration of the melt grows by $\Delta C = \Delta M_g / M_{melt} = C_0 \rho_m \Delta\beta / (\rho_m(1 - \beta))$, or in other terms $\Delta C / C_0 \sim \Delta\beta / (1 - \beta)$. Assuming $\beta \sim 0.5$, further crystallization of 10% leads to a considerable increase of volatile concentration by 20%. The potential contribution due to crystallization is demonstrated in Figure 4a which shows the effect of the initial water content on degassing. Crystallization shifts the melt to higher curves at it proceeds.

Couch *et al.* (2003*b*) have determined the time-scales of crystal nucleation and growth in melt with composition similar to that of SHV. They carried out experiments where samples were decompressed in steps to simulate the episodic magma ascent that would be expected in the stick-slip extrusion cycles. The experiments suggest that during decompression and degassing, the actual crystal growth

lags somewhat behind equilibrium crystallization. We assume a similar behaviour and model the growth of the crystal mass during the slip phase as exponential. The equilibrium crystal fraction, β_e, follows $\beta_e(P_g) = 0.115\exp((1.2 \times 10^8 - P_g)/6.67 \times 10^7)$, while the dynamic crystallization lags behind: $\beta_d(P_g) = 0.115\exp((1.2 \times 10^8 - P_g)/8 \times 10^7)$. During the stick stage, the crystal content caches up with the equilibrium value according to: $\beta_{stick} = \beta_d + (\beta_e(P_{slip}) - \beta_d) \cdot (1 - e^{-t/\tau})$. Finally, at each step of the numerical solution, the water content, C_S in equations 9 and 15 is adjusted by factor $1/(1 - \beta)$ according to the volume decrease of the residual melt during crystallization (ignoring the small water content of crystallizing assemblages).

Depressurization

Plug motion. The pressurization stage comes to its end with the onset of plug motion, which is when the stress conditions meet the criterion for the onset of slip. During the slip stage, the motion of the plug is modelled using the rate- and state-dependent friction approach (e.g. Dieterich 1979; Ruina 1983) that is commonly used to describe stick-slip motion along pre-defined sliding surfaces. The model accounts for the evolution of frictional strength as a function of slip, slip-velocity and state variables that characterize the properties of the sliding surfaces. This simplified approach is widely used in seismology to simulate important aspects of earthquake cycles (e.g. Ben-Zion & Rice 1993; Burridge & Knopoff 1967).

Various laboratory experiments have been conducted in order to study variations of the dynamic friction during block sliding (for review see Marone 1998). An abrupt change in the slip velocity from v_0 to v results in an almost instantaneous change of the dynamic friction coefficient. The friction changes from its steady-state value, f_0, to its dynamic value with an amplitude proportional to the logarithm of the velocity ratio, $a \cdot \ln(v/v_0)$ (Marone 1998). This change is followed by an evolutionary effect with an opposite sign and amplitude $b \cdot \ln(v/v_0)$ over a characteristic slip distance L, which is less than a millimetre (Marone 1998). The overall change in the quasi-static friction, f (also termed steady-state friction), is controlled by the value of $a - b$:

$$f = f_0 + (a - b)\ln\left(\frac{v}{v_0}\right). \qquad (16)$$

Stesky (1978), Blanpied *et al.* (1991, 1995) Kilgore *et al.* (1993), Dieterich & Kilgor (1996) and others measured the values of a and b for granite samples at various pressures and temperatures. If $(a - b)$ is negative, there is an overall change of velocity-weakening and the frictional response favours unstable sliding (e.g. Scholz 1998). This regime is used to model earthquake dynamics (e.g. Ben-Zion & Rice 1993). On the other hand, if $(a - b)$ is positive, the overall change is referred to as velocity-strengthening and the frictional response favours stable aseismic sliding. Blanpied et al. (1991) reported changes of the $(a - b)$ sign from negative to positive for wet crystalline granite at about 300 °C and its increase up to 0.04 at 600 °C and 100 MPa. Positive $(a - b)$ values for semi-brittle behaviour of granites at high temperatures are attributed to the onset of crystal plasticity of quartz (Scholz 1998). Similar semi-brittle behaviour of sufficiently crystalline magma in SHV noted by Sparks et al. (2000) and other dacite domes (Smith et al. 2001), suggests that positive $(a - b)$ value should be adopted for the SHV conditions. Extrapolating the experimental results of Blanpied et al. (1991) to the typical temperatures in an active dome, high positive values of about $(a - b) = 0.1$ are expected.

The ratio between the driving stress (equation 3) and the normal stress (equation 4) is now related to the plug velocity using the steady-state friction (equation 16):

$$\frac{\tau_{dr}}{\sigma_N} = \frac{P_g - P_{ms}}{K_N P_{ms}} = f_0 + (a - b)\ln\left(\frac{v}{v_0}\right). \quad (17)$$

Rearranging, we obtain the rate of motion:

$$v = v_0 \exp\left[\frac{1}{a - b}\left(\frac{P_g - P_{ms}}{K_N P_{ms}} - f_0\right)\right]. \quad (18)$$

The criterion for plug arrest is the decrease of τ_{dr}/σ_N below the dynamic friction, which in a seismological modelling of the tectonic earthquake source is taken as 0.2–0.3 (Ben-Zion & Rice 1993).

$$\frac{P_{ar} - P_{ms}}{K_N P_{ms}} = f_{ar}. \quad (19)$$

Motion continues as long as $P_g > P_{ar}$. Lower dynamic friction leads to a lower arrest pressure, larger pressure amplitude and longer relaxation period.

Plug motion with velocity v leads to magma expansion which reduces gas pressure and introduces an additional term to equation 15. The decompression due to expansion is equal to $dP_g/dt = -vP_g/\alpha H$; this relation reflects the change in

gas volume due to plug motion, similar to equations 11 and 14. Accounting for this motion and for the diffusive flux, magma flow and conduit deformation (equation 15), the rate of pressure change is:

$$\frac{dP_g}{dt} = \frac{3\rho_m D}{R^2}\frac{GT}{M}(C_S - C_R)$$
$$+ K_{mf}(P_{op} - P_g)\frac{P_g}{\alpha H}$$
$$- \frac{\Delta V_{em}}{V_{co}\alpha}\frac{P_g}{P_{mo} - P_{ar}}\frac{dP_g}{dt} - v\frac{P_g}{\alpha H}. \quad (20)$$

Gas escape. Magma degassing in a conduit is generally attributed to permeable gas flow laterally into the conduit wall (Eichelberger 1995; Jaupart & Allegre 1991), or vertically through the magma column (Boudon et al. 1998). Gonnermann & Manga (2003) estimated the timescale for permeable gas flow through the vesicular magma within the conduit of the order of weeks to years. They also speculated that shear-induced fragmentation may create an interconnected fractured network of high permeability. This assumption is consistent with the observed correlation between gas emission and tilt cycles (Watson et al. 2000) and the suggested mechanism of crack opening related to plug movement in lava domes (Tuffen & Dingwell 2005; Tuffen et al. 2003). Hence, we assume that gas escape is ineffective during the stick stage and is very efficient during slip stage. Two different regions along the conduit should be considered: permeable gas flow horizontally through the magma below the plug and vertical gas flow along plug walls. There are also two permeabilities related to each region, namely the permeability of magma and the plug margin. It is unclear which mechanism might be rate limiting for overall gas escape. However, in both regions the gas flux is described by the Darcy's law:

$$q = -\frac{k}{\eta}\nabla P_g, \quad (21)$$

where q is flux of the gas, η is the gas viscosity, and k is the effective permeability representing the whole system. If the rate of the gas escape is limited by the horizontal flow through the magma below the slipping plug, the permeability is vesicularity dependent: $k(\alpha) = 5 \times 10^{-12}\alpha^{3.5}\,\mathrm{m}^2$ (Costa 2006). Permeability of the plug walls strongly depends on the aspect ratio of the open cracks and could hardly be estimated. Using the conservation of gas mass, the rate of gas volume change is: $e = -div(q)$. Finally, we use the Biot relation between effective pressure and fluid volume change:

$$\frac{dP_g}{dt} = M_b e = \frac{M_b k}{\eta} \nabla^2 P_g, \qquad (22)$$

where M_b is the Biot modulus of magma (approximately the bulk modulus of melt). We approximate the second derivative of P_g by the ratio between pressure drop and cross-sectional area, characterized by a squared length scale. The pressure drop is the difference between the gas and atmospheric pressure; the latter is negligible (relative to P_g). Introducing some representative length scale L (e.g. Edmonds *et al.* 2003) for the whole system leads to the approximation $\nabla^2 P_g \cong P_g/L^2$. This length scale is related to the radius of the conduit below the plug, or to the height of the plug, or both, depending of which mechanism is rate-limiting. With these, equation 22 is simplified to:

$$\frac{dP_g}{dt} \cong B P_g, \qquad (22a)$$

where B $(1/s)$ is the pressure diffusion coefficient $B = M_b k/\eta L^2$ controlling the timescale of the gas loss due to permeability. Equation 22a is the generalization of the complex problem of the gas escape from the vesiculating magma, which detailed description is out of the scope of this study. Combining equations 22a and 20 yields:

$$\frac{dP_g}{dt} = \frac{3\rho_m D}{R^2} \frac{GT}{M}(C_S - C_R) + K_{mf}(P_{op} - P_g)\frac{P_g}{\alpha H}$$
$$- \frac{\Delta V_{em}}{V_{co}\alpha}\frac{P_g}{P_{mo} - P_{ar}}\frac{dP_g}{dt} - \frac{v}{\alpha H}P_g - B P_g. \quad (23)$$

Before we discuss the coupled solution of equations 18 and 23, we search for the steady-state solution of equation 23, for which $dP_g/dt = 0$:

$$\frac{3\rho_m D}{R^2}\frac{GT}{M}\left(C_S - K_H\sqrt{P_{sr}}\right) + K_{mf}(P_{op} - P_{sr})\frac{P_g}{\alpha H}$$
$$- B P_{sr} = \frac{v_0}{\alpha H}P_{sr}\exp\left(\frac{\dfrac{P_{sr} - P_{ms}}{K_N P_{ms}} - f_0}{a - b}\right), \qquad (24)$$

where P_{sr} is the steady-state pressure and C_R is related to P_{sr} through the solubility law (equation 5). The value of the steady-state pressure, obtained from equation 24, distinguishes between two modes of the system: steady-state extrusion and cyclic motion. If depressurization is not efficient, $P_{sr} > P_{ar}$, then a steady motion of the plug is expected. However, under conditions leading to

efficient decompression, the gas pressure meets the arrest value before the steady-state conditions are achieved, $P_g = P_{ar} > P_{sr}$. In this case, the depressurization stage is arrested and a new cycle of pressurization begins. The efficiency of depressurization is mostly controlled by frictional parameter $(a - b)$ and the initial slip velocity. Figure 5 allows distinguishing between steady-state extrusion and cyclic motion for a given values of $(a - b)$ and v_0. For reasonable range of the initial slip velocity $v_0 = 0.1-10$ m/s, corresponding to the plug acceleration $10^{-2} - 10^2$ g over a few mm slip, the cyclic motion is realized for the $(a - b)$ values of about 0.1 or higher (middle curve). In the case of effective diffusion, higher $(a - b)$ values are required for the onset of cyclic motion (upper curve). On the other hand, gas loss due to permeability enhances the depressurization and lower $(a - b)$ values are sufficient for the onset of cyclic motion (lower curve). As discussed above, high positive values of about $(a - b) = 0.1$ are expected. With this value, cyclic activity is expected at initial slip velocity $v_0 > 0.15$ m/s.

Under conditions of the transition between steady-state motion and cyclic activity, the duration of the depressurization cycle is extremely sensitive to the friction parameter $(a - b)$. Figure 6 presents the depressurization during the plug motion for $(a - b) = 0.09-0.11$. Higher $(a - b)$ values shorten the duration of depressurization from of c. 4 hrs for $(a - b) = 0.1$ to c. 1 hr $(a - b) = 0.11$. Permeable gas escape contributes to depressurization and further reduces the duration of the

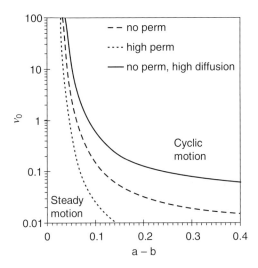

Fig. 5. Steady-state solution distinguishes between steady-state extrusion and cyclic motion for a given values of $(a - b)$ and v_0.

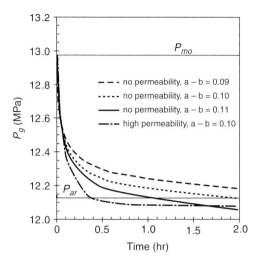

Fig. 6. Depressurization due to plug motion and permeability — see text.

depressurization. With high permeability ($B = 5–7 \times 10^{-5}\ s^{-1}$) the relaxation time is $c.\ 0.5$ hr.

Figure 7 combines together the pressurization and depressurization stages into a single cycle for reasonable values of the model parameters ($a - b = 0.1$, $v_0 = 0.1$ m/s, tilt $= 10^{-5}$ rad). Accounting for permeability provides efficient mechanism for the depressurization in addition to the plug motion. We should note here that the physical parameters (permeability, length scale and others forming B-value in equation 22a) controlling the gas lost by permeable flow along the conduit walls in the slip phase are not well constrained.

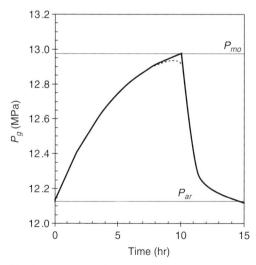

Fig. 7. A pressure cycle.

Low gas flux associated with low B-values (low permeability and long pathways for gas escape) leads to negligible permeability-controlled pressure release. On the other hand, enhanced permeability-controlled pressure release may significantly shorten the depressurization stage, as shown above.

Discussion and conclusions

We presented a model of cyclic volcanic activity that includes pressure accumulation and relaxation during stick and slip stages. Below, we discuss the role of the various processes controlling the rate of pressure change during these stages.

Pressurization mechanisms

The present model takes into account two pressurization mechanisms: diffusion-controlled exsolution of water from a supersaturated melt into bubbles and magma inflow from a deeper source into the conduit. The supersaturation is the result of non-equilibrium decompression and crystallization of the magma. As long as the pressure in the conduit does not exceed the static friction, it builds up towards the equilibrium pressure which is controlled by the saturation pressure (under diffusion control, Fig. 4a) or the magmatic overpressure (under magma flow control, Fig. 4b).

If pressurization is driven only by diffusion from an initially supersaturated melt, then the pressure asymptotically approaches equilibrium pressure, P_0, depending on the number density of bubbles, the initial supersaturation and conduit deformation. The timescale of this process follows from equations 9 and 14 and is:

$$\tau = \frac{3}{2}\frac{R^2}{D}\frac{\rho_m}{\rho_{g,0}}K_H\sqrt{P_0}$$
$$\times\left[1 + \Delta V_{em}/V_{co}\alpha \cdot P_g/(P_{mo} - P_{ar})\right].$$

The duration of the diffusive controlled pressurization from $P_g = P_{ar}$ to $P_g = P_{mo}$ may be estimated as $t_{mo} = \tau\ln(P_0 - P_{ar})/(P_0 - P_{mo})$.

Using the reasonable values for the various parameters (Table 1), t_{mo} is of the order of an hour for an undeformable conduit and is about twice as long, once wall deformation is included. The minimum water content in the melt needed for the onset of the plug motion in the deformable conduit by a diffusive mechanism as presented in Figure 8. The calculations are done based on the water mass balance considerations. For the typical range of water contents estimated for the SHV magma (1.5–1.6%, Rutherford & Devine 2003; Sparks

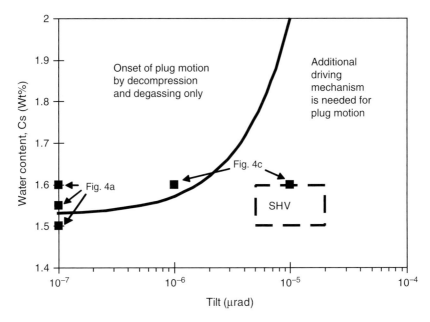

Fig. 8. The minimum water content in the melt needed for the onset of the plug motion in the deformable conduit by diffusive mechanism. For the typical range of water content in the SHV, the critical pressure for the onset of motion could be achieved by diffusive mechanism only for small values of tilt ($<10^{-6}$ μrad). For the observed tilt values (above 5×10^{-6} μrad), the water content should be 1.75% and an additional source of pressure is needed for onset of plug motion.

et al. 2000) the critical pressure for the onset of motion could be achieved by diffusive mechanism only for small values of tilt ($<10^{-6}$ μrad). For the observed tilt values (above 5×10^{-6} μrad), the required water content is 1.75%. Higher water contents may in fact be plausible, since estimates of bulk water content based on melt inclusions and phase equilibrium are typically minimum values as they do not include any excess exsolved water in the magma chamber. Alternatively, an additional source of pressure is needed for onset of plug motion.

Crystallization of the magma is a source for extra degassing. It acts to maintain supersaturation, to increase the release time and to enhance pressure build-up until P_{mo} is reached. Couch *et al.* (2003*a, b*) reported timescales of a few hours to days for crystal growth under continuous decompression compatible with the SHV decompression rate. When pressure was kept constant, the crystals approached equilibrium within a few hours. Hence, the crystallization process ensures that enough water is released and extends the time for pressurization and surpassing P_{mo} to a few hours.

The other pressurization mechanism is related to the magma flow from the feeding source region into the conduit, which is blocked at its top by the plug. Magma flux is assumed to be proportional (with

flow rate coefficient) to the pressure difference between the magmatic overpressure in the source and the ambient pressure. In the case of efficient diffusion-controlled pressurization, when high ambient pressure builds up in the conduit, the model accounts for backflow into the source, reducing the ambient pressure to the magmatic overpressure. The characteristic time of the flow-controlled pressurization is the ratio between the gas volume in the conduit $H\alpha$, and the flow rate of the overpressurized magma, $\tau = H\alpha / K_{mf}P_g$, where P_g is a typical value of the gas pressure. For typical SHV conditions ($H = 10^3$ m, $\alpha = 0.2$, $K_{mf} = 10^{-8}$ m/Pa/s, $P_g = 10^7$ Pa), the pressurization timescale for a rigid conduit is *c.* 1 hr, compatible with the numerical solution in Figure 4b. This timescale is extended by an order of magnitude when the deformation of the SHV conduit is taken into account (Fig. 4d). In agreement with the derived timescale of the flow-driven pressurization, the numerical solutions show that the increase in vesicularity reduces the rate of pressurization. We note that when α approaches zero, the pressurization time reduces significantly. This may explain the very short pressurization time that Iverson *et al.* (2006) achieved in their model, which ignores vesicularity of magma.

Figure 9 presents the pressure evolution by diffusion (upper curve), magma flow (lower curve) and their coupled solution (middle curve with maximum). When the two processes are taken into account, the equilibrium pressure is controlled by the magmatic overpressure. At the initial stage, the two processes act to increase pressurization rate, with the timescale for both diffusion and flow of c. 1 hr (assuming tilt of only 10^{-6} μrad). After one hour, the pressure achieves its maximum and from then it decreases and asymptotically approaches the magma overpressure within a few hours. In other words, the long-term pressure is dictated by the magma overpressure and in the short term (hour) the diffusion can contribute to pressure build-up. If decompression-induced degassing is sufficient for reaching P_{mo}, the time needed to set the plug in motion may be approximated by the diffusion timescale discussed above. In the SHV, the observed tilt is c. 10^{-5} μrad, timescales should be longer than an hour and the relative contribution of decompression-induced degassing becomes less important. Pressurization is controlled by the rate of crystallization or by the flow rate from the feeding source region, which act over extended timescale.

Depressurization mechanisms

The presented model utilizes the frictional framework assuming sliding on pre-existing surfaces. When the difference between magma pressure and ambient pressure exceeds the static friction between the plug and the host rock, the plug starts to extrude, and consequently bubbles expand and overpressure is relaxed. Figure 7 extends the pressurization stage up to the onset of plug motion with abrupt transition to the depressurization stage. This simplified approach leaves the detailed description of the micro-crack nucleation and very initial stage of slip out of the scope of the presented study. These processes are expected to smooth the transition from stick- to slip-mode, shown schematically in Fig. 7.

The detailed description of the very initial stage of slip is out of the scope of the presented model. During this stage, we assume a short-term (<1 second) inertia controlled acceleration over a distance of a few millimetres (these conditions were analysed by Iverson et al. 2006). The present model accounts for the decay of the accelerated motion followed by a steady-state slip during which the velocity slowly decreases. In reality, the onset of the motion may be more gradual, and involve competition between the formation of slip surfaces, their healing and the failure and generation of asperities. These processes lead to smoothing of the transition between the stick and the slip stages. For example, plug motion is associated with low-frequency earthquakes (Hydayat et al. 2000; Neuberg et al. 2006; Tuffen & Dingwell 2005; Tuffen et al. 2003), which are probably generated at the stage of the formation of the slip surfaces and asperities failure during the plug motion. In the case of lava domes, temperatures are high and fracture healing is fast (Hydayat et al. 2000; Neuberg et al. 2006; Tuffen & Dingwell 2005; Tuffen et al. 2003), so it is clear that a model of motion on smoothed slip surfaces is a simplification. While macroscopic motion is continuous, it may consist of many microscopic slip events (e.g. the repetitive drumbeats described by Iverson, 2006). Such geometrical complexities and detailed earthquake mechanism are not considered in this model. However, it does account for the dependence of frictional slip on pressure and is used here to express the pressure release during dome extrusion and plug motion in order to get a first order model approximation of the process.

As the plug extrudes, the bubbly magma expands and the development of permeability in the magma allows gas escape from the system, probably along connected shear planes along the plug walls (Tuffen & Dingwell 2005; Tuffen et al. 2003). Both plug motion and gas escape lead to gas pressure relaxation with a timescale that is highly sensitive to the frictional parameter $(a - b)$ and permeability (see Fig. 6). When magma overpressure drops to the dynamic strength of the slip surfaces, the plug sticks to the conduit walls and blocks the vent.

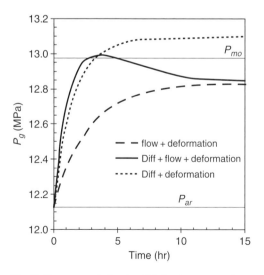

Fig. 9. Pressure evolution by diffusion (upper curve), magma flow (lower curve) and their coupled solution (middle curve with maximum).

Different regimes

Three different dynamic regimes of volcanic eruptions are described by the model: stick solution, constant slip, and the intermediate stick-slip cyclic activity. Additional cases not discussed here are failure of the wall-rock during pressurization that may lead to either gas escape or to dyke and sill injection, and the case where P_{mo} exceeds the internal strength of the magma. In this latter case, the initiation of motion and gas escape along the plug-wallrock surfaces may lead to sudden decompression, a rarefaction wave, magma disruption and vulcanian explosions.

The style of eruption is controlled by the relative values of four pressures. Two of them are the mechanical limits for plug motion and arrest, P_{mo} and P_{ar}; the other two are the steady states for gas pressurization during the stick stage and for relaxation during the slip stage, P_{sp}, P_{sr}.

In the stick solution, if $P_{sp} < P_{mo}$ then the conduit remains sealed, no magma extrudes and the dome does not grow.

For constant slip: $P_{sp} > P_{mo}$, $P_{sr} > P_{ar}$ the gas pressure increases along the pressurization curve until P_{mo} is achieved and the plug is set in motion (Fig. 10a). Cycle starts with $P_g = P_{ar}$ and high pressurization rate (point a). When $P_g = P_{mo}$, the plug is set in motion and the fast transition from the pressurization to relaxation regime is presented by the dotted line. Plug motion releases the pressure, but contributions from continued crystallization and water transfer to the bubbles, magma flow into the conduit and relaxation of the chamber walls may compensate and lead to a steady-state extrusion with $P_{sr} > P_{ar}$. The extrusion rate for the steady-state plug motion is expressed by equation 18 with $P_g = P_{sr}$. Periods of steady extrusion with no discernible pulsations at SHV (Sparks *et al.* 1998;

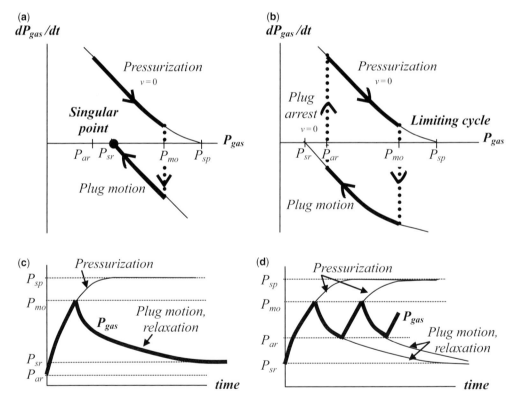

Fig. 10. Gas pressure dynamics of a volcano during continuous flow plug (**a, b**) and during cyclic activity (**c, d**). The rate of pressurization decreases with increasing pressure (a) and time (b) according to equation 5. Pressurization may proceed until equilibrium is achieved (P_{sp}). If the strength of the plug is lower ($P_{mo} < P_{sp}$), the plug slips and pressure descends according to equation 22 (the lower solid curve in a). Decompression may proceed (equations 13, 18, 22) until a steady-state is achieved at $P_g = P_{sr}$ and flow continues as long as magma is supplied from below (equations 14, 19, 23). Figures (c) and (d) describe cyclic activity which takes place when $P_{mo} < P_{sp}$ (as before) and the dynamic friction are high enough so that P_{ar} exceeds P_{sr}.

Wadge *et al.* 2006) can be attributed to this flow regime.

For cyclic activity when $P_{sr} < P_{ar}$, stick-slip behaviour is expected (Fig. 10c, d). When $P_g = P_{ar}$, the plug arrests and a new pressurization stage begins. The dotted line at P_{ar} in Figure 10c represents the transition from pressure decrease along the relaxation curve to its increase along the pressurization curve. Thus, cyclic pattern is now expected (Fig. 10d) with gas pressure oscillating between P_{mo} and P_{ar}; dP_g/dt is never zero and the steady-state solutions are never reached, as long as $P_{sp} > P_{mo}$. This condition demands recharge of the conduit with fresh magma.

Limitations of the model

The presented model is an heuristic one and considers certain processes. It splits the whole system into two cells: the first stands for the magma evolution in the conduit, while the second represents the plug and the brittle failure associated with its motion. The model ignores any kind of thermal effects or heat exchange between these virtual cells and surroundings. In this sense, it is isothermal and the only interaction with the host rock is mechanical response of edifice to the magma pressure change. The adopted relation between conduit deformation and magma pressure mimics linear elasticity of the edifice and leaves out any irreversible deformation components (e.g. Cayol 2003; Widiwijayanti *et al.* 2005) or effect of shear stress along the conduit wall (discussed by Green *et al.* 2006). The amount of deformation, or coefficient of proportionality between pressure and volumetric deformation, is not well constrained.

The sharp transition between the ductile magma and brittle plug disregards the semi-brittle behaviour observed in the laboratory experiments, where dome materials at low strain rate act as non-Newtonian fluids while acting as brittle solids at high strain rates. (e.g. Dingwell 1998; Lavalle *et al.* 2007) and discussed by Neuberg *et al.* (2006) and Tuffen & Dingwell (2005) in context of SHV seismic activity. In this context, a low rate of pressure build-up might result in aseismic flow of the magma without reaching the brittle-ductile transition. It remains to be established whether such an extrusion regime is capable of causing cycles, and the occurrence of seismic swarms during inflation cycles indicates a brittle mechanism of the kind we have modelled. The model assumes frictional plug sliding along pre-defined surfaces and uses rate- and state-dependent friction to produce a first order approximation for average rate of plug motion. It does not account for local details such as asperity failures and gradual nucleation and formation of slip

surfaces and associated generation of low-frequency seismicity. After the arrest of plug motion, the model considers fast healing and full recovery of strength during the stick. This is similar to Tuffen *et al.* (2003) and reasonable under the high temperature conditions typical for SHV.

Detailed kinetics of the fracture nucleation, development and healing is expected to introduce smoothed transition from stick to slip stages. In order to simulate the details of the volcanic activity, including stochastic temporal cyclic behaviour, evolving seismicity pattern, as well as details of magma flow from the source region into the conduit, one has to consider a 2- or 3-D model incorporating the physical processes discussed in this paper.

The formation of the plug itself is beyond the scope of this model. A plausible mechanism is that gas escape from the foamy magma below the overlying plug results in densification of the magma by bubble collapse as the internal pressure in the bubbles reduces below the magma pressure itself. This collapse involves volume reduction and will act against prolonged movement of the plug. Rapid loss of gas from the melt to the under-pressured bubbles in this stage could also induce further microlite formation, large increases in magma viscosity and strength. This densified and rheologically stiffened magma will then add a zone of impermeable magma to the base of the plug, which will assist in the start of a new cycle of pressure build-up. A more advanced model of the cycles would seek to include the compaction and gas pressure evolution in bubbles in the relaxation stage of the stick-slip cycles.

Concluding remarks

The present model demonstrates that the timescale of water diffusion is too fast to account for the pressurization of the dome. Edifice deformation increases the pressurization time, but it is still of the order of an hour or two and shorter than the 6–10 hours of the Montserrat Dome. When edifice deformation is included, the amount of water released due to decompression and degassing is not sufficient to set the plug in motion. Crystallization and magma flow into and out of the conduit may lengthen the time and produce higher pressures. Gas escape during the slip stage ensures that pressure drops and the plug sticks. Together, they produce a physical model that accounts for the major processes that must take place and reproduces the right timescale for pressurization and relaxation.

We acknowledge the reviewers H. Tuffen and A. Longo for their constructive reviews which greatly improved the manuscript. We thank O. Melnik for fruitful discussions. R. S. J. Sparks would like to acknowledge a Royal Society Wolfson Merit award and NERC research grants.

References

BEN-ZION, Y. & RICE, J. R. 1993. Earthquake failure sequences along a cellular fault zone in a three dimensional elastic solid containing asperity and nonasperity regions. *Journal of Geophysical Research*, **98**, 14,109–14,131.

BLANPIED, M. L., LOCKNER, D. A. & BAYERLEE, J. D. 1991. Fault stability inferred from granite sliding experiments at hydrothermal conditions. *Geophysical Research Letters*, **18**, 609–612.

BLANPIED, M. L., LOCKNER, D. A. & BYERLEE, J. D. 1995. Frictional slip of granite at hydrothermal conditions. *Journal of Geophysical Research*, **100**, 13,045–13,064.

BLUTH, G. J. S. & ROSE, W. I. 2004. Observations of eruptive activity at Santiaguito volcano, Guatemala. *Journal of Volcanology and Geothermal Research*, **136**, 297–302.

BOUDON, G., VILLENMANT, B., KOMOROWSKY, J.-C., ILDEFONSE, P. & SEMET, M. P. 1998. The hydrothermal system at Soufriere Hills Volcano, Montserrat (West Indies): characterization and role in the on-going eruption. *Geophysical Research Letters*, **25**, 3693–3696.

BURNHAM, C. W. 1975. Water and magmas; a mixing model. *Geochimica et Cosmochimica Acta*, **39**, 1077–1084.

BURRIDGE, R. & KNOPOFF, L. 1967. Model and theoretical seismicity. *Bulletin of the Seismological Society of America*, **57**, 341–371.

BYERLEE, J. D. 1967. Frictional characteristics of granite under high confining pressure. *Journal of Geophysical Research*, **72**, 3639–3648.

CASHMAN, K. V., THORNBER, C. R. & PALLISTER, J. S. 2007. From dome to dust: shallow crystallization and fragmentation of conduit magma during 2004–2006 dome extrusion of Mount St. Helens, Washington. *In*: SHERROD, D., SCOTT, W. E. & STAUFFER, P. H. (eds) *A Volcano Rekindled: The First Year of Renewed Eruption at Mount St. Helens, 2004–2006.* US Geological Survey Professional Paper, Ch. 19.

CAYOL, V. (2003). Anelastic deformation on Montserrat. *EOS Transactions of AGU*, **84**, V52E-03.

CLARKE, A. B., STEPHENS, S., TEASDALE, R., SPARKS, R. S. J. & DILLER, K. 2007. Petrological constraints on the decompression history of magma prior to vulcanian explosions at the Soufriere Hills volcano, Montserrat. *Journal of Volcanology and Geothermal Research*, **161**, 261–274.

CONNOR, C. B., SPARKS, R. S. J., MASON, R. M., BONADONNA, C. & YOUNG, S. R. 2003. Exploring links between physical and probabilistic models of volcanic eruptions: the Soufriere Hills Volcano, Montserrat. *Geophysical Research Letters*, **30**, 1697–1701, doi:10.1029/2003 GLO17384.

COSTA, A. 2006. Permeability-porosity relationship: a re-examination of the Kozeny-Carman equation based on fractal pore-space geometry. *Geophysical Research Letters*, **33**, L02318, doi:10.1029/2005GL025134.

COSTA, A., MELNIK, O., SPARKS, R. S. J. & VOIGHT, B. 2007. The control of magma flow in dykes on cyclic lava dome extrusion. *Geophysical Research Letters*, **34**, L02303, doi:10.1029/2006GL027466.

COUCH, S., HARFORD, C. L., SPARKS, R. S. J. & CARROLL, M. R. 2003a. Experimental constraints on the conditions of formation of highly calcic plagioclase microlites at the Soufriere Hills Volcano, Montserrat. *Journal of Petrology*, **44**, 1455–1475.

COUCH, S., SPARKS, R. S. J. & CARROLL, M. R. 2003b. The kinetics of degassing-induced crystallization at Soufriere Hills Volcano, Montserrat. *Journal of Petrology*, **44**, 1477–1502.

DENLINGER, R. P. & HOBLITT, R. P. 1999. Cyclic eruptive behavior of silicic volcanoes. *Geology*, **27**, 459–462.

DIETERICH, J. H. 1979. Modeling of rock friction 1. Experimental results and constitutive equations. *Journal of Geophysical Research*, **84**, 2161–2168.

DIETERICH, J. H. & KILGORE, B. 1996. Imaging surface contacts; power law contact distributions and contact stresses in quartz, calcite, glass, and acril plastic. *Tectonophysics*, **256**, 219–239.

DINGWELL, D. B. 1998. Recent experimental progress in the physical description of silicic magma relevant to explosive volcanism. *In*: GILBERT, J. S. & SPARKS, R. S. J. (eds) *The Physics of Explosive Volcanic Eruptions.* The Geological Society, London, Special Publications, **145**, 1–7.

DRUITT, T. H., YOUNG, S. R., BAPTIE, B. *ET AL.* 2002. Episodes of cyclic vulcanian explosive activity with fountain collapse at Soufriere Hills Volcano, Montserrat. *In*: DRUITT, T. H. & KOKELAAR, B. P. (eds) *The Eruption of Soufriere Hills Volcano, Montserrat, from 1995 to 1999.* Geological Society of London, London, **21**, 281–306.

EDMONDS, S., OPPENHEIMER, C., PYLE, D. M., HERD, R. A. & THOMPSON, G. 2003. SO_2 emissions from Soufriere Hills Volcano and their relationship to conduit permeability, hydrothermal interaction and degassing regime. *Journal of Volcanology and Geothermal Research*, **124**, 23–43.

EICHELBERGER, J. C. 1995. Silicic volcanism — ascent of viscous magmas from crustal reservoirs. *Annual Review of Earth and Planetary Sciences*, **23**, 41–63.

FORMENTI, Y. & DRUITT, T. H. 2003. Vesicle connectivity in pyroclasts and implications for the fluidisation of fountain-collapse pyroclastic flows, Montserrat (West Indies). *Earth and Planetary Science Letters*, **214**, 561–574.

FORMENTI, Y., DRUITT, T. H. & KELFOUN, K. 2003. Characterisation of the 1997 Vulcanian explosions of Soufriere Hills Volcano, Montserrat, by video analysis. *Bulletin of Volcanology*, **65**, 587–605.

GONNERMANN, H. M. & MANGA, M. 2003. Explosive volcanism may not be an inevitable consequence of magma fragmentation. *Nature*, **426**, 432–435.

GREEN, D. & NEUBERG, J. 2006. Waveform classification of volcanic low-frequency earthquake swarms and its implication at Soufriere Hills Volcano, Montserrat. *Journal of Volcanology and Geothermal Research*, **153**, 51–63.

GREEN, D. N., NEUBERG, J. & CAYOL, V. 2006. Shear stress along the conduit wall as a plausible source of tilt at Soufriere Hills volcano, Montserrat. *Geophysical Research Letters*, **33**, L10306, doi:10.1029/2006GL025890.

HARFORD, C. L. & SPARKS, R. S. J. 2001. Recent remobilisation of shallow-level material on Montserrat revealed by hydrogen isotope composition of amphiboles. *Earth and Planetary Science Letters*, **185**, 285–297.

HAUTMANN, S., GOTTSMANN, J., SPARKS, R. S. J., COSTA, A., MELNIK, O. & VOIGHT, B. 2008. Modelling ground deformation response to oscillating overpressure in a dyke conduit at Soufriere Hills Volcano, Montserrat. *Geophysical Research Letters*, in press.

HYDAYAT, D., VOIGHT, B., LANGSTON, C., RATDOMO-PURBO, A. & EBELING, C. 2000. Broadband seismic experiment at Merapi Volcano, Java, Indonesia: very-long-period pulses embedded in multiphase earthquakes. *Journal of Volcanology and Geothermal Research*, **100**, 215–231.

IVERSON, R. M., DZURIZIN, D., GARDNER, C. A. *ET AL.* 2006. Dynamics of seismogenic volcanic extrusion at Mount St. Helens in 2004–05. *Nature*, **444**, 439–443.

JAEGER, J. C. & COOK, N. G. W. 1969. *Fundamentals of Rock Mechanics*. Chapman & Hall, London.

JAQUET, O., SPARKS, R. S. J. & CARNIEL, R. 2006. Magma memory recorded by statistics of volcanic explosions at the Soufriere Hills Volcano, Montserrat. *In*: MADER, H. M., COLES, S. & CONNOR, C. (eds) *Statistics Volcanology*. Special Publications of IAVCEI, **1**. Geological Society of London, 175–184.

JAUPART, C. & ALLEGRE, C. J. 1991. Gas content, eruption rate and instabilities of eruption regime in silicic volcanoes. *Earth and Planetary Science Letters*, **102**, 413–429.

KILGORE, B. D., BLANPIED, M. L. & DIETERICH, J. H. 1993. Velocity-dependent friction of granite over a wide range of conditions. *Geophysical Research Letters*, **20**, 903–906.

LAVALLE, Y., HESS, K. U., CORDONNIER, B. & DINGWELL, D. B. 2007. Non-Newtonian rheological law for highly crystalline dome lavas. *Geology*, **35**, 483–486.

LENSKY, N. G., NAVON, O. & LYAKHOVSKY, V. 2004. Bubble growth during decompression of magma: experimental and theoretical investigation. *Journal of Volcanology and Geothermal Research*, **129**, 7–22.

LENSKY, N. G., NIEBO, R. W., HOLLOWAY, J. R., LYAKHOVSKY, V. & NAVON, O. 2006. Bubble nucleation as trigger for xenolite entrapment in mantle melts. *Earth and Planetary Science Letters*, **245**, 278–288.

LYAKHOVSKY, V., HURWITZ, S. & NAVON, O. 1996. Bubble growth in rhyolitic melts: Experimental and numerical investigation. *Bulletin of Volcanology*, **58**, 19–32.

MARONE, C. 1998. Laboratory-derived friction laws and their application to seismic faulting. *Annual Reviews of Earth and Planetary Sciences*, **26**, 643–696.

MELNIK, O. & SPARKS, R. S. J. 1999. Nonlinear dynamics of lava dome extrusion. *Nature*, **402**, 37–41.

MELNIK, O. & SPARKS, R. S. J. 2005. Controls on conduit magma flow dynamics during lava dome building eruptions. *Journal of Geophysical Research*, **110**, B02209, doi:10.1029/2004JB003183.

MELNIK, O. E. & SPARKS, R. S. J. 2002. Modeling of conduit flow dynamics during explosive activity at Soufriere Hills Volcano, Montserrat. *In*: DRUITT, T. H. & KOKELAAR, B. P. (eds) *The Eruption of Soufriere Hills Volcano, Montserrat, from 1995 to 1999.* Geological Society, London, Memoirs, **21**, 153–172.

MILLER, A. D., STEWART, R. C., WHITE, R. A. *ET AL.* 1998. Seismicity associated with dome growth and collapse at the Soufriere Hills Volcano, Montserrat. *Geophysical Research Letters*, **25**, 3401–3404.

MURPHY, M. D., SPARKS, R. S. J., BARCLAY, J., CARROLL, M. R. & BREWER, T. S. 2000. Remobilisation of andesite magma by intrusion of mafic magma at the Soufriere Hills Volcano, Montserrat, West Indies. *Journal of Petrology*, **41**, 21–42.

NAKADA, S. & MOTOMURA, Y. 1999. Petrology of the 1991–1995 eruption at Unzen: effusion pulsation and groundmass crystallization. *Journal of Volcanology and Geothermal Research*, **89**, 173–196.

NAVON, O., CHEKHMIR, A. & LYAKHOVSKY, V. 1998. Bubble growth in highly viscous melts: theory, experiments, and autoexplosivity of dome lavas. *Earth and Planetary Science Letters*, **160**, 763–776.

NEUBERG, J., BAPTIE, B., LUCKETT, R. & STEWART, R. C. 1998. Results from the broadband seismic network on Montserrat. *Geophysical Research Letters*, **25**, 3661–3664.

NEUBERG, J. & O'GORMAN, C. 2002. The seismic wave-field in gas-charged magma. *In*: DRUITT, T. H. & KOKELAAR, B. P. (eds) *The Eruption of Soufriere Hills Volcano, Montserrat, from 1995 to 1999.* Geological Society of London, London, **21**, 603–609.

NEUBERG, J. W., TUFFEN, H., COLLIER, L., GREEN, D., POWELL, T. & DINGWELL, D. 2006. The trigger mechanism of low-frequency earthquakes on Montserrat. *Journal of Volcanology and Geothermal Research*, **153**, 37–50.

PROUSSEVITCH, A. A., SAHAGIAN, D. L. & ANDERSON, A. T. 1993. Dynamics of diffusive bubble growth in magmas: Isothermal case. *Journal of Geophysical Research-Solid Earth*, **98**, 22283–22307.

ROBERTSON, R., COLE, P., SPARKS, R. S. J. *ET AL.* 1998. The explosive eruption of Soufriere Hills Volcano, Montserrat, West Indies, 17 September, 1996. *Geophysical Research Letters*, **25**, 3429–3432.

RUINA, A. L. 1983. Slip instability and state variable friction laws. *Journal of Geophysical Research*, **88**, 10359–10370.

RUTHERFORD, M. J. & DEVINE, J. D. 2003. Magmatic conditions and magma ascent as indicated by hornblende phase equilibria and reactions in the 1995–2002 Soufriere Hills magma. *Journal of Petrology*, **44**, 1433–1453.

SCHOLZ, C. H. 1998. Earthquakes and friction laws. *Nature*, **391**, 37–42.

SMITH, J. V., MIYAKE, Y. & OIKAWA, T. 2001. Interpretation of porosity in dacite lava domes as ductile-brittle failure textures. *Journal of Volcanology and Geothermal Research*, **112**, 25–35.

SPARKS, R. S. J. 1997. Causes and consequences of pressurisation in lava dome eruptions. *Earth and Planetary Science Letters*, **150**, 177–189.

SPARKS, R. S. J., MURPHY, M. D., LEJEUNE, A. M., WATTS, R. B., BARCLAY, J. & YOUNG, S. R. 2000.

Control on the emplacement of the andesite lava dome of the Soufrière Hills volcano, Montserrat by degassing-induced crystallization. *Terra Nova*, **12**, 14–20.

SPARKS, R. S. J. & YOUNG, S. R. 2002. The eruption of Soufriere Hills Volcano, Montserrat: overview of scientific results. *In*: DRUITT, T. H. & KOKELAAR, B. P. (eds) *The Eruption of Soufriere Hills Volcano, Montserrat, from 1995 to 1999*. Geological Society, London, London, **21**, 45–69.

SPARKS, R. S. J., YOUNG, S. R., BARCLAY, J. *ET AL.* 1998. Magma production and growth of the lava dome of the Soufriere Hills Volcano, Montserrat, West Indies: November 1995 to December 1997. *Geophysical Research Letters*, **25**, 3421–3424.

STESKY, R. 1978. Mechanisms of high temperature frictional sliding in Westerly granite. *Canadian Journal of Earth Sciences*, **15**, 361–375.

STIX, J., TORRES, R., NARVAEZ, L., CORTES, G. P., RAIGOSA, J., GOMEZ, D. & CASTONGUAY, R. 1997. A model of vulcanian eruptions at Galeras volcano, Colombia. *Journal of Volcanology and Geothermal Research*, **77**, 285–303.

TUFFEN, H. & DINGWELL, D. 2005. Fault textures in volcanic conduits: evidence for seismic trigger mechanisms during silicic eruptions. *Bulletin of Volcanology*, **67**, 370–387.

TUFFEN, H., DINGWELL, D. B. & PINKERTON, H. 2003. Repeated fracture and healing of silicic magma generate flow banding and earthquakes? *Geology*, **31**, 1089–1092.

VOIGHT, B., HOBLITT, R. P., CLARKE, A. B., LOCKHART, A., MILLER, A. D., LYNCH, L. & MCMAHON, J. 1998. Remarkable cyclic ground deformation monitored in real time on Montserrat and its use in eruption forecasting. *Geophysical Research Letters*, **25**, 3405–3408.

VOIGHT, B., LINDE, A. T., SACKS, I. S. *ET AL.* 2006. Unprecedented pressure increase in deep magma reservoir triggered by lava dome collapse. *Geophysical Research Letters*, **33**, LO3312, doi:10.1029/2005GLO24870.

VOIGHT, B., SPARKS, R. S. J., MILLER, A. D. *ET AL.* 1999. Magma flow instability and cyclic activity at Soufriere Hills Volcano, Montserrat, British West Indies. *Science*, **283**, 1138–1142.

VOIGHT, B., YOUNG, K. D. & HIDAYAT, D. 2000. Deformation and seismic precursors to dome-collapse and fountain-collapse nu'es ardentes at Merapi Volcano, Java, Indonesia, 1994–1998. *Journal of Volcanology and Geothermal Research*, **100**, 261–287.

WADGE, G., MATTIOLI, G. S. & HEARD, R. A. 2006. Ground deformation at Soufriere Hills volcano, Montserrat during 1998–2000 measured by radar interferometry and GPS. *Journal of Volcanology and Geothermal Research*, **152**, 157–173.

WATSON, I. M., OPPENHEIMER, C., VOIGHT, B. *ET AL.* 2000. The relationship between degassing and ground deformation at Soufriere Hills Volcano, Montserrat. *Journal of Volcanology and Geothermal Research*, **98**, 117–126.

WATTS, R. B., SPARKS, R. S. J., HERD, R. A. & YOUNG, S. R. 2002. Growth patterns and emplacement of the andesitic lava dome at the Soufriere Hills Volcano, Montserrat. *In*: DRUITT, T. H. & KOKELAAR, B. P. (eds) *The Eruption of Soufriere Hills Volcano, Montserrat, from 1995–1999*. Geological Society of London, London, **25**, 115–152.

WIDIWIJAYANTI, C., CLARKE, A., ELSWORTH, D. & VOIGHT, B. 2005. Geodetic constraints on the shallow magma system at Soufriere Hills Volcano, Montserrat. *Geophysical Research Letters*, **32**, L11309, doi:10.1029/2005GLO22846.

WYLIE, J. J., VOIGHT, B. & WHITEHEAD, J. A. 1999. Instability of magma flow from volatile-dependent viscosity. *Science*, **285**, 1883–1885.

ZHANG, Y. & BEHRENS, H. 2000. H_2O diffusion in rhyolitic melts and glasses. *Chemical Geology*, **169**, 243–262.

Source mechanisms of vulcanian eruptions at Mt. Asama, Japan, inferred from volcano seismic signals

TAKAO OHMINATO

Volcano Research Center, Earthquake Research Institute, University of Tokyo, Yayoi 1-1-1, Bunkyo-ku, Tokyo 113-003, Japan (e-mail: Takao@eri.u-tokyo.ac.jp)

Abstract: During the 2004 Asama volcanic activity, five summit eruptions, accompanied by the broadband seismic signals, were observed. We re-analyse the broadband waveform data analysed by Ohminato *et al.* (2006) using relaxed restrictions. The results are essentially the same as those shown in the previous study. The results of the waveform inversions that assume a point source show that the force system is dominated by vertical single-force components. The source depths with dominant single-force components are 200–300 m beneath the summit crater. In the source-time history of the vertical single-force component, two downward forces separated by an upward force lasting for 5–6 s are clearly seen. We conduct a grid search for the best combination of two point sources, each consisting of a single-force component. The best waveform-match solution was obtained when one of them is positioned near the top of the conduit, and the other source is positioned 2000 m below the upper source. When a combination of single-force and moment-tensor components is assumed for the two-point source model, the moment source is located out of the vertical hypocentre distribution, suggesting a steeply inclined conduit.

After a quiescent period lasting 21 years, Mt. Asama erupted on 1 September, 2004. This initial explosion was followed by four eruptions. Mt. Asama is now relatively quiet in terms of volcanic activity. Ohminato *et al.* (2006) analysed very-long-period (VLP) seismic waveforms accompanying these eruptions, and showed that the corresponding force systems are all dominated by a vertical single-force component. However, these authors restricted their waveform analyses to point sources located along a vertical line extending below the centre of the summit crater. They also limited their investigation to two point sources. One of the two point sources had a mechanism consisting of single-force components, and the other had a mechanism consisting of moment-tensor components.

In this paper, we re-analyse the same data as Ohminato *et al.* (2006). We conducted a grid search for the source location not only in the vertical direction but also in the horizontal direction. A more extensive combination of source mechanisms, such as a combination of two point sources with mechanisms consisting of single-force components, is also tested. The dates and times used throughout this paper are Japan Standard Time.

Setting of Asama volcano

Mt. Asama, located in central Japan, is one of the most active andesitic volcanoes in Japan (Fig. 1). The summit elevation is 2560 m above sea level, and the active summit crater is 450 m in diameter and 150 m in depth. To the west of Mt. Asama is a row of older Quaternary volcanoes collectively known as Eboshi volcano. The volcanism near Mt. Asama appears to have progressed eastward, with Asama volcano representing the eastern end and youngest member of the row.

The first eruption recorded in historic documents occurred in 685 AD. After a 400-year period of apparent dormancy, a large-scale Plinian eruption in 1108 produced more than 1 km³ of volcanic ejecta. In 1783, another Plinian eruption wiped out four villages and killed several hundred people (Aramaki 1963). Most of the eruptions since then have been of the vulcanian type. In the early 1900s, vulcanian and sub-plinian eruptions were frequently observed. From 1960 to 2004, Mt. Asama was relatively calm except for 1973 and 1984. The last major eruptions were vulcanian and occurred during the periods 1973–1974 and 1982–1983.

Volcanic activity in 2004

GPS observations indicated magma injection beneath Mt. Asama from the middle of July to the end of August 2004 (Aoki *et al.* 2005; Takeo *et al.* 2006). The source of the ground deformation inferred from the GPS data was modelled as the opening of a nearly vertical dyke at 2–3 km beneath the Eboshi–Asama volcanic row, west of the summit of Mt. Asama. From late 1999, the seismic activity started demonstrating a clear upward trend, and the eruption in 2004 occurred in the middle of this increasing trend. Starting in 2002, VLP seismic signals (*c.* 10 s) with unique waveforms were

From: LANE, S. J. & GILBERT, J. S. (eds) *Fluid Motions in Volcanic Conduits: A Source of Seismic and Acoustic Signals*. Geological Society, London, Special Publications, **307**, 189–206.
DOI: 10.1144/SP307.11   0305-8719/08/$15.00 © The Geological Society of London 2008.

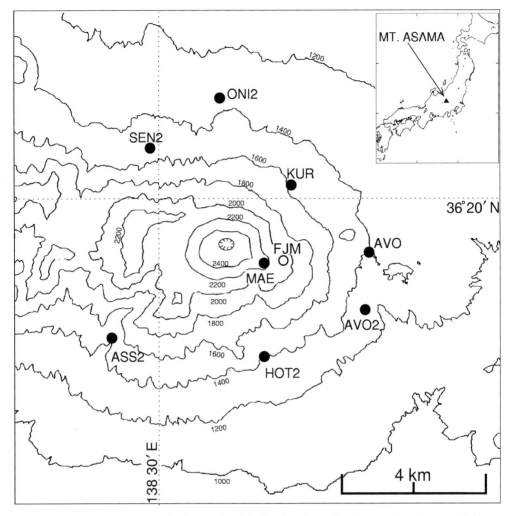

Fig. 1. Map of Mt. Asama and the seismic network. Eight closed circles are broadband stations. An open circle is a short-period station, whose waveform records are shown in Figure 2. Contours represent 200 m elevation intervals. The inset map shows the location of Mt. Asama in Japan.

frequently observed and maximum temperatures at the bottom of the crater in excess of 200 °C were recorded (Takeo *et al.* 2006).

An intense seismic swarm started at about 3 pm on 31 August, 2004 (Nakada *et al.* 2005), and the first vulcanian eruption occurred at 20:02 on 1 September. From 15 to 17 September, Mt. Asama continuously emitted volcanic ash that reached as far as metropolitan Tokyo about 130 km away. Following this ash emission stage, four more eruptions occurred on 23, 29 September, 10 October and 14 November. The estimated amounts of ash deposits are 4.9 × 10^7 kg, 8.5 × 10^6 kg, 1.3 × 10^7 kg, 2.8 × 10^6 kg, and 2.5 × 10^7 kg, for eruptions on 1, 23, 25 and 29 September, 10 October, and 14

November, respectively (Yoshimoto *et al.* 2005). The explosion on 1 September was the largest in terms of the amount of ash deposits, while the explosion earthquake on 23 September was the largest of the five explosion earthquakes that occurred during the 2004 activity (Ohminato *et al.* 2006).

GPS observations indicate that the propagation of the dyke beneath the Eboshi–Asama volcanic row, which started in mid-July, continued during the volcanic activity (Aoki *et al.* 2005; Murakami 2005). The VLP seismic activity with unique waveforms disappeared a few days before the eruption on 1 September, suggesting a change in the volcanic edifice as a precursor to the eruptions that followed (Yamamoto *et al.* 2005).

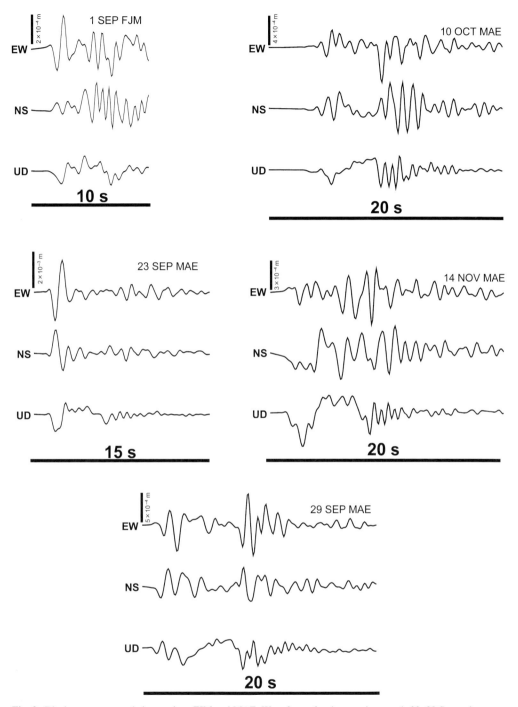

Fig. 2. Displacements recorded at stations FJM and MAE. Waveforms for the eruptions on 1, 23, 29 September, 10 October, and 14 November recorded at stations FJM and MAE are shown. Horizontal bars indicate timescales. Vertical bars indicate amplitude scale. For all the traces, sensor responses are corrected and the band-pass filter is applied in the 0.1–2 Hz band using a six-pole zero-phase Butterworth filter.

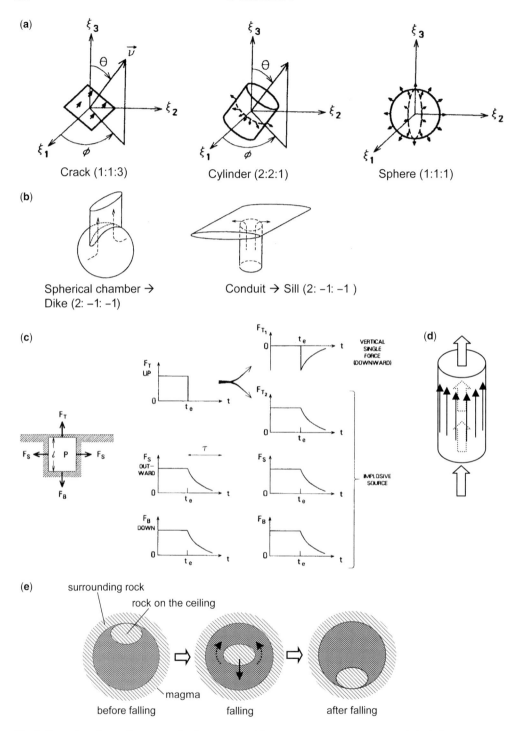

Fig. 3. Examples of several source mechanisms expressed by moment-tensor components and single-force components in the context of the source of volcanic earthquakes. (**a**) Examples of source geometries including volume change. Crack, cylinder and sphere are shown. The ratio under each panel represents the amplitude ratio of three eigenvalues of the corresponding moment tensor (reproduced from Chouet 1996). (**b**) Examples of source geometries

The hypocentre distributions before, during and after the volcanic activity in 2004 fall into two groups (Takeo *et al.* 2006). One group consists of a vertical distribution of sources ranging from the bottom of the summit crater to a depth of 1 km below sea level, outlining the vertical extension of the volcanic conduit. The other group includes a horizontal distribution of sources extending 2 km in the west-northwest direction from the bottom of the vertical distribution of the sources. The latter hypocentre distribution coincides with the top edge of the vertical dyke inferred from GPS observations.

Seismic network around Mt. Asama

The seismic network in operation before the eruption on 1 September consisted of short-period seismic sensors with narrow dynamic range data loggers and one broadband seismic sensor (STS-2) positioned 4 km from the summit. Owing to the narrow dynamic range of these sensors and also due to the extremely large amplitude of the accompanying air shock, many of the seismic records from this eruption are saturated and only five stations are available for analysis.

We installed temporary broadband stations featuring seven broadband sensors around Mt. Asama after the ash-emission activity on 15–17 September (Fig. 1). Three stations were equipped with STS-2 sensors with a natural period of 120 s and four stations were equipped with CMG-3T sensors with natural periods ranging from 100 to 360 s. The second eruption occurred on 23 September, one week after the installation of the temporary network. Eventually, we recorded four of the five eruptions that occurred during the activity of 2004 with eight broadband seismic stations. We checked the broadband sensor responses and sensor orientations using teleseismic signals dominated by 20 s surface waves. We also checked the sensor orientations using a gyro-compass with 1° accuracy. In Figure 2, waveforms of five eruptions recorded at stations FJM and MAE are shown. Since the broadband station MAE was not in operation when the event on 1 September occurred, we show the waveforms recorded at FJM, the second closest station to the summit, instead of the waveforms recorded at MAE.

Data analyses

In the following, we will focus on the results of analyses for the event on 23 September as this event had the largest amplitude among the five eruptions. For this event, the broadband seismic records from eight stations were available.

Our search for source location and source mechanism uses the inversion method of Ohminato *et al.* (1998). This method is based on a very simple and straightforward idea. Assuming a point source, the observed waveforms are expressed by the linear combination of Green's functions convolved with the appropriate source-time functions for three single-force components and six moment-tensor components. The combination of the source-time functions that best explains the observed waveforms is then determined by solving a set of linear equations. Our search for the best-fitting signal is carried out for trial point sources distributed in a uniform mesh sampling the inferred source volume. The source volume is defined around the initial source location based on the results of particle motion analyses. For details of the procedure, see Ohminato *et al.* (1998).

The Green's functions used in our waveform analyses are calculated by the 3-dimensional finite-difference method (Ohminato & Chouet 1997) assuming a homogeneous medium with realistic topography. The P- and S-wave velocities and density values used in our calculations are 3280 m/s, 1660 m/s, and 2400 kg/m^3, respectively. These velocity values are determined so that the observed travel time residuals are minimized (Ohminato *et al.* 2006). The density value is based on the gravity survey around Mt. Asama (Onizawa *et al.* 1996).

The onsets of the particle motions for the eruption on 23 September point to the summit crater. As stated above, earthquake hypocentres are distributed vertically from the bottom of the crater to 1 km below sea level. Considering the results of particle motion analyses together with the hypocentre distribution, we consider a source volume of 1.6 km in width in both the north–south and east–west directions, and 3.2 km in the vertical directions. This volume includes almost the entire vertical hypocentre distribution beneath the summit. In the uniform mesh sampling the volume, the grid size

---

Fig. 3. (*Continued*) without volume change. These solutions are expressed by a combination of two volume sources with opposite signs. One of the volume sources corresponds to a volume expansion and the other corresponds to a shrinkage of the same volume (reproduced from Chouet 1996). (**c**) A model of vertical single-force excitation proposed by Kanamori *et al.* (1984). A force system representing the sudden removal of a lid of a pressurized gas pocket near the ground surface is equivalent to a combination of a vertical single force and an implosive source. (**d**) A drag force due to viscous flow through a narrow path is a candidate for the source of the single force. (**e**) An example of source mechanisms with single-force components proposed by Takei & Kumazawa (1994). An exchange of linear momentum between the source region and the surrounding rock is observed as the source of a vertical single force. For all possible sources with single force components, the total linear momentum must be conserved.

of 50 m is used, and thus the number of grid points searched for is approximately 70 000.

Possible source models expressed by the moment tensors and single forces

Here, we briefly summarize several source-models relevant to volcanic processes so that the concepts being used in this paper can be understood easily. The following explanation mainly follows Chouet (1996).

The solutions obtained from waveform analyses yield information about the force system exerted in the source region. The derived force system does not image the real forces but represents an equivalent system of force (Aki & Richards 1980). We must then interpret the physical meaning of these equivalent forces.

Figure 3 shows some of the force systems that represent seismic sources in volcanoes. For example, a tension crack opening or closure is represented by $1:1:3$ amplitude ratios among the three diagonal components of the diagonalized moment tensor. If the ratio is $2:2:1$, it is interpreted as a radial oscillation of a cylinder. If the ratio is $1:1:1$, it is an isotropic volume change. For these ratios, $\lambda = \mu$ is assumed, where λ and μ are the Lamé coefficients of the rock matrix. For rock near liquidus temperature, $\lambda = 2\mu$ may be more appropriate, yielding the ratios $1:1:2$ and $3:3:2$ for oscillations of crack and cylinder, respectively. These three examples represent the volume change at the source (Fig. 3a). The two examples shown in Figure 3b correspond to physical processes that produce no net volume change and are called 'Compensated Linear Vector Dipole' or CLVD.

Some force systems are not necessarily expressed by the moment-tensor components. Single-force components are also observed in nature. One well-studied example of single-force component in nature is the model introduced by Kanamori et al. (1984) to quantify the explosive

eruption at Mt. St. Helens in 1980. In this model, a pressurized portion of the volcanic conduit capped by a lid is assumed. When the lid is removed suddenly, a vertical single force and an isotropic implosive source are generated (Fig. 3c). The drag force due to viscous flow through a narrow channel is another possible mechanism for a single force (Fig. 3d).

As proposed by Takei & Kumazawa (1994), a single force can be generated by an exchange of linear momentum between the source region and the rest of the Earth. Suppose a block of rock is attached to the ceiling of a magma chamber. When the block detaches from the ceiling, it starts falling through the magma chamber with increasing velocity. The block gradually looses the downward velocity as it approaches the bottom of the chamber. Finally, the block reaches the bottom of the chamber. During this process, there is a downward acceleration of the centre of mass of the source volume, followed by a downward deceleration of the centre of mass of the source volume. The resulting reaction force in the surrounding medium is an upward force followed by a downward force (Fig. 3e).

Note again that what we obtain from seismic waveform analyses is a force system relevant to volcanic processes but not the physical process itself.

Results of the waveform analyses for cases assuming a point source

Ohminato et al. (2006) assumed that the source locations are beneath the centre of the summit crater, and they searched in the vertical direction guided solely by particle motion analyses. In this study, in contrast to the previous study, we search for source locations in a volume that include most of the hypocentre distribution of the volcanic earthquakes as stated above. As a first step, we test seven equivalent force cases assuming a single point source (Table 1). The waveform matches for the

Table 1. Residual errors and corresponding Akaike's Information Criterion (AIC) for the seven source mechanisms considered in the waveform analyses assuming a single point source

Mechanisms (Mom: moment, SF: single force ISO: isotropic volume source)	Error	AIC
Case 1: 'Fz'	0.512	−4671
Case 2: '6Mom'	0.317	−7382
Case 3: '6Mom + 2SF (Horizontal)'	0.238	−9138
Case 4: '3 SF'	0.260	−9244
Case 5: '6Mom + Fz'	0.202	−10 466
Case 6: 'ISO+3 SF'	0.205	−10 810
Case 7: '6Mom+3 SF'	0.146	−12 504

'Fz' represents the vertical single-force components. '6Mom' refers to the six moment-tensor components. 'SF' and 'ISO' refer to the single-force and isotropic components, respectively. Case 7 shows the best solution for single point source.

vertical components at the closest station MAE, the closest station, are shown in Figure 4. The source locations for the seven cases have slight offsets in the horizontal direction. The offsets are 200–300 m southwest (Cases 1, 4, 5, 6 and 7) or southeast (Cases 2 and 3) of the centre of the summit crater, and thus the source locations obtained in this study are close to that obtained by

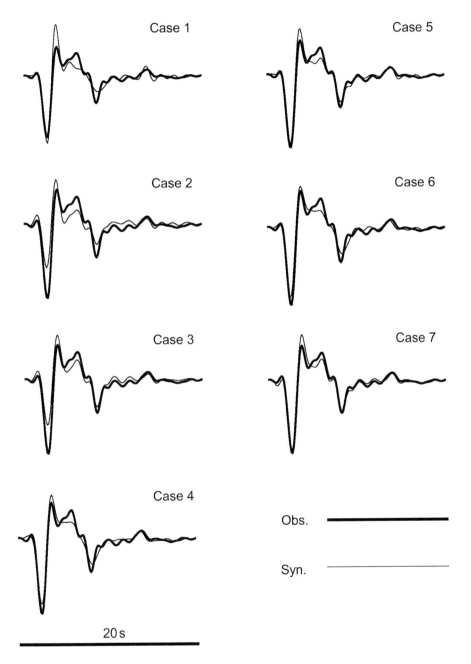

Fig. 4. Differences in waveform match for different source mechanisms corresponding to Cases 1 to 7. Waveforms are displacements and are band-passed at 0.1–2 Hz. A single point source is assumed in each case. The vertical component recorded at station MAE during the event on 23 September is shown. Waveform matches around the initial downward phase for mechanisms including the 'Fz' component (Cases 1, 4–7) are good, while waveform matches for cases without 'Fz' (Cases 2 and 3) are significantly worse.

Ohminato *et al.* (2006). The errors are smaller than that obtained in the previous study because of the relaxed condition for the horizontal source location.

In order to evaluate the trade-off between the data fit and the number of free parameters, we use Akaike's Information Criterion (AIC) (Akaike 1974) that is defined as $AIC = N_s \ln(e) + 2N_m$ in which the constant term is omitted. N_s is the number of data fitted by the model, and is calculated by the number of data traces multiplied by the number of samples in each trace. e represents the squared error. N_m is the number of free parameters, and is calculated by the number of assumed single-force and moment-tensor components multiplied by the number of samples of each source-time history. The preferred model is the one with the lowest AIC value. To have a lower AIC is better than to have a lower residual.

The AIC values obtained in this study (Table 1) are larger than those obtained in the previous study (Ohminato *et al.* 2006, Table 2) because the number of samples in each seismic trace and the number of elementary source pulses used to represent the source time function are 150 and 75, and are half of those used in the previous study. Details of the results of the seven cases are as follows.

Case 1: The source mechanism includes a vertical single-force component only (abbreviated as 'Fz'). In this simplest case, the vertical components of the waveforms are well explained, but the horizontal components are poorly explained and thus the residual error is large. The source location is 200 m west and 50 m south of the centre of the summit crater. The source elevation is 2450 m above sea level. Hereafter, we use the Cartesian coordinate with origin set at sea level beneath the centre of the summit crater to indicate source locations. The source location for Case 1 is expressed as (−200, −50, 2450).

Case 2: The source mechanism consists of six moment-tensor components and no single force (abbreviated as '6Mom'). Although the number of free parameters is six times larger than that for Case 1, the AIC value for this case is smaller than that for Case 1 owing to an improved waveform match, especially in the horizontal components. Note, however, that the waveform match in the vertical components is degraded compared to Case 1. The source location for Case 2 is (200, −200, 2100).

Case 3: The source mechanism consists of six moment-tensor components and two horizontal single-force components (abbreviated as '6Mom + 2SF'). Additional horizontal single-force components improves both error and AIC. In this case, however, the vertical components of the waveforms are not well explained as in Case 2. The source location for Case 3 is (200, −300, 2100).

Case 4: The source mechanism consists of three single-force components (abbreviated as '3SF'). Although the error reduction is worse than that for Case 3, a smaller AIC value is obtained owing to the smaller number of free parameters compared to Case 3, and thus Case 4 is better than Case 3 in terms of AIC. In this case, the vertical components of the waveforms are well explained. The source location for this case is (−200, −100, 2250).

Case 5: The assumed source mechanism consists of six moment-tensor components and one vertical single-force component (abbreviated as '6Mom + Fz'). Comparison between this case and Case 2 clearly shows that the addition of the vertical single-force component dramatically improves waveform match, and thus results in a smaller AIC value. This observation supports the idea that the 'Fz' is an indispensable component for the mechanism of the eruptions at Mt. Asama. The source location for Case 5 is (−200, −100, 2150).

The results of the above five cases clearly show that the 'Fz' component is indispensable for explaining the vertical components of the observed waveforms. The question arises, then, of which components are important for explaining the horizontal component of the waveforms. In order to clarify this, we test additional cases.

Case 6: In this case, the source mechanism consists of an isotropic component and three single-force components (abbreviated as 'ISO + 3SF'). We replace the 6 moment-tensor component in Case 5 with an isotropic component, and we add two horizontal single-force components. These changes reduce the AIC slightly. The effect of the reduction of the number of free parameters overcomes the small increase in the residual error compared to Case 5. For both Cases 5 and 6, the vertical components of the waveforms are already well explained by the 'Fz' component that is included in both cases. Thus, a comparison between Case 5 and Case 6 suggests that the contribution of the non-isotropic (deviatoric) components of the moment tensor and the contribution of the two-horizontal single-force components to the horizontal components of the waveforms are almost the same. In other words, the horizontal components of the observed waveforms are explained similarly by both a deviatoric component of the moment-tensor or two horizontal single-force components. The source location for Case 6 is (−250, −150, 2200).

Case 7: Finally, we tested the source mechanism composed of six moment-tensor components and three single-force components (abbreviated as '6Mom + 3SF'). This case resulted in the best error residual and best AIC solution. A comparison between Cases 5, 6 and 7 indicates that both the deviatoric components of moment tensor and the two horizontal single-force components are

Table 2. *Residual errors and AIC for the three cases assuming two point sources*

Mechanisms (Mom: moment, SF: single force ISO: isotropic volume source)	Error	AIC	Location of the unfixed source (ISO, 3SF, 6Mom)
Case 8: '3SF (fixed) + ISO'	0.205	−10 810	ISO: same as the fixed one
Case 9: '3SF (fixed) + 3SF'	0.166	−12 041	3SF: deep and slightly west
Case 10: '3SF (fixed) + 6Mom'	0.119	−14 000	6Mom: deep and east

One of the two point sources with three single forces is spatially fixed. The locations of the other point forces with various source mechanisms are grid-searched. The same abbreviations as used in Table 1 are used for the source mechanisms. The locations of unfixed sources are also shown. Case 10 shows the best solution of all 10 cases.

indispensable for explaining the observed waveforms, and they cannot compensate for each other's contributions. The source location for Case 7 is (−100, −100, 2200).

Figure 5 shows the distribution of the residual error in the volume where the grid search for the best source location was conducted. The two horizontal axes in the north–south and east–west directions represent distances from the centre of the summit crater. The vertical axis shows the elevation above sea level. In this figure, six horizontal slices are shown. The residual value increases from blue to red. The colour scale on the right is from 100% to 160% of the smallest residual. The distribution corresponding to Case 7 is shown. The residual distribution is simple and has only one global minimum. Similar simple residual distributions are obtained for all seven cases stated above. One-point-source solutions are stable and robust.

Results of the waveform analyses for cases assuming separated sources

In Ohminato *et al.* (2006), it was assumed that the explosion source was composed of two point sources; one of them corresponded to three single-force components, and the other corresponded to six moment-tensor components. They fixed the horizontal source location at the centre of the summit crater and searched for the best combination of depths for these two sources in the vertical direction. For the best AIC solution, the depth of the single-force source was at the bottom of the crater, and the depth of the six moment-tensor source was at 400 m beneath the single-force source. The source mechanism of the moment-tensor component was found to be an almost isotropic volume change.

For the next step in our study, we investigate sources consisting of two point sources. In this investigation, one of the two sources has a fixed location and a search for the other source location is made grid by grid. We search not only in the vertical

direction but also in the horizontal directions. As discussed above, the results of Cases 1 to 4 show that a source composed of three single-force components is indispensable. The source locations for Cases 4 to 7, all including 'Fz' component, are almost the same. Accordingly, we fix the source location of the single-force component at the location (−200, −100, 2250) that is obtained for Case 4, and search for the location of another point source. We test three cases for the two-point-source model. We assume 'ISO', '3SF', and '6Mom' for the source mechanisms of the second point source, whose location is grid-searched. They are abbreviated as '3SF + ISO', '3SF + 3SF', and '3SF + 6Mom', and are referred to as Case 8, Case 9 and Case 10, respectively. Errors and the AIC value for Cases 8, 9 and 10 are shown in Table 2.

The global minimum for Case 8 is obtained for an isotropic source located 1 km east-southeast of the summit crater. This location is beneath station MAE, which is the station closest to the summit. Near this station, there is nothing that suggests shallow hydrothermal or magmatic activity such as fumaroles, vents or cracks on the ground surface. Judging from the distance from the summit crater and the surface condition around MAE station, this location is highly inappropriate as a source of the explosion signal. We find a second local minimum at (−250, −150, 2200). This position is the same as the source location obtained for Case 6. We therefore discard the former position and adopt the latter.

Although we assumed separate sources composed of '3SF' and 'ISO' in this case, the source location of the 'ISO' component found to be very close to the location of the fixed point source of '3SF'. The separated source of '3SF + ISO' can be regarded as a point source. The residual for this solution is much smaller than that of the solution composed of '3SF' and 'ISO' obtained at (0, 0, 2200) by Ohminato *et al.* (2006). It is highly likely that the source locations obtained by these authors were inaccurate positions resulting from their imposed restriction of horizontal source locations fixed beneath the centre of the summit crater.

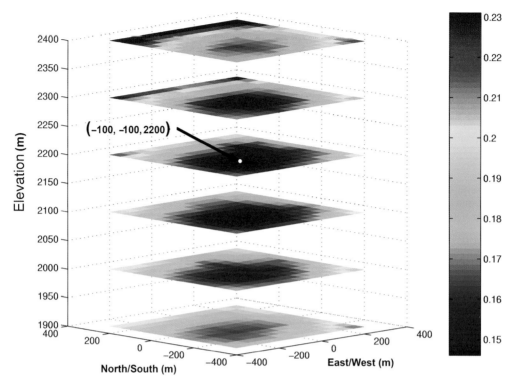

Fig. 5. Residual distribution in the volume in which the grid search for the best source location is conducted. The horizontal axes show the distance from the centre of the summit crater. The vertical axis shows elevation above sea level. Six horizontal slices with a vertical interval of 100 m are shown. The residual value increases from blue to red. The bar range on the right is from 100% to 160% of the smallest residual. Only the distribution corresponding to Case 7 is shown.

In Case 9, we again observe that a minimum residual error is obtained nearly 1 km east of the summit crater. Such a solution is unrealistic for the same reason stated for Case 8, and so we discard it. We find a second-best local minima beneath the summit at two distinct elevations, namely at (−400, −100, 1500) and (−300, −100, 0) (Fig. 6). The former is 1500 m above sea level and the latter is at sea level. The latter solution gives us a slightly better AIC value, although the difference in AIC is very small. In Table 2, the result for the latter is shown. The source-time functions corresponding to the deep '3SF' source at sea level and to the fixed '3SF' source are illustrated in Figure 7.

In Case 10, one point source of '3SF' is fixed at the location for Case 4 and the other point source of '6Mom' is grid-searched. The best source location for a source of '6Mom' is (450, −200, 1250). The global minimum is obtained at the southeast of the summit crater, and the location is out of the vertical hypocentre distribution. For other two cases, Cases 8 and 9, the locations of the 'unfixed'

components, 'ISO' and '3SF', are within the hypocentre distribution, showing that the locations for Cases 8 and 9 are not attracted to MAE station.

Discussion

Characteristic source-time history of 'Fz'

The solution obtained for the 23 September event assuming a point source is essentially the same as that obtained by Ohminato et al. (2006). The source-time history of the vertical single-force component 'Fz' is characteristic. Figure 8 shows the source-time histories of 'Fz' for all five events. These are vertically aligned so that their first downward phases coincide with each other. We can summarize the characteristics seen in the source-time function of 'Fz' as follows.

For all the source-time histories of 'Fz', (i) the initial motion of the vertical force is always downward; (ii) it is followed by an upward force; (iii) there is another downward force 5–6 sec after the first downward force.

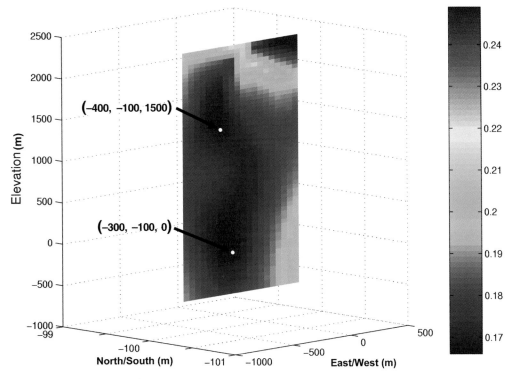

Fig. 6. Residual distribution for Case 9 in the volume in which the grid search for the best source location is conducted. One vertical section sliced at 100 m south of the centre of the crater is shown. Horizontal extent of the slice is 1500 m in the east–west direction and 2500 m in the vertical direction. Two local minima are seen at (−400, −100, 1500) and (−300, −100, 0). The horizontal axes show the distance from the centre of the summit crater. The vertical axis shows elevation above sea level. The residual value increases from blue to red. The bar range on the right is from 100% to 160% of the smallest residual.

The above-mentioned characteristic source-time histories of 'Fz' are interpreted by Ohminato *et al.* (2006) as follows.

(1) Before an eruption, the source region is pressurized. This region is located at the top of the conduit. (2) A lid at the top of the pressurized conduit is suddenly removed, and downward and implosive forces are excited. This corresponds to the source process proposed by Kanamori *et al.* (1984) as explained earlier in Figure 3c. (3) The depressurized magma increases its volume due to vesiculation and starts ascending in the conduit. The highly viscous andesitic magma exerts an upward force on the conduit wall. This is the source process shown schematically in Figure 3d. During this stage, the observed vertical single force is expressed by the superposition of two vertical forces. One is the downward force that is gradually increasing after the sudden removal of the lid as expected from Kanamori's model. The other is the upward force originated from the viscous ascending magma. Since the observed

vertical force is upward during this stage, the amplitude of the upward force is larger than the downward force. (4) Finally, explosive fragmentation starts when the magma reaches some critical depth. (5) Magma viscosity disappears due to magma fragmentation, and thus the upward viscous force also disappears. Since the vertical force component is the superposition of two force components, the sudden disappearance of the upward viscous force results in the appearance of the second downward force.

Conservation of total linear momentum

Ohminato *et al.* (2006) attributed the origin of the upward single force to the viscous drag force exerted on the wall of the volcanic conduit by the ascending viscous magma. The way linear momentum is exchanged between the ascending magma column and the rest of the Earth is slightly complicated and requires some explanation.

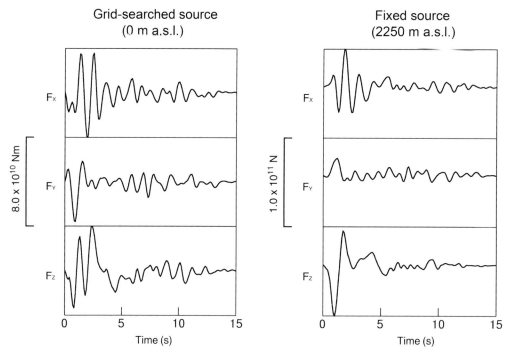

Fig. 7. Source time functions for Case 9 ('3SF + 3SF'). The '3SF' source located at (−300, −100, 0) is on the left and the '3SF' source fixed at (−200, −100, 2250) is on the right. Horizontal axes represent time and vertical axes represent amplitude. From the top to the bottom of each panel, east–west, north–south, and vertical components are shown.

Magma ascends in the conduit because it gains upward linear momentum from the rest of the Earth. In return, the rest of the Earth receives downward linear momentum. The exchange of the linear momentum mainly takes place at a certain depth in the conduit. A part of the upward linear momentum that the ascending magma column receives from the rest of the Earth is returned to the Earth through the viscous drag force exerted on the shallow portion of the conduit. The total linear momentum that the rest of the Earth receives from the ascending magma column consists of the momentum transferred at certain depths and momentum transferred through the upward viscous drag force exerted near the top of the conduit.

The shallow force should be observed as an upward force. Naturally, if the viscosity of the magma is low enough, the shallow force is not detectable. In contrast, the excitation of the seismic signals of observable amplitude by the force corresponding to the exchange of linear momentum at the deep portion in the conduit depends on the depth where the momentum exchange takes place. Note that the upward linear momentum of the magma column must balance the downward linear momentum gained by the rest of the Earth. The

force exerted on the shallow portion of the conduit and the force exerted at the deep portion of the conduit do not necessarily balance. The deep force must be stronger than the shallow force for the magma column to gain upward linear momentum. What matters here is the depth of the zones where the single forces are applied.

The upward force may act mainly on a relatively short segment of conduit near the top of the magma column, where the pressure gradient is the largest (e.g. Sparks 1997), and thus the traction force on the conduit wall is the largest. In contrast, the portion of the conduit where the downward force is exerted probably extends over a significant depth range. It is difficult to specify the extent of the portion of the magma column that exerts the downward force on the conduit wall. Such a portion of the magma column can be extended from the vesiculation depth of 2–3 km to the greater depth of a few kilometres down to the magma chamber. The contribution of the portion of conduit, which receives the downward force from the magma column, to the excitation of seismic signals may not be well observed because of the large hypocentral distance to the seismic stations (Fig. 9a).

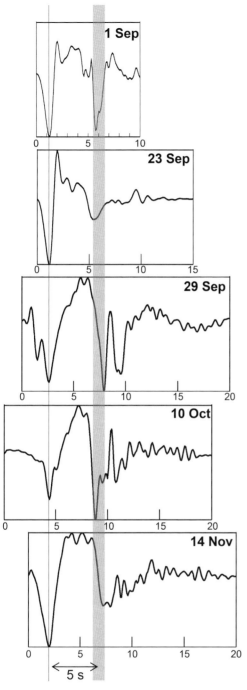

Fig. 8. Source-time histories of the 'Fz' component for the five explosions on 1, 23, 29 September, 10 October and 14 November. Horizontal axes represent time and vertical axes represent amplitude. The five panels are vertically aligned so that initial downward phases coincide with each other. For all five cases, a positive signal corresponding to an upward force follows the

There may be another shape for the distribution of downward single force over the conduit wall. Suppose there is a discontinuity in the conduit at a certain depth. This discontinuity may be a constriction or a sharp bend in the conduit. In this case, the downward force necessary to conserve the linear momentum would concentrate at the discontinuity, and the distribution of the vertical single force in the conduit may be as shown in Figure 9b. In this case, the portion where the downward force is exerted is much shallower than in the case where there is no discontinuity, and the seismic signals associated with the downward force can be observed.

In the rest of this section, we investigate whether a source composed of two point sources, each with '3SF', can explain the observed waveform better than a single point source. If a source consisting of one point source composed of '3SF' explains the observed waveform better than a source consisting of two point sources, then the force distribution shown in Figure 9a is preferable. In contrast, if a second point source composed of '3SF' is discovered at a depth that is not extremely great, then the model of Figure 9b would be better.

A model of one or two point sources may seem to be too simple as the actual source must be a fully continuous extended source. This simplistic model consisting of a few point sources represents an intermediate step from the traditional single point source model to the extended source model. A fully extended source is beyond the scope of this study, however.

In Case 9, two source locations are obtained. The source location of the shallower source at 1500 m is in the middle of the vertical hypocentre distribution beneath the summit crater, and the source location of the deeper source at sea level coincides with the depth where the hypocentre distribution changes its spatial density significantly (Fig. 10). The sudden change in density of the hypocentre distribution at sea level suggests the existence of a certain change in the physical condition around the conduit at this depth. If there is a certain condition change in the conduit at sea level, then the observation that the deep '3SF' source is slightly better than the shallow '3SF' source may suggest that the model with conduit constriction (Fig. 9b) is preferable to the model with no constriction (Fig. 9a).

To interpret the source-time functions of the six force components illustrated in Figure 7 is not easy. The horizontal components have the amplitudes almost equivalent to that of the vertical component. The behaviour of the horizontal components of the

Fig. 8. (*Continued*) initial negative phase (thin line). Five to six seconds later, another downward phase (shaded area) appears for all five explosions.

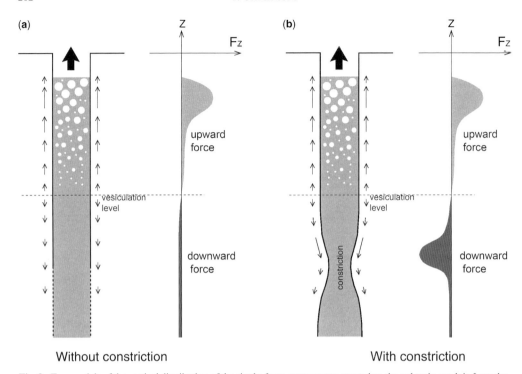

Fig. 9. Two models of the vertical distribution of the single-force components exerted on the volcanic conduit. Lengths of the arrows indicate the strength of the drag force. (**a**) The magma column has no bottom. In this case, the downward single force is not concentrated. (**b**) The magma column has a constriction. Any structure that can effectively support the vertical component of the drag force in order for the magma column and the rest of the Earth to exchange total linear momentum would work as conduit discontinuity. A constriction or a sharp bend in the conduit is a good candidate for such discontinuity.

shallow and deep sources looks as if they are compensating each other, and are mimicking the horizontal dipoles distributed along the vertical conduit. Similar behaviour as the horizontal components is seen for the vertical components. On the source-time function similar to that of the one point-source case (see Fig. 11), time functions of relatively high frequency are superimposed. These additional components at shallow and deep sources have opposite sign, and thus they are behaving as a vertical dipole. These waveforms strongly suggest that in addition to the three single-force components, the six-moment-tensor components are necessary to explain the observed waveforms. Although the necessity of both single-force and moment-tensor components have been already suggested above, an inclusion of '6Mom' component in addition to two '3SF' will make the inversion unstable.

Location and interpretation of '6Mom' component

We summarize the source locations for the ten cases investigated in this study (Fig. 10). The source

locations for Cases 1 to 10 are compared with the hypocentre distribution of the volcanic earthquakes before, during and after the 2004 Mt. Asama volcanic activity determined by Takeo et al. (2006) using the double-difference method (Waldhauser & Ellsworth 2000). For all of the cases that include the vertical single-force component 'Fz', the source locations are within the vertical hypocentre distribution. In contrast to this, sources including '6Mom' components are liable to be located outside of the hypocentre distribution (an open circle for Cases 2 and 3, and a closed circle for Case 10). The results for Cases 2, 3 and 7 indicate that the solution including the '6Mom' component and excluding vertical single force 'Fz' at the same time, are likely to be located east of the vertical hypocentre distribution of volcanic earthquakes. In contrast, other solutions including 'Fz', regardless of other source mechanisms included, are located within the hypocentre distribution. The point source solution for Case 7 includes both '6Mom' component and 'Fz' component, but the source location is inside of the vertical hypocentre distribution. This observation indicates that the tendency that the '6Mom' source is attracted to the east is overwhelmed by the tendency

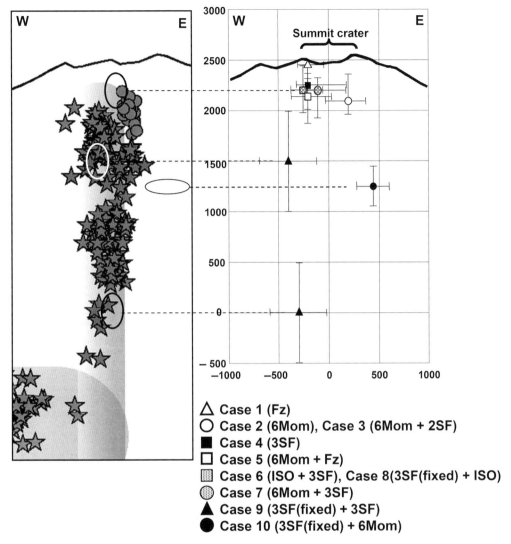

Fig. 10. Comparison between source locations (right) obtained by waveform analyses and the hypocentre distribution of volcanic earthquakes (left) determined by the double-difference method. Source locations for 10 cases are shown with error bars superimposed on the E–W section. Error bars correspond to 10% increment above the minimum error. Cases 8, 9 and 10 are the two-point-source cases, and for these cases, one of the point sources is fixed at the source location for Case 4 (closed square). The other source is searched. For Case 9, two locations (closed triangles) give almost the same residual errors and AIC values.

that the 'Fz' source stays close to the vertical hypocentre distribution.

These observations suggest that the '6Mom' components are sensitive to waveform uncertainty, which can be attributed to unknown velocity and density structures. The reason that the solutions including the '6Mom' component are attracted to the closest station MAE may be that the amplitude of the waveforms observed at MAE is the largest due to the proximity of this receiver to the

source, and thus the portion of the waveforms attributable to the uncertainty is also the largest. A source location near the station with the largest uncertainty is advantageous to reduce the overall waveform mismatch. Such a source can reduce the overall waveform mismatch only by explaining the waveform discrepancy of the waveforms at the largest-amplitude station. This explanation would mainly work for the source of '6Mom,' which has the largest number of free

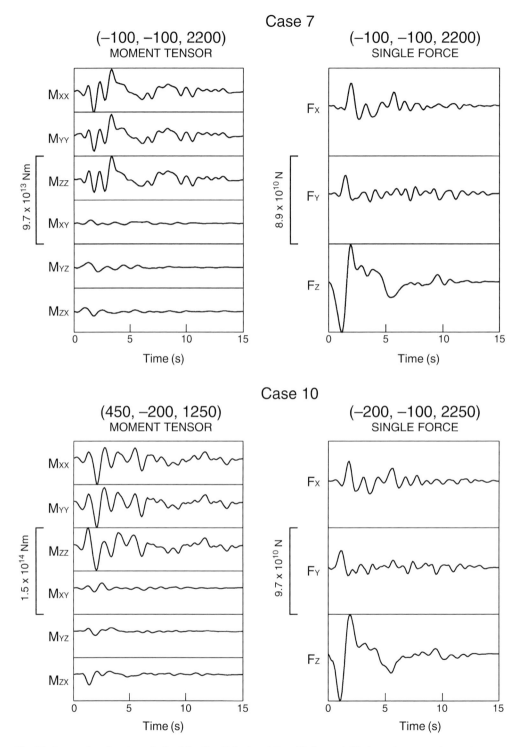

Fig. 11. Source-time functions obtained for Case 7 (top) and Case 10 (bottom). Horizontal axes are time and vertical axes are amplitude. For Case 7, the locations of point sources of six moment-tensor components and three single-force components are the same, while they are located at different positions for Case 10. Source locations are shown at the top of each panel.

parameters. For other sources, such as 'Fz', '3SF', and 'ISO' that have smaller numbers of free parameters, the explanation would not work effectively.

Figure 11 illustrates the source-time functions of the six components of the moment tensor and three components of single force for Cases 7 and 10. The source-time functions for three single-force components are almost the same for these two cases, while the source-time functions for six moment-tensor components are significantly different for these two cases. This suggests that the six moment-tensor component does not necessarily reflect the physical process in the source region.

According to Kanamori's lid removal model, the amplitude of the moment-tensor component is obtained by the amplitude of the downward vertical single force multiplied by the extent of the pressurized portion of the conduit capped by a lid. Since the size of such a portion does not exceed the diameter of the summit crater of 400 m, the amplitude of the moment-tensor component is at most 2×10^{13} Nm, which is much smaller than the amplitude of the moment-tensor components for both Cases 7 and 10. This estimation shows that the contribution of the volume change at the source related to the mass withdrawal from the system to the source-time functions for six moment-tensor component is small.

Furthermore, the source-time functions of three diagonal components for Case 10 are not necessarily in phase. This also makes it difficult to give a physical interpretation to the source-time functions of the moment-tensor component.

If we accept the fact that the solution displays a significantly lower residual error and clear minimum AIC compared to the other models, we need to interpret the source location in the east. Guided by the steeply dipping crack model obtained at Popocatépetl Volcano, Mexico (Chouet et al. 2005), it is conceivable that the upper conduit under Asama may be steeply inclined to the southeast and slightly offset from the distribution of earthquakes at a depth of 1.2 km below the summit crater. The fact that there is no hypocentre distribution that connects the summit crater and the '6Mom' source nor the hypocentre distribution connecting the '6Mom' source and the horizontal dyke inferred from GPS observations (shaded region in Fig. 10) seems to be against the idea of a steeply inclined conduit model. However, hypocentre distributions do not necessarily coincide with the magma path beneath volcanoes (e.g. Brancato & Gresta 2003), and thus, solely from the hypocentre distribution, we should not discard the inclined conduit model, which is favourable to the solution for Case 10.

Summary

We re-analyzed the waveform data associated with the vulcanian eruptions that occurred during the 2004 Mt. Asama activity. We loosened the restrictions on source locations and source mechanisms used in the previous study. The results are essentially the same as those obtained with more restrictions in the previous study. The source locations are slightly shifted in both the horizontal and vertical directions, but the characteristics of the source-time histories of the vertical single-force component do not change.

The source mechanisms include dominant vertical single-force components. The source-time history of the vertical single force starts with a large downward phase followed by an upward phase. After several seconds, another downward component appears. The initial downward force can be explained by the sudden removal of a lid capping the pressurized conduit. The following upward force can be interpreted as a drag force due to ascending viscous magma. The secondary downward force appears when the fragmented magma stops exerting an upward viscous drag force on the conduit wall, and the downward force due to the cap removal reappears.

The results of the analyses assuming two point sources suggest that the conduit has a constriction at sea level that works as a deep source of the secondary downward single-force component. The results of the analyses assuming one point source of single-force components and another point source of six moment-tensor components suggest an inclined conduit model, although the inclined conduit is not necessarily suitable for the hypocentre distribution beneath the summit crater.

The author greatly thanks Steve Lane, who organized the wonderful WS and the Special Publication of the GSL on 'Fluid motion in volcanic conduits: A source of seismic and acoustic signals'. The author also thanks Jeff Johnson and Bernard Chouet, who gave us valuable and intense comments which improved the manuscript greatly. We also thank Jennie Gilbert, who patiently encouraged us to complete this work.

References

AKAIKE, H. 1974. A new look at the statistical model identification. *IEEE Transactions on Automatic Control*, **AC-9**, 716–723.

AKI, K. & RICHARDS, P. 1980. *Quantitative Seismology*, Vol. I. Freeman, New York.

AOKI, Y., WATANABE, H., KOYAMA, E., OIKAWA, J. & MORITA, Y. 2005. Ground deformation associated with the 2004–2005 unrest of Asama Volcano, Japan. *Bulletin of the Volcanological Society of Japan*, **50**, 585–584 (in Japanese with English abstract).

ARAMAKI, S. 1963. Geology of Asama Volcano. *Journal of the Faculty of Science, University of Tokyo*, **14**, 229–443.

BRANCATO, A. & GRESTA, S. 2003. High precision relocation of microearthquakes at Mt. Etna (1991–1993 eruption onset): a tool for better understanding the volcano seismicity. *Journal of Volcanology and Geothermal Research*, **124**, 219–239.

CHOUET, B. 1996. New methods and future trends in seismological volcano monitoring. *In*: SCARPA, R. & TILLING, R. I. (eds) *Monitoring and Mitigation of Volcano Hazards*, Springer-Verlag, Berlin.

CHOUET, B., DAWSON, P. & ARCHNIEGA, A. 2005. Source mechanism of vulcanian degassing at Popocatepetl Volcano, Mexico, determined from waveform inversion of very long period signals. *Journal of Geophysical Research*, **110**, doi: 10.1029/2004JB003524.

KANAMORI, H., GIVEN, J. & LAY, T. 1984. Analysis of seismic body waves excited by the Mount St. Helens eruption of May 18, 1980. *Journal of Geophysical Research*, **89**, 1856–1866.

MURAKAMI, M. 2005. Magma pluming system of the Asama volcano inferred from continuous measurement of GPS. *Bulletin of the Volcanological Society of Japan*, **50**, 347–361 (in Japanese with English abstract).

NAKADA, S., YOSHIMOTO, M., KOYAMA, E., TSUJI, T. & URABE, T. 2005. Comparative study of the 2004 eruption with old eruptions at Asama Volcano and activity evolution. *Bulletin of the Volcanological Society of Japan*, **50**, 303–313 (in Japanese with English abstract).

OHMINATO, T. & CHOUET, B. 1997. A free-surface boundary condition for including 3D topography in the finite difference method. *Bulletin of the Seismological Society of America*, **87**, 494–515.

OHMINATO, T., CHOUET, B., DAWSON, P. & KEDAR, S. 1998. Waveform inversion of very-long-period impulsive signals associated with magmatic injection beneath Kilauea volcano, Hawaii. *Journal of Geophysical Research*, **103**, 23839–23862.

OHMINATO, T., TAKEO, M., KUMAGAI, H. ET AL. 2006. Vulcanian eruptions with dominant single force components observed during the Asama 2004 volcanic activity in Japan. *Earth Planets Space*, **58**, 583–593.

ONIZAWA, S., MATSUSHIMA, T., OIKAWA, J. ET AL. 1996. GPS and gravity measurement at Asama volcano. *Abstracts of the Volcanological Society of Japan*, Fall Meeting, B19 (in Japanese).

SPARKS, R. S. J. 1997. Causes and consequences of pressurisation in lava dome eruptions. *Earth and Planetary Science Letters*, **150**, 177–189.

TAKEI, Y. & KUMAZAWA, M. 1994. Why have the single force and torque been excluded from seismic source models? *Geophysical Journal International*, **118**, 20–30.

TAKEO, M., AOKI, Y. & OHMINATO, T. 2006. Magma supply path beneath Mt. Asama volcano, Japan. *Geophysical Research Letters*, **33**, doi:10.1029/2006 GL026247.

WALDHAUSER, F. & ELLSWORTH, W. L. 2000. A double difference earthquake location algorithm: method and application to the Northern Hayward Fault, California. *Bulletin of the Seismological Society of America*, **90**, 1353–1368.

YAMAMOTO, M., TAKEO, M., OHMINATO, T. ET AL. 2005. A unique earthquake activity preceding the eruption at Asama volcano in 2004. *Bulletin of the Volcanological Society of Japan*, **50**, 393–400 (in Japanese with English abstract).

YOSHIMOTO, M., SHIMANO, T., NAKADA, S. ET AL. 2005. Mass estimation and characteristics of ejecta from the 2004 eruptions of Asama volcano. *Bulletin of the Volcanological Society of Japan*, **50**, 519–533 (in Japanese with English abstract).

Dome-building eruptions: insights from analogue experiments

S. J. LANE[1], J. C. PHILLIPS[2] & G. A. RYAN[3]

[1]*Department of Environmental Science, Lancaster University, Lancaster, LA1 4YQ, UK*
(e-mail: s.lane@lancaster.ac.uk)

[2]*Department of Earth Sciences, University of Bristol, Wills Memorial Building,*
Queen's Road, Bristol, BS8 1RJ, UK

[3]*Montserrat Volcano Observatory, Flemmings, Montserrat, West Indies*

Abstract: Laboratory flows, self-capped by high-viscosity fluid, exhibit vertical pressure gradients similar to those postulated within conduits feeding dome-building eruptions. Overpressure and pressure cycles exhibited at laboratory scale provide insight into the mechanism of tilt cycles at volcanic scale. Experimental pressure cycles correlated with the rate of gas escape, with pressure rise being controlled by diffusion of volatile into bubbles during times when gas escape from the flow was negligible. The increase in pressure continued until margin decrepitation created preferential pathways for rapid gas escape from permeable foam, thereby reducing pressure within the flow. As pressure reduced, the gas escape pathways sealed and diffusion repressurized the system. This implies that tilt cycles, such as those exhibited by the Soufrière Hills Volcano, Montserrat, result from a diffusively pumped process that oscillates around the viscous-elastic transition within the outer regions of the flow. Phases of open- and closed-system degassing result, with gas escaping through fractures created and maintained by the flow process itself. Fluid-dynamically, this mechanism generates an oscillation between Poiseuille flow and plug flow, with a 1-D model of plug motion giving a reasonable representation of observation in both experimental and volcanic cases.

The existence of overpressure (also known as superstatic pressure or excess pressure) within conduits feeding dome-building eruptions is most dramatically illustrated by the common occurrence of explosive activity (Newhall & Melson 1983), and variation in overpressure may be measured as ground deformation (Voight *et al.* 1998). Overpressure may be defined as there being a positive value to the difference between the absolute pressure in the flow and that in the surrounding country rock. Pressure in the flow is determined by gravity acting on the static overburden, dynamic effects, and the pressure gradient driving the flow against the fluid rheology. During dome growth, dynamic effects are negligible because flow velocities are small. However, rheological heterogeneity has a significant effect on changing the pressure from the static value. Sparks (1997) developed a theoretical model of pressure within magma-filled conduits and overlying domes where magma viscosity increased non-linearly with decreasing depth (i.e. pressure) due to loss of water from magma (Eichelberger *et al.* 1986; Hess & Dingwell 1996) and the growth of crystals. This suggested the build-up of substantial overpressure within a few hundred metres of the surface in volcanic conduits feeding domes.

The models of Sparks (1997) and Melnik & Sparks (1999), as applied to the Soufrière Hills Volcano, Montserrat (SHV), show rapid decline in overpressure approaching the surface. Physically, this is because the flow rate is independent of viscosity in regions of low viscosity, which forces the steepest pressure gradients to occur where viscosity is highest; therefore the region of rapid pressure decline defines a rheologically stiffened plug that controls flow in the conduit. Here we use a series of analogue experiments, using a degassing fluid whose viscosity is a function of volatile content, to mimic flow in conduits feeding volcanic domes and give insight into the subterranean processes accompanying dome growth.

The experimental flow is driven by an exsolving and expanding gas phase and the liquid phase of the flow has a viscosity that varies with temperature and dissolved volatile content in a manner analogous to hydrated silicate melts. The experimental foams continued to expand and flow for many hours after the initial explosive event. It was, therefore, possible to investigate the experimental flows for much longer than in previous studies such as Lane *et al.* (2001). Phenomenological similarity between experimental system behaviour and documented dome-building behaviour was observed in

From: LANE, S. J. & GILBERT, J. S. (eds) *Fluid Motions in Volcanic Conduits: A Source of Seismic and Acoustic Signals.* Geological Society, London, Special Publications, **307**, 207–237.
DOI: 10.1144/SP307.12   0305-8719/08/$15.00 © The Geological Society of London 2008.

the form of a highly non-linear pressure gradient, on which was superimposed long timescale (seconds to minutes) cyclic pressure variations. The possibility of a relationship between the experimental phenomena and conduit overpressure (Sparks 1997) combined with volcanic tilt cycles (Voight *et al.* 1998, 1999) is explored.

Experimental methods

The experimental fluid used (GRDEE) was a solution of thermally vacuum-degassed gum rosin (GR, predominantly abietic acid, $C_{19}H_{29}COOH$) and $19.0 \pm 0.2\%$ w/w diethyl ether (DEE, $C_4H_{10}O$). The rapid decompression of this solution produces expanding gas-liquid flows that can fragment and produce solid, vesicular, pumice-like products. The applicability of gum rosin as an analogue for aluminosilicate melt, and an organic solvent for water dissolved in that melt, has been discussed by Cobbold & Jackson (1992), Phillips *et al.* (1995), Lane *et al.* (2001), Blower *et al.* (2001*a*), Mourtada-Bonnefoi & Mader (2001), Ryan (2002), and Bagdassarov & Pinkerton (2004). We measured the viscosity, η (Pa s), of experimental GRDEE as a function of volatile concentration, [*DEE*] (mol m^{-3}), and temperature, T (K), with $\log_{10}\eta = -0.0534T + 16.03$ ($R^2 = 0.986$, [*DEE*] $= 2560$ mol m^{-3}, 275 K $< T < 305$ K), and $\log_{10}\eta = 10774/[DEE] -3.9262$ ($R^2 = 0.993$, $T = 293$ K, $2000 < [DEE] < 5000$ mol m^{-3}).

Viscosity of GRDEE solutions was measured using a falling-ball method, and volatile concentration in mol m^{-3} is given by (% w/w volatile $\times$ solution density (kg m^{-3}))/(100 $\times$ volatile molar mass (kg)). In terms of moles per unit volume, 19.0% w/w diethyl ether in gum rosin is equivalent to about 2% w/w water in magma, both being about 2500 mol m^{-3}.

Diethyl ether (DEE) was chosen as the volatile phase in preference to acetone, although gum rosin mixed with acetone (GRA) has been used in most other gum rosin shock-tube experimental studies (Phillips *et al.* 1995; Blower *et al.* 2001*a*; Mourtada-Bonnefoi & Mader 2001; Lane *et al.* 2001). The decision to use DEE was made because it was found to form a solution that was more temporally stable than GRA, the properties of which change significantly after approximately 24 hours; a timescale comparable to source-liquid preparation time and experiment duration. Diethyl ether also has a higher saturated vapour pressure (*c.* 60 kPa at 20 °C) than acetone (*c.* 20 kPa at 20 °C); GRDEE flows are, therefore, inherently more explosive than GRA flows for a given molar volatile content and depressurization (Lane *et al.* 2001).

Lane *et al.* (2001) developed an argument for geometric, thermodynamic and dynamical similarity between the experimental and volcanic systems. Based on this similarity argument, they proposed that the experimental model be used as a means of interpreting seismo-acoustic signals produced by sustained explosive volcanic eruptions, enabling the translation of seismic data into information about flow processes. Lane *et al.* (2001) also demonstrated that pressure oscillations are inherent to degassing foam-flow systems and are linked to changes in flow pattern. Changes in conduit geometry (constrictions, widenings etc.) or elastic coupling to the confining medium may also trigger oscillations. The experimental apparatus used here (Fig. 1) was similar to that used by Lane *et al.* (2001), namely the GRDEE solution was initially isolated from the vacuum by a polymer diaphragm. The vacuum was created by a continuously running vacuum pump evacuating a cylindrical chamber that acted as a vacuum reservoir. Connected to the vacuum chamber was the experimental shock tube of borosilicate glass, 1.5 m long and of 38 mm internal diameter, with the shorter source tube attached via the rupture collar (Fig. 1).

The polymer diaphragm was electrically ruptured by the logging and control software (LabView), rapidly decompressing the GRDEE solution in the source tube from atmospheric pressure to the pressure of the vacuum chamber (*c.* 10 Pa). This rapid decompression caused bubbles of DEE vapour to form in the GRDEE solution, which turned the solution into first a bubbly fluid and then foam. Initially, the foam accelerated up the shock tube and then fragmented into smaller pieces several centimetres in scale, which were then ejected into the vacuum chamber (Fig. 1b). After a few seconds, this fragmentation behaviour (documented in Lane *et al.* 2001) ceased. The foam then slowly ascended the shock tube with 'explosions' at the head of the flow involving small amounts of material being occasionally observed. This previously unstudied slow-flow behaviour forms the focus of our paper.

The experimental shock tube was instrumented using a variety of sensors mounted in transducer blocks (Fig. 1). Piezoelectric (Pz) pressure transducers measured high frequency (>10 Hz) pressure changes. Active strain gauge (ASG) pressure transducers measured absolute pressures over timescales longer than 1 ms. Optical gates measured the positions of foam fragments, and hence their velocities and accelerations. A Barocel absolute pressure gauge measured pressure in the vacuum chamber, which in combination with calibrated pump-speed for diethyl ether, the vacuum system volume, and measuring the amount of volatile remaining

(a)

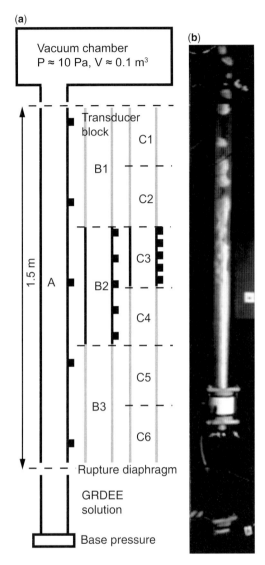

Fig. 1. (a) Shows a schematic diagram of the experimental apparatus. The 19.0% w/w solution of diethyl ether in gum rosin (GRDEE) is loaded into the base section of tube below the rupture diaphragm. The tube above the diaphragm is evacuated to a pressure of *c.* 10 Pa, and the *c.* 0.1m³ chamber acts as a vacuum reservoir to assist the 40 m³/hr rotary vane vacuum pump. The 1.5-m-long flow tube was configurable in a number of ways. Configuration A allowed broad-scale features to be measured and consisted of a single 1.5-m-long tube with five transducer blocks arranged along its length. Configuration B arranged the transducer blocks along a 0.5-m-long tube that could be positioned in one of three places (B1, B2 or B3). For spatially detailed measurement, configuration C arranged the five transducer blocks along a 0.25-m-long tube placed in one of six possible positions. Experimental flow was initiated by rupturing the diaphragm and data was

(Fig. 2), allowed an estimate of the flux of gas being exsolved into the chamber as a function of time. The flows were also recorded with a data-synchronized 25 fps video camera operating at 0.25 ms frame exposure, which allowed flow-front position to be measured as well as providing flow visualization. Post-experiment vesicularity and permeability were obtained by making mass and volume measurements, together with gas flow rate and pressure drop across specific sections of the flow (Fig. 1). These should be considered as approximate, but give indications of trends. For greater details see Ryan (2002).

A single experimental condition (reproducing run 3 of Lane *et al.* 2001) was studied, and Table 1 shows the variability of starting parameters, and Table 2 compares experimental and volcanic parameters. Samples of each source liquid were degassed at 80 °C for 48 hours, giving a measured $17.1 \pm 0.4\%$ w/w diethyl ether. This indicates that $1.8 \pm 0.3\%$ w/w diethyl ether is quite strongly bound with the gum rosin and will not be removed during a decompression experiment. The transducers were placed in different positions for different runs in order to maximize the information obtained about the experimental flow (Fig. 1). The range and number of sensors and the number of run repetitions gave information about all parts of the flow and increased the potential for a more detailed understanding of the behaviour of this experimental system. Estimates of the molar flux of gas escaping from the flow were made. This flux is related to the amount of open system degassing from the flow and gas loss through permeable pathways in the foam. These are both processes that are thought to be important in determining the evolution of volcanic eruptions (Jaupart & Allègre 1991).

The approach used in this study of building up a picture of flow behaviour by combining data from a number of different runs assumes that flows with nominally identical initial conditions behave similarly. This assumption is borne out to large extent by the data, although flow parameters do vary between runs. Analyses of run parameters where transducers were in the same positions show good similarity between experiments (Table 1). This makes it possible to generalize behaviours from one run to all runs. Mader *et al.* (1996) took a similar approach, where video footage of different

collected to a logging system and a synchronized video camera. Flow structure about 0.8 s after decompression is shown in (b). Rapid acceleration at the onset of inverse annular flow results in the separation of degassing foam fragments from the foam core. Fragmenting flow (documented in Lane *et al.* 2001) is initiated by rapid decompression and precedes the formation of the head–body–source structure of the extruding flow that forms the focus of this study.

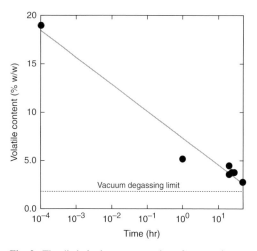

Fig. 2. The diethyl ether concentration of gum rosin remaining in the experimental tube is shown as a function of time. The dashed line represents the concentration of strongly bound diethyl ether (c. 1.8% w/w) inaccessible to the degassing process. There is a suggestion that volatile content decreases logarithmically with time.

parts of the experimental flow from five nominally identical runs was combined to give information about the motion of the flow; good reproducibility of experimental flow behaviour was found.

Results

To set the scene, we summarize the essential phenomenological behaviour of two experiments whose duration was 10^5 s (c. 28 hrs), and give detailed accounts in the following section. The diaphragm ruptured 1.697 s after logging started, resulting in rapid reduction of the pressure above the source solution from atmospheric to c. 10 Pa (Fig. 3). Within 0.02 s, pressure at the tube base fell to within measurement error of the fluid-static value (c. 2360 Pa). Base and chamber pressure (i.e. volatile flux) then increased rapidly (Fig. 3)

on the degassing of diethyl ether from the surface of the source solution, driving a vigorous fragmenting flow (Fig. 1b) that demonstrated flow patterns described by Lane *et al.* (2001). About 0.2 s after flow initiation, bubble nucleation began from the base of the source tube (Figs 4 & 5). The height of the fragmentation level remained relatively stable until bubbles filled the source tube (about 1 s after decompression, Figs 4 & 5), at which point the fragmentation level ascended. Simultaneously, the volatile flux (chamber pressure) and base pressure both increase, peaking about 1.5 s after decompression (Figs 3 & 5). This point represents the maximum pressure drop across the flow (Fig. 3b). After about 3.5 s (1.8 s after flow initiation, see Lane *et al.* 2001, Plate 4) the volatile source is exhausted (Fig. 5) and vigorous fragmentation ended resulting in a rapid decline in escaping volatile flux (Fig. 3c).

The post-fragmentation flow front developed from the fragmentation level and slowly ascended the shock tube as an expanding flow. Base and chamber pressures continued to decline, but about 3.5 s after flow initiation, unanticipated pressure cycles commenced (Fig. 3b). During the period over which pressure cycles occurred, the gas escape rate appears approximately proportional to the inverse of time, implying that the mass of volatile escaping from the flow during this period was proportional to log [time] (Figs 2 & 3c). A small number of foam plugs separated from the flow front after 3.5 s (Fig. 6), adding to the mass of material ejected during vigorous fragmenting flow and representing a transition flow behaviour between vigorous explosive and non-explosive patterns. After about 100 s, driving pressure stabilized with the pressure cycles increasing in period (Fig. 3b). Another phase of behaviour began here, with the flow front being occasionally disrupted by impulsive removal of a thin carapace of foam (Fig. 7) that then settled back down onto the flow front as loose material. The smallest of the loose material was lofted by gas escaping the flow front. With time, this impulsive behaviour declined.

Table 1. *Reproducibility of pre-, syn- and post-experimental parameters*

Variable	Average ± range or 1 sd error %
Source-liquid volume	267.5 ± 0.5 cm^3 (19 runs)
Temperature	23.0 ± 0.5 °C (19 runs)
Density	1007 ± 7 kg m^{-3} (19 runs)
Viscosity	1.32 ± 0.51 Pa s (19 runs)
Foam column height after 30 mins	156 cm ± 9% (18 runs)
Average pressure at base after 500 s	18 700 Pa ± 5% (9 runs)
Average stable pressure-cycle height	1035 Pa ± 13% (5 runs)
Percentage of gum-rosin mass erupted	38.8 ± 7 (14 runs)

Table 2. *Comparison of parameters between the non-fragmenting phase of the experimental GRDEE flow and a typical effusive rhyolitic eruption*

Parameter	GRDEE flow	Effusive rhyolite eruption
r_b (bubble radius)	10^{-5} m	10^{-6}–10^{-2} m
		Lane *et al.* (2001)
U (average flow velocity)	0.2–1 mm s^{-1}	0.1 ms^{-1}
		Gardner *et al.* (1999)
σ (liquid surface tension)	0.029 N m^{-1}	0.3–0.2 N m^{-1}
	Phillips *et al.* (1995)	Phillips *et al.* (1995)
$\dot{\gamma}^a$ (strain rate at conduit wall)	8×10^{-2}–0.2 s^{-1}	8×10^{-2}–8×10^{-3} s^{-1}
ρ (average density of flowing fluid)	220–140 kg m^{-3}	2200–550 kg m^{-3}
		Proussevitch & Sahagian (1998)
η (liquid phase viscosity)	>1.32 Pa s	10^6 Pa s
		Lane *et al.* (2001)
η_{app} (apparent bulk fluid viscosity)	1600–600 Pa s	10^6 Pa s
L (conduit length)	1.8 m	500–5000 m
		Lane *et al.* (2001)
D (conduit diameter)	0.038 m	10–100 m
		Lane *et al.* (2001)
Ca (capillary number)	10^{-5}–10^{-4}	10^{-2}–10^3
Re (Reynolds number)	10^{-6}–10^{-8}	10^{-2}–10^{-4}
L/D	c. 50	c. 50
R_C/r_b (R_C = conduit radius)	1900	500 to 5×10^7

$^a\dot{\gamma} = 4U/R_c$ (shear rate at the conduit wall assuming Poiseuille flow; gives order of magnitude estimate for shear rate).

After about 4000 s, pressure cycles changed in nature and flow behaviour became less reproducible. In one experiment (Fig. 3a, b), pressure cycles ceased, driving pressure declined, and although the estimated volatile flux became undetectable, chamber pressure remained slightly elevated indicating volatile was escaping from the flow. Driving pressure then recovered between about 17,000 s and 30,000 s, accompanied by a sequence of small-amplitude, rapid base-pressure drops and slower rises. In another experiment (Fig. 3d), irregular pressure cycles of large amplitude and period continued as average pressure declined. Although there was no detectable volatile release over this period in either case, between 30,000 and 35,000 s the chamber pressure was observed to decrease by 2–3 Pa to the c. 14 Pa measured before flow initiation This suggests that slight volatile release was occurring up to c. 35,000 s, but was not evident thereafter, unless there was a large and rapid drop in base pressure (Fig. 3d). The average volatile contents of the remaining material in the tube for the experiments shown in Figure 3 was 3.8 ± 0.3% w/w, with 1.8 ± 0.3% w/w being inaccessible to drive the flow. The relatively steady nature of the base pressure also suggests that diffusion is keeping pace with any gas loss and flow expansion.

Interpretation and analysis

Boiling flow or exsolving flow

Average base pressure (c. 20 kPa) was about one-third the saturated vapour pressure of diethyl ether (c. 60 kPa at 20 °C) during the entire flow. Combining this with the observation that about 1.8% w/w diethyl ether is strongly bound to gum rosin, suggests that gum rosin and diethyl ether form a solution, rather than a mixture, at least at lower volatile concentrations. This is consistent with the behaviour of the GRA system (Phillips *et al.* 1995, Fig. 3a). At higher acetone concentrations (>c. 20% w/w), there is evidence (Phillips *et al.* 1995, Fig. 3a; Mourtada-Bonnefoi & Mader 2004, Fig. 2) that chemical interaction reduces, and acetone vapour pressure approaches that of the pure solvent. Therefore, above about 20%, w/w acetone flows can be considered as boiling, and below about 20% w/w flows are exsolving. This boundary represents an approximate 1:1 molar ratio of gum rosin to acetone (more exactly, c. 19.2% w/w), giving physical reason to a change of behaviour above about 20% w/w acetone. We hypothezie that the oxygen atom in acetone or diethyl ether interacts with the hydrogen atom in the OH group of the abietic acid molecule, possibly as a hydrogen bond. Once the 1:1 ratio is

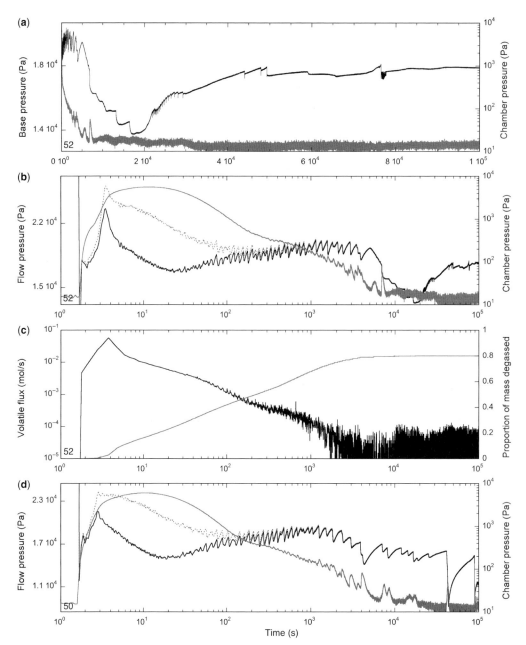

Fig. 3. Two experiments were carried out over a period of 10^5 s. (**a**) Shows pressure measured at the base of the tube (black line) and above the flow in the vacuum chamber (grey line, logarithmic scale) as a function of time. For much of the experiment, these pressures remain stable, with most variability occurring at relatively short time periods. Plotting time logarithmically (**b**) reveals the range of timescales over which pressure is changing. The pressure across the flow (black line) was obtained by subtracting the chamber pressure (grey line, logarithmic scale) from the base pressure (dashed grey line). (**c**) Shows the volatile flux (black line, logarithmic scale), derived from the chamber pressure data in (**b**), and the estimated mass-proportion of volatile lost from the experimental fluid (grey line). Repeatability of experiments is illustrated in (**d**), which can be directly compared with (**b**) for flow pressure (black line), derived from chamber pressure (grey line, logarithmic scale) and base pressure (dashed grey line) as before. The overall pressure trends, as well as the variability over a range of timescales, show considerable similarity with the experiment in (**b**). There are significant differences at longer timescales with large and rapid pressure changes being evident in (**d**).

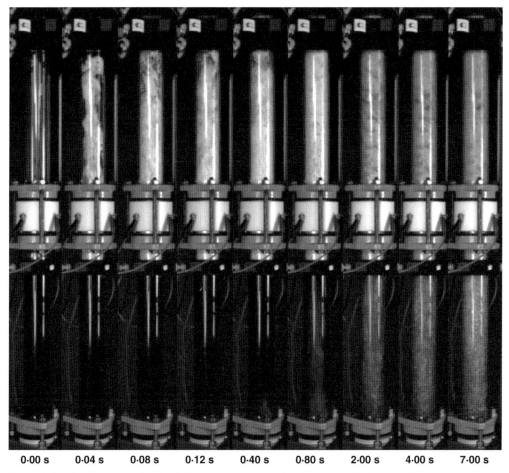

0·00 s   0·04 s   0·08 s   0·12 s   0·40 s   0·80 s   2·00 s   4·00 s   7·00 s

Fig. 4. Selected video frames showing fluid source, the housing for the polymer diaphragm and the lower section of the flow tube to the first transducer block of configuration A (Fig. 1). At **0.00 s**, the haze in the flow tube indicates that this is the first frame exposed after the polymer diaphragm has ruptured. One frame (**0.04 s**) later, rapidly accelerating gum rosin foam escapes from the liquid surface as fragmenting flow stabilizes by **0.08 s**. Before **0.40 s**, bubble nucleation is from the liquid surface, but by **0.40 s**, bubbles are growing from the tube base and within the source liquid. By **0.80 s**, the supersaturated liquid source is nearly exhausted and vigorous fragmenting flow is waning, but the foam at the wall of the flow tube is still a pale yellow, indicating a high degree of degassing. By **2.00 s**, continuous inverse annular flow is not stable and fragmenting flow has ceased. The colour of the foam in the flow tube has darkened, indicating an increase in volatile concentration at the wall. Features evident in the flow margin at **2.00 s** are blurring at **4.00 s** and all but gone by **7.00 s**. This suggests that liquid viscosity was significantly less at this height than in fragmenting flow and represents increasing pressure as a flow-front plug forms. Flow is characterized by the detachment of sizeable foam plugs (Fig. 6) from the flow front, which can be considered as intermittent inverse annular flow. The last of the flow-front plugs attains sufficient wall shear strength to resist driving pressure and forms the rheologically stiff plug that controls subsequent extrusive flow behaviour.

reached, additional solvent does not interact preferentially with gum rosin molecules and behaves more as it would in pure solvent. The 1 : 1 molecular ratio for diethyl ether represents about 24.5% w/w, therefore experiments carried out here (19.0% w/w) are considered as degassing by exsolution (in similarity with dehydrating magma), rather than boiling.

Flow structure

Figure 8 shows the upper *c*. 12 cm of an established (>100 s old) expanding flow. The upper *c*. 7 cm, termed the 'flow head', was pale yellow, and visibility of the tube wall indicated that wall wetting was not occurring. Individual bubbles were too small to be resolved, but structure was visible.

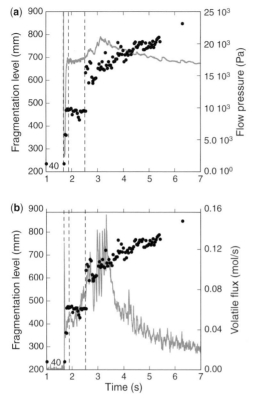

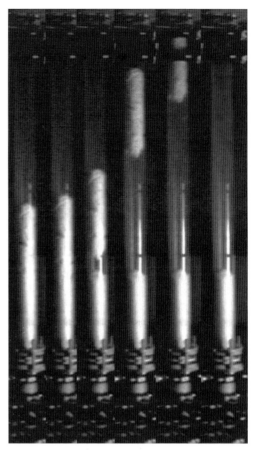

Fig. 5. Fragmentation level (black dots) as a function of time is compared to (**a**) flow driving pressure, and (**b**) calculated volatile flux. The time of diaphragm rupture (1.69 s); the time at which bubble nucleation first starts at the base of the source tube (1.90 s); and the time at which bubbles fill the source tube (2.54 s) is also indicated on the plot.

Fig. 6. The separation of a foam plug is shown in video frames 0.5 s apart. Separation results in a large downward change in flow-front position and precedes slower extrusive motion (Fig. 10a). Although images are of low resolution, there is visible detail at the margin of the detaching plug, but very little is visible in the remaining fluid. This suggests that high-viscosity degassed material (at least at the wall) in the plug separates from lower-viscosity fluid in the underlying flow. Compare this to the higher-resolution image of steady extrusive flow in Figure 8.

Post-experiment analysis revealed the upper flow had a strong radial structure, with a heavily degassed, essentially solid foam 'shell' at the wall, and liquid material relatively rich in dissolved volatile at the centre; a structure not observed when liquid viscosity remains constant or independent of volatile content (e.g. Namiki & Manga 2005, 2006, 2008; Taddeucci *et al.* 2006). The visible structure partly represented lumps of degassed foam from small impulsive explosions (Fig. 7) that were incorporated at the flow margins as it expanded. These observations suggest that flow progressed by extrusion of material from the flow front, with the foam ascending through the centre of the flow and spreading out to the wall on reaching the flow front to form the degassed 'shell'. The ejection of foam plugs (Fig. 6) after vigorous fragmentation, but before extrusive flow became established, possibly represented separation of embryonic 'dry' foam

heads that then slid up the tube driven by gas pressure from beneath. Once established, the flow head appears to plug the tube and control the rate of expansion of the flow, preventing the rapid expansion of more volatile-rich fluid below.

Deeper in the flow, termed the 'flow body', the volatile content is higher, resulting in a darker colour and evident wall wetting. Individual bubbles are visible here, and the base of the degassed shell is consumed as volatile diffuses into the foam that was once degassed at the flow front. Beneath the flow body is the source region extending to the tube base, where bubbles became

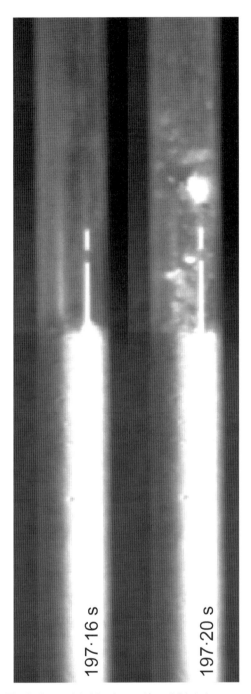

197·16 s

197·20 s

Fig. 7. Sequential video frames ($\Delta t = 0.04$ s) show impulsive removal of the thin flow cap generating foam fragments. Fine fragments were lofted by escaping organic vapour and larger fragments became trapped at the wall by extrusive flow.

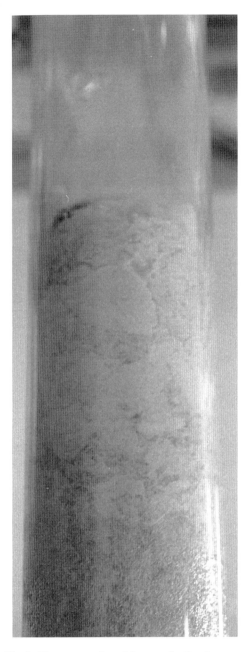

Fig. 8. The upper section of the extrusive flow has a carapace of pale yellow foam fragments from impulsive events. Fragments can be seen trapped at the walls by fluid emerging from the top of the flow and are apparent some distance below the flow front, but do not wet the glass wall. As the flow expands, degassed fragments are reinvaded by volatile because pressure and volatile concentration rise. This darkens their yellow colour and allows wall wetting. Some small pale-yellow remnant fragments can be seen surrounded by amber material. The amber material at the base of the image can be seen to be foam. The outside diameter of the tube is 46 mm.

large with time and pressure remains high (Fig. 3). This structure is similar to that for unfragmenting GRA flows shown in Figure 3 of Lane *et al.* (2001).

Vesicularity and permeability

Overall, flow vesicularity was estimated to be 90–93% from knowing the volume of source liquid, the final flow volume and the amount of material ejected during fragmentation. In more

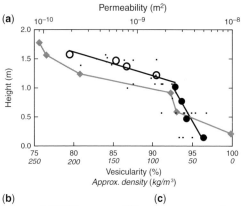

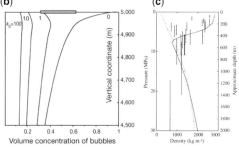

Fig. 9. Post-experiment vesicularity (*density*) of Run 52 (source, filled circles; body and head, open circles), with a linear curve fit to each segment, is shown in (**a**). Measurements from eight other experiments are shown with small black squares. The post-experiment specific permeability of Run 50 is shown on a logarithmic scale by the grey diamonds. Note that the permeability measurement technique is likely to alter foams, especially those with high volatile content in the lower regions of the tube. (**b**) Reproduces Figure 2 from Melnik & Sparks (1999) (reprinted by permission of Macmillan Publishers Ltd, *Nature* copyright © 1999), which shows that the volume concentration of bubbles is likely to decline in the high-viscosity plug of a dome-building eruption when vertical magma permeability is within the measured range. Models that also incorporate gas escape through the conduit wall ((**c**), solid line in Fig. 8 of Clarke *et al.* (2007), reproduced with kind permission of Elsevier) allow the dense plug to develop to greater depth, showing greater similarity to experiment.

detail, Figure 9a shows that bubble volume fraction decreased with increasing height. The source region had vesicularities in the region of 95%, and was discriminated from the flow body and head by a significant change in the vesicularity gradient, with gas volume fractions falling to $<80\%$ at the top of the flow. This suggests that foam is compressed and increases in density as it degasses towards the flow front. Axial permeability for the whole flow cross-section (Fig. 9a) was in the region of $>2 \times 10^{-9}$ m^2 in the source region, and then decreased rapidly by an order of magnitude on entering the flow head. Flow permeability of any foam-column segment incorporating the head was in the range of 8×10^{-11} to 1.3×10^{-10} m^2 (five experiments), suggesting that the flow head/cap acted to control the overall flow permeability and, hence, the rate of degassing due to passive permeability processes. These axial permeability values are very similar to those found in non-fragmenting expanding flows of corn syrup with $>70\%$ v/v chemically generated CO_2 ($10^{-11}-10^{-9}$ m^2, Fig. 9 of Namiki & Manga 2008) where bubbles were elongated. In corn syrup, a capillary tube model of *permeability* = $\delta^2/32$ (δ is the bubble diameter (10^{-4} m) and less porosity) provides reasonable representation of experimental data. The same model predicts permeability in the flow head for gum rosin, but breaks down in the high vesicularity of the flow source where bubbles are not elongated.

Flow expansion

During the fragmentation stage, degassed high-viscosity fluid was removed from the flow. However, once the source liquid was unable to supply fresh material into the expanding flow at sufficient rate to sustain the flow pattern changes leading to fragmentation (Lane *et al.* 2001), then the degassed fluid acts to restrict the flow and control the rate of expansion by 'plugging' the tube. Figure 5 illustrates fragmentation-front position data, together with pressures and volatile escape flux, for the first 7 s of an experiment. On initial decompression, fluid expansion probably moved the flow front with at least constant acceleration (Phillips *et al.* 1995), but more likely exponential displacement with time (t) (Navon *et al.* 1998; Mourtada-Bonnefoi & Mader 2001). Flow was being driven by high volatile supersaturation and retarded by viscosity and/or inertia, but diffusion of volatile was unlikely to be controlling the flow propagation on these timescales. This is, however, difficult to confirm when the flow is fragmenting and no stable flow front exists. Were the flow not fragmenting, then the position of the flow front (h) is likely to have been approximated by

$h \propto exp[100t]$ (Mourtada-Bonnefoi & Mader 2001, Fig. 5). Figures 5 and 10 show the presence of two fragmentation stages. From 1.69 s to 2.54 s in Figure 5, fragmentation is vigorous and the fragmentation front stable. Between 2.54 s and 3.54 s, a small number of foam plugs (possibly embryo degassed flow-heads) are ejected and the volatile flux reaches a maximum whilst showing distinct peaks. The peaks are thought to represent rapid volatile loss as individual foam plugs break from the flow beneath and reopen a clear degassing path to the vacuum chamber. After about 4 s, volatile flux has declined and the flow front has become more stable.

Analysis of video images allowed measurement of the position of the flow head as a function of time. Figure 10 shows the flow front stabilizing after about 10 s following separation of a c.

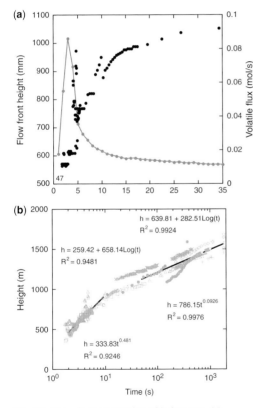

Fig. 10. Flow front progression with time was, (**a**), initially indeterminate during fragmentation (see also Fig. 5), then erratic as foam plugs continued to separate (Fig. 6). After about 15 s, flow front position stabilized and progressed more smoothly up the tube. On average, the relationship between flow front position and time (**b**) shows two regimes between 3 and 10 s, and between 20 and 2000 s. These can be related with both logarithmic or fractional power-law relationships (black lines).

70 mm-long foam plug during one experiment (Fig. 6 shows example). This foam plug slid against the tube wall before separating from the fluid beneath. An overall view of post-fragmentation flow-head position for eight runs as a function of time for 2000 s is shown in Figure 10b. Flow expansion shows two regimes, both of them decelerating. Between about 5 s and 10 s, flow expansion is approximately proportional to the square root of time, which may indicate diffusive control (Navon *et al.* 1998). Behaviour can alternatively be described logarithmically (Proussevitch & Sahagian 1998), also suggesting diffusive control. During this time, the flow head slides against the tube wall and represents the final expression of the inverse annular or detached flow pattern (gas at wall with liquid or foam core in centre) required for fragmentation in this degassing system (Lane *et al.* 2001). Interestingly, detached flow precludes fragmentation where flow is driven by bubble expansion alone (Namiki & Manga 2005). After about 20 s, flow expansion becomes approximately proportional to $t^{0.09}$, again also describable as a logarithmic relationship, and possibly representing increasing viscous control as the sliding of the flow head against the tube wall ceased and extrusive flow becomes established (Fig. 8). Figure 11 shows the declining velocity of the flow head, with a dependence on $c. t^{-0.9}$.

Impulsive events

Figure 7 illustrates impulsive events observed during the early stages (<1000 s, Fig. 12) of the

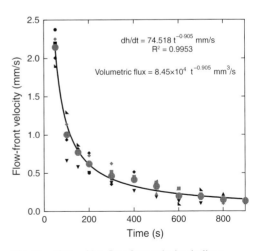

Fig. 11. After c. 20 s, flow front velocity declines almost inversely with time. Data are from six experiments, with a power-law line of best-fit to the average velocity at each time (large grey circles).

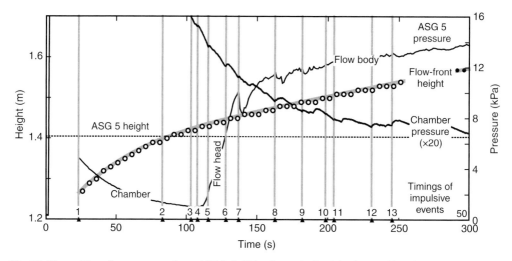

Fig. 12. The position of pressure transducer ASG 5 (1.404 m from tube base) is shown with a dashed line. This is compared to the position of the flow front (open circles, grey curve fit) as a function of time. Pressure measured by ASG 5 (labelled black line) remains at chamber level as the first few centimetres of flow pass the transducer, then rises steeply (flow head), finally rising more slowly (flow body) as ASG 5 measures deeper into the flow. Impulsive events (black triangles and grey lines labelled 1 to 13, see also Fig. 7) sometimes coincide with pressure drops in the flow (e.g. 7), and often pressure rises in the vacuum chamber (labelled black line with 20x pressure exaggeration), but sometimes happen with no detectable pressure effect in the flow (e.g. 6).

slow expansion of the foam. These events appear to be superficial, removing only the top few milli-metres of the flow front. Fine particles generated by each event were lofted on gas escaping the flow front, and larger pieces of foam were moved to the walls as the flow progressed (Fig. 8). Figure 12 shows pressure at the tube wall as the flow front passes a transducer whilst impulsive events are taking place. A pressure drop in the flow-front of *c.* 1.5 kPa is coincident with event 7, and events 8 and 9 generate smaller pressure decreases of about 400 and 300 Pa respectively (Fig. 12). None of the other impulsive events generates mea-surable pressure change elsewhere in the flow, suggesting that impulsive events cause depressuri-zation within < 10 cm of the flow front. Events 5 and 6 occur when the flow is pressurizing the transducer, but no change is measured (Fig. 12). This indicates that the top *c.* 5 cm of the flow did not respond significantly to the pressure decrease, possibly because of rigidity and low radial per-meability of degassed gum rosin material in the shell at the wall. Once the transducer is deeper in the flow, pressure change is communicated to the tube wall, possibly because radial permeability is higher or the volatile content of the gum rosin is increased at the wall and viscous flow can occur. Chamber pressure indicates that some events (e.g. 6, 8 and 13) coincide with significant releases of volatile, whilst others (e.g. 7, 9 and 10) cause

only small or undetectable change in chamber pressure (Fig. 12). Note that there are significant changes in chamber pressure in the absence of impulsive events.

We hypothesize that impulsive events result from the brittle bending-failure of a highly degassed flow 'cap' that represents the upper extreme of the flow head. Impulsive events commence once extru-sive flow becomes established, with pressure under the low-permeability flow head/cap increased by diffusion from below, and decreased by gas escape through the flow cap; the flow is in dynamic equilibrium. The steadily rising nature of the pressure-drop across the flow (Fig. 3) after 20 s suggests that these gas fluxes are approxi-mately in balance, with the rise in pressure drop being attributable to the increase in flow length to maintain pressure gradient. Degassing from the flow cap will cause it to become increasingly brittle at a particular deformation rate, and cap failure occurs once ductile deformation can no longer be maintained. Release of stored elastic energy within the cap creates the impulsive event. Rapid volatile loss from the newly exposed GRDEE foam quickly halts any increase in expan-sion, limits the propagation depth of the pressure decrease, and creates the new flow cap ready for the next event. Deformation rates decline with time (Fig. 11) and the amount of degassing required for the ductile-brittle transition (Papale 1999) increases until it no longer occurs on the experimental timescale, if at all, and impulsive events stop. The occurrence of rises in chamber pressure in the absence of impulsive events, and the presence of small and large changes, suggests the operation of other, possibly cyclic, degassing mechanisms.

Pressure gradients

The pressure change illustrated in Figure 12 as the flow front passes transducer ASG 5, suggests that the vertical pressure gradient in the flow is non-linear. The flow front (cap) passes ASG 5 (1.404 m) at 89 s. Pressure starts to rise at 112 s, when the flow front is at 1.430 m. This indicates that the flow front did not apply radial pressure to the wall and that the top $c.$ 2.6 cm may comprise a domed rigid cap overlain with a layer of debris (Fig. 8). As the flow progresses past ASG 5, pressure rises steeply by $c.$ 10 kPa over a distance of $c.$ 2.3 cm. This gives a pressure gradient of $c.$ 400 kPa/m across the flow cap/head, indicating that wall shear stress (τ_w) in the region of 4 kPa can be sustained without the flow cap breaking down (estimated from $R_c \Delta P / 2 \Delta z$, where R_c is tube radius and $\Delta P / \Delta z$ is pressure gradient (Hanselmann & Windhab 1996)). By this measure, the rheological flow head is $c.$ 5 cm in length at 100 s, in broad agreement with the visually identified flow head (Fig. 8) of 9 ± 4 cm after about 0.5 hours. Over the next 12 cm of expansion, pressure rises by about 3 kPa (Fig. 12), giving a 25 kPa/m gradient (about 200 Pa wall shear stress), and considered to represent the flow body. By comparison, the static pressure gradient is $c.$ 1.5 kPa/m and dynamic effects negligible. The large pressure gradient in the flow head implies much higher apparent viscosity than elsewhere in the flow, and this is likely to control the flow expansion. This is similar to the control of experimental sugar-syrup flows by their high-viscosity skins described in Stasiuk *et al.* (1993), where viscosity increased as temperature decreased. Here, viscosity increases as volatile concentration decreases.

Composite pressure data from a number of experiments using different arrays of transducers reveal pressure gradients at specific times. Once extrusive flow becomes established (>10–20 s), the flow pressure gradient remains stable in form with time, but varies in height as the flow expands. Figure 13a shows composite data from eight experiments at 100 s, together with the flow-head pressure gradient derived from Figure 12 between 112 s and 159 s. Figure 13a shows that the pressure gradient can be seen to define the flow head, body and source. The source region maintained near-constant pressure in both time and height ($c.$ 1.5 kPa/m static) over the duration

of the experiments, and maintained an average height of about 0.78 m after the first 50 s. This suggests that there was little flow in the source region and that viscosity was relatively low. This

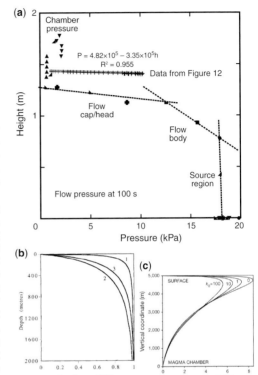

Fig. 13. (**a**) Shows pressure, as a function of height from the base of the experimental tube, measured 100 s after flow initiation. Data were taken at the flow base, along the tube wall and in the vacuum chamber for seven experiments. Regions of the flow are demarked by their pressure gradients. Pressure data from Figure 12 are shown with + symbols, and represent the pressure gradient of the flow head over the time period 112 s to 159 s. The grey line represents a linear fit to this data. Also shown is Figure 1 from Sparks (1997) (**b**), reproduced with kind permission of Elsevier and RSJ Sparks, where normalized overpressure is plotted against depth. Note the similarity in form of the pressure gradients measured experimentally in this work and calculated numerically by Sparks (1997). Figure 2 from Melnik & Sparks (1999) (**c**), reprinted by permission from Macmillan Publishers Ltd: *Nature* copyright © 1999, shows overpressure as a function of vertical conduit position. Overpressure increases with height in the conduit because static pressure is large in comparison; experimentally, static pressure is always small compared with superstatic pressure. Note that the terms 'overpressure', 'excess pressure' and 'superstatic pressure' all refer to the difference between absolute and static pressures at any point.

is consistent with the source region retaining the highest volatile content within the flow, and with the stationary lower boundary.

The flow body maintains a pressure gradient of about 16.5 kPa/m from the 100 s composite data (Fig. 13a), compared to about 25 kPa/s in Figure 12. Using the flow-front velocity at 100 s (*c.* 1.3 mm/s from Fig. 11), and assuming Newtonian rheology and uniform steady flow of viscous liquid (Poiseuille flow), we estimate the effective viscosity of the flow body as approximately 700 Pa s. This suggests a volatile content of about 10% w/w. Interestingly; a more rigorous approach (Ryan 2002) taking account of non-Newtonian foam behaviour yields viscosities in the range 700 to 800 Pa s at low strain rates. Radial heterogeneity and non-Newtonian rheology are not accounted for in Poiseuille flow, but do not appear to play a significant role in the flow body, where these are not as pronounced as in the flow head. The apparent applicability of Poiseuille behaviour to the flow body may result from the controlling role of the flow head.

The flow head shows a pressure gradient of about 90 kPa/m at 100 s. However, the pressure gradient derived from Figure 12 (*c.* 335 kPa/m, Fig. 13) is considered more representative because of the much higher data density, although being measured at the wall means this is a maximum value. This gradient is much higher than in either the flow body or the static value, and its development allows the lower regions of the flow to become significantly super-pressured compared to the static value. The same naïve assumption of Poiseuille flow gives a viscosity estimate of *c.* 10 kPa s, but strong radial heterogeneity, foam yield strength and viscoelastic behaviour in the flow head/cap make a single value unlikely to be realistic. At 10 kPa s, the volatile content is about 7% w/w (Fig. 2 of Lane *et al.* 2001). This appears unrealistically high, as the average volatile content is at least 5% w/w (Fig. 2) and the flow head will have much less volatile concentration than the average; the assumption of Poiseuille flow is probably inadequate here.

Force, applied by pressure in the flow source to the low-permeability flow head, is balanced by force exerted back by the yield stress of the confining flow head. This gives an estimate of the yield stress of the low-permeability foam as

$$\tau_y = \frac{\Delta P R_c}{2B}, \tag{1}$$

where B is the length of the yielding section of flow (estimated to be about 0.5 m, Figs 9 & 13), R_c is the tube radius and ΔP is the pressure drop across the yielding foam column (*c.* 10 kPa, Fig. 13). This gives a flow-head yield stress of *c.* 200 Pa. Foam in the body and source regions are also likely to have a yield stress. Using the whole flow length, the calculated yield stress reduces to *c.* 70 Pa.

A theoretical estimate of foam yield stress is given by (Gardiner *et al.* 2000) as

$$\tau_y = \frac{0.51(\phi - \phi_c)^2 \sigma}{r_b}, \tag{2}$$

where ϕ is gas volume fraction, σ is surface tension, r_b is mean bubble radius and ϕ_c is the critical packing fraction for randomly packed spheres, which has a value of 0.62. ϕ of the flow source has an average value of 0.95, and 2.5×10^{-5} m was used for r_b and a value of 0.029 Nm^{-1} for the surface tension (Phillips *et al.* 1995). This gives a yield stress for the foam of about 65 Pa, in broad agreement with the whole-flow value calculated from experiments. This analysis suggests that overall flow expansion may be resisted by surface tension during the interaction and deformation of numerous small bubbles.

Pressure cycles

Cyclic pressure fluctuations were recorded at the base of all flows (Fig. 3), and at higher transducers located in the flow source, but not those at the level of the flow head. Cycles became resolvable once fragmenting flow waned after about 3 s, and stable in waveform after about 60 s when extrusive flow is established and the head-body-source configuration has fully developed. Figure 14 shows that a typical base (i.e. flow source) pressure cycle comprises a pressurization phase followed by a more rapid depressurization phase resulting in a sawtooth type waveform. The cycle height varied within a limited range, initially being about 250 Pa then increasing to the average stable pressure-cycle height of 1035 Pa ± 13% (five experiments) after about 100 s.

The pressure-cycle period (T_p) increased linearly with time (t). The empirical relationship $T_p = mt + t_0$, where m and t_0 are constants, was found to be a good representation of the experimental data. Over five experiments, the average value of m was 0.11 ± 13%, and the average t_0 was −1.3 s ± 25%, with linear regression correlation coefficients exceeding 0.95 for individual experiments. Video imagery showed small but wholesale up-and-down movement of the bubbly fluid in the source region, which was related to the depressurization phase of the pressure cycles (Fig. 14c). Given the relatively constant cycle height, the period can

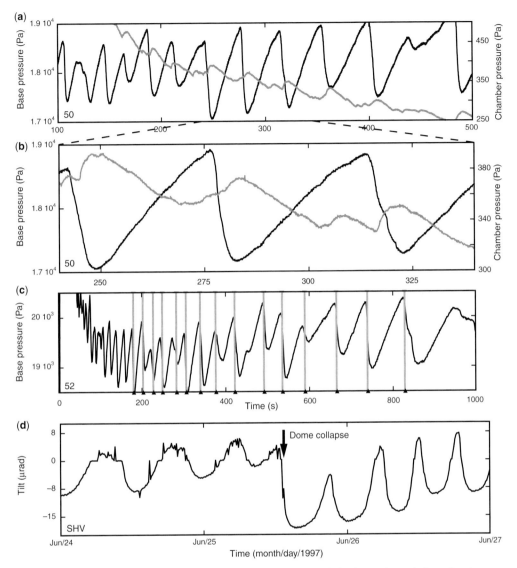

Fig. 14. Pressure cycles measured at the flow base showed, (**a**) stable height but increasing period as a function of time (black line, see also Fig. 3d). Chamber pressure (grey line) also showed cyclical behaviour. (**b**) In general terms, base pressure increase (black line) was correlated to chamber pressure decrease (grey line), and the more rapid base pressure decrease to chamber pressure increase. This suggested that base pressure drop was the result of gas escape from the flow. Visible disturbances within the flow source correlate (**c**) with the rapid base-pressure drops. Tilt cycles measures at SHV in 1997 (**d**) show similar waveforms to the pressure cycles measured at the base of the experiments (Barry Voight is gratefully acknowledged for access to tilt cycle data).

only increase if the rates of pressure change decrease. Figure 15a shows that the rate of flow pressurization declines steadily from about 150 Pa/s to about 10 Pa/s between 80 s and 1000 s when cycles are most stable. Rates of depressurization are more rapid (several 100 Pa/s) and variable in time, but generally also decline.

This suggests that different mechanisms may control these two processes.

Pressure cycles were also detected within the vacuum chamber, and correlated with the base pressure cycles (Fig. 14b). Relatively rapid rises in chamber pressure (*c.* 5–10 Pa/s, degassing rate exceeded pump rate, Fig. 15b, c) correlated with

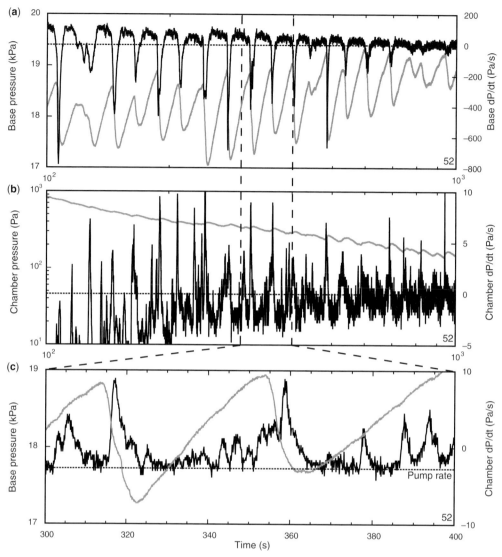

Fig. 15. Between 100 s and 1000 s, pressure cycle height remained stable but period increased. For base pressure (**a**), this resulted from a steadily declining rate of pressurization. Depressurization rates were an order of magnitude higher, more variable, and did not decline steadily with time. For chamber pressure (**b**), pressure fall was related to pump rate and absolute pressure as well as gas escape rate. Again, rates of pressure rise showed no consistent pattern with time, and were considerably higher than rates of pressure fall. The relationship between base pressure and gas escape rate into the chamber (**c**) is complicated by the background gas removal of the pump. The dashed line in (c) shows the gas-removal rate, and when the rate of chamber pressure change (black line) is above this level, gas is escaping from the flow at detectable rates. Gas escape rates are minimal during the first half of the base pressurization phase, and peak during the maximum rate of base pressure drop.

the rapid fall of base pressure, and the slower rise in base pressure was mirrored by a fall in chamber pressure ($c.$ -2 Pa/s, degassing rate less than pump rate). This indicates that during flow depressurization, gas was decoupling and escaping from the flow in an open-system manner. However, gas escape was much slower during the pressurization phase. The average chamber pressure increase of about 15 Pa represents (ideal gas assumption) about 6×10^{-4} moles of gas. An average source depressurization of about 1000 Pa at 296 K yields a depressurization volume of $c.$ 1.5×10^{-3} m^3.

Given a high gas volume fraction in the flow (Fig. 9a), this suggests that a 1.2-m-length of the flow loses gas during the depressurization phase. This would be consistent with the high-permeability flow source (Fig. 9a) being periodically vented through the low-permeability flow head/cap, and explains the motion seen in the flow source during depressurization.

We hypothesize that once fragmentation has ceased after about 5 s, liquid in the source region is still in a supersaturated state and diffusion is acting to increase bubble pressure. Once cycles stabilize after about 60 s, the rate of base pressure rise during the pressurization phase declines by about 25% within each cycle (Fig. 15a). This could represent falling supersaturation and hence diffusion rate, but may also be explained by expansion of the source region. Here, we make the simplification that over one cycle, the rate of pressure rise is effectively constant, meaning that $\Delta(C_l - C_{eq})/(C_l - C_{eq}) \ll 1$, where C_l is liquid volatile concentration and C_{eq} is the equilibrium volatile concentration.

Bubble pressure in the source is able to increase because of the confining yield stress and low permeability of the flow head/cap, and continues to increase until the foam yield stress is exceeded; at which point the flow shears. Bubbles in any shear zone may coalesce and collapse (Stasiuk *et al.* 1996; Jaupart 1998) to form a preferential gas-escape pathway from the source/body region to the vacuum chamber. Gas escapes because the source/body is relatively permeable (Fig. 9a) and the rapid depressurization phase occurs because the rate of gas escape greatly exceeds that of diffusive replenishment.

As pressure declines, stress decreases and the shearing foam self-seals, closing the preferential pathway for gas escape. Diffusion then increases source-bubble pressure and the cycle repeats. As gas is lost with time, the volatile concentration declines, and rates of diffusion decrease, resulting in decreasing pressurization rate and increasing cycle period. Figure 14 also indicates that both the pressurization and depressurization phases show events on shorter timescales than the main cycles. These may represent partial attempts at a complete cycle and could represent incremental compression of the flow head; however, they are likely to result from the same basic balance of diffusive gas input and gas escape during structural failure of the flow.

Physical motion of the flow front was not observed to significantly coincide with pressure cycles (Fig. 12), even though pressure cycles represent a major degassing mechanism. However, the flow cap was regularly subject to impulsive events (Fig. 12) and establishing such a link was experimentally difficult. This absence of coincidence may indicate that flow front motion was dominated by steady extrusive flow driven by the superstatic source pressure of *c.* 20 kPa, with experimentally unobserved movement due to foam-failure during the *c.* 1 kPa pressure cycles. The decrease in vesicularity in the flow head (Fig. 9a) may indicate foam compression during the cycle pressurization phase, effectively isolating the flow front from significant motion coincident with the pressure cycles.

Diffusion pumping

We hypothesize that the rate of pressure rise during cycles is likely to be diffusion controlled in the source and body sections of the flow. A shell model (Proussevitch *et al.* 1993; Lensky *et al.* 2001) was used in which each bubble was assumed to be a sphere of radius r_b surrounded by a thin shell of fluid of radius S_b. For simplicity, bubble size was assumed to be constant for each cycle, and video images of the source region indicated only a narrow range of bubble sizes at the tube wall. Within each quasi-linear pressurization phase of a cycle, the diffusion coefficient (D) was assumed to remain constant, as was C_l, the volatile concentration at radius S_b. C_{eq}, the concentration of volatile at radius r_b, is the equilibrium concentration of volatile in the fluid at internal bubble pressure, P_b. Since pressure varies by *c.* 5% (1 kPa in 20) during a pressure cycle, equilibrium concentration was assumed to remain constant. The volatile flux into a bubble (adapted from Crank, 1975), when the fluid shell is thin is given by

$$\frac{dN_b}{dt} = \frac{4\pi r_b^2 D(C_l - C_{eq})}{(S_b - r_b)}, \qquad (3)$$

where N_b is the number of volatile molecules inside the bubble, t is time and D is diffusion coefficient. The relationship between bubble pressure and bubble volume is given by the relationship

$$P_b\left(\frac{4}{3}\pi r_b^3\right) = N_b kT, \qquad (4)$$

where k is the Boltzmann constant (Toramaru 1995). This assumes that surface tension effects are negligible (in fact about 4%). If S_b, r_b and D are assumed to be constant (over one pressure cycle), it follows that

$$D = \frac{\Delta P r_b (S_b - r_b)}{3(C_i - C_{eq})kT\Delta t} \qquad (5)$$

ΔP and Δt were measured from each pressure cycle. Starting values of the other parameters, estimated from images (and other measurements), are listed below. Figure 16a shows the diffusion coefficient for diethyl ether in gum rosin calculated from pressure cycles as a function of estimated volatile concentration.

The chamber pressure gauge measured the balance between pump rate and the sum of all the flow degassing processes. Characterization of the pump rate allowed estimation of flow degassing rates. Assuming diffusion is the dominant control on gas loss, once fragmentation has ceased, allows use of the rate of pressure change in equation 5, assuming all other parameter values remain constant for each pressurization phase. Figure 16b shows the diffusion coefficient calculated from chamber pressure data with $r_b = 2.8 \times 10^{-5}$ m, $S_b = 3.02 \times 10^{-5}$ m, initial $C_l = 1.53 \times 10^{27}$ m^{-3} and $C_{eq} = 3.23 \times 10^{26}$ m^{-3}. Data from both flow pressure cycles and chamber pressure yield diffusion coefficients in the same range, suggesting that gas escape rates were, overall, diffusion controlled. Values of the diffusion coefficient for the GRA system were measured as 2.55×10^{-11} m^2/s and 7.6×10^{-12} m^2/s for acetone concentrations of 30 and 25% w/w (Blower pers. comm.), which correspond to molecular concentrations of 3.1×10^{27} m^{-3} and 2.0×10^{27} m^{-3} respectively. The molecular concentration of volatile in the GRDEE system under consideration has a theoretical maximum of 1.53×10^{27} m^{-3}. Given the likely

high dependence of diffusion coefficient on molecular concentration suggested by Figure 16 and Blower (pers.comm.), the values calculated for diffusion coefficient appear to be reasonable on an order-of-magnitude scale. This supports the hypothesis that cycle pressurization was diffusion controlled and the dominant gas loss process was periodic generation of preferential pathways (Fig. 17), with diffusion balancing degassing rates (on average) to yield a flow in dynamic equilibrium.

Volcanological Implications

Broad comparisons

Jenkins *et al.* (2007) investigated the statistics of explosive volcanic activity. 'Episodes' incorporating explosive volcanic activity have a timescale of decades to centuries, and comprise one or more volcanic 'events', generally with timescales of days to years. Each 'event' may be further subdivided into 'stages' (hours to weeks of one style of activity) and 'pulses' (minutes to hours of continuous activity). This gives approximate median timescale ratios of $10^6 : 10^4 : 50 : 1$. If we consider 35,000 s as the point at which the experimental flow became dormant (volatile escape became undetectable, Fig. 3), then in the phraseology of Jenkins *et al.* (2007), this represents a single-event episode. Within this experimental event, vigorous explosive activity lasted about 1 s and ejected about 39%

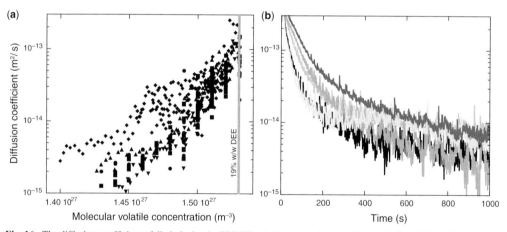

Fig. 16. The diffusion coefficient of diethyl ether in GRDEE solution was calculated from a shell model combined with two different data sources. (**a**) The rate of increase in pressure for individual pressure cycles was assumed to be diffusion controlled. The calculated diffusion coefficient for each cycle in five experiments is plotted as a function of the calculated equilibrium volatile concentration. The grey line indicates the volatile concentration of initial source liquid. (**b**) Assuming an overall diffusive control over degassing allows calculation of diffusion coefficient from chamber pressure measurements of five experiments. There is considerable scatter in the data, but these two approaches agree at order-of-magnitude level suggesting that diffusion plays a significant role in gas loss from this system once fragmenting flow has ceased.

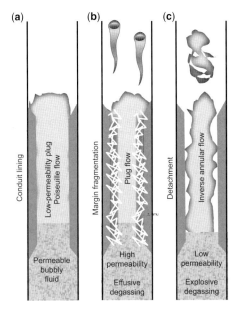

Fig. 17. Schematic cartoons of cyclic flow behaviour based on our experiments and dome-growth literature (see text for details) illustrate behavioural regimes. (**a**) Poiseuille flow of the low-permeability plug impedes gas escape and allows pressure to rise by diffusive pumping. (**b**) Increasing strain rate at the plug-lining interface results in margin decrepitation, the development of degassing pathways and plug flow. Gas decouples rapidly from the high permeability fluid beneath the plug, pressure reduces and Poiseuille flow re-establishes. If the foam beneath the plug is insufficiently permeable to degas efficiently, then (**c**) coupled gas escape in the form of inverse annular flow may develop, leading to explosive behaviour. Pressure then decreases rapidly and Poiseuille flow re-establishes.

of the gum rosin mass. This continuous activity could be considered as a pulse, with an approximate $1:3 \times 10^4$ timescale ratio to its associated event. This meshes with the timescale ratios for explosive volcanic activity (Jenkins *et al.* 2007).

The experimental pressure cycles occupied at least 80%, but probably nearer 95%, of the flow duration (Fig. 3), and resulted in a flow volume expansion by a factor of about 1.8. The pressure cycles were continuous within an event and, therefore, represent a pulse whose duration is close to that of the event. However, volcanically these may be interpreted as stages, therefore the timescales between event and stage just mesh with those found by Jenkins *et al.* (2007). It is interesting to note that even though the experiment appears dormant after 35,000 s, gas release and pressure change may still occur at longer timescales from a system that is not replenished by fresh source fluid. It is also interesting that the ratios of timescales of the

different stages of activity are broadly comparable between laboratory and volcanic systems. This may indicate some fundamental behaviour of degassing systems worthy of further investigation.

Pressure gradients

Rapid decline in overpressure is predicted approaching the surface (see Fig. 1 of Sparks (1997), and Fig. 2 of Melnik & Sparks (1999), reproduced here in Fig. 13b, c) because a rheologically stiffened plug controls flow. The experiments show similar rheological control (Fig. 13a), with the flow cap/head sustaining the largest pressure gradient and being analogous to the plug in dome-building eruptions. Figure 1 of Sparks (1997), which shows dimensionless superstatic pressure as a function of depth, is directly comparable to the experimental pressure gradient because, experimentally, fluid-static pressure is negligibly small (<5%) compared to measured pressure. These approaches demonstrate that substantial superstatic pressures are generated within and below a rheologically stiffened plug. Our experimental work supports the numerical approach taken by Sparks (1997) and Melnik & Sparks (1999), as well as field measurements of ground deformation attributed to pressure sources a few 100 m beneath lava domes (e.g. Voight *et al.* 1998, 1999; Green *et al.* 2006).

Experimentally, the plug is created by diethyl ether escape causing liquid viscosity to rise, but in magmas, microlite growth (due to water loss causing solidus migration, and possibly cooling) is also likely to play a role (Sparks & Pinkerton 1978), especially in static systems (Sparks 1997). However, the experiments suggest that microlite growth is not essential for such pressure gradients to develop, provided flow is occurring. Nevertheless, microlite growth does significantly enhance conduit plugging (Melnik & Sparks 1999). The plug in both experimental and volcanic systems is also likely to have developed some yield strength due to high bubble and crystal volume fractions (Voight *et al.* 1999), as well as the high-viscosity liquid phase encountering the viscoelastic transition within high-shear regions of the flow (Gonnermann & Manga 2003; Taddeucci *et al.* 2006). Such non-Newtonian behaviour may act to increase the plugging effect and result in steeper pressure gradients that allow significant superstatic pressures to approach closer to the surface.

In the lower-viscosity magma below the plug, pressure is elevated above magmastatic levels, and may approach that of the magma chamber overpressure (Sparks 1997, Fig. 1; Melnik & Sparks 1999, Fig. 2, reproduced in Fig. 13). The low-viscosity water-rich magma acts to communicate pressure up the conduit (Voight *et al.* 1999) because the pressure gradient required to maintain flow rate controlled by the water-depleted plug is small compared to that required in the absence of the plug. If the plug were removed, the increase in pressure gradient within the lower-viscosity magma would increase flow rate, possibly to the point of fragmentation.

Experimentally, pressure in the source region is some 20 times the static value and is maintained by the vapour pressure of diethyl ether in the source. This suggests that the vapour pressure of water will contribute significantly to magma overpressure in the presence of a low-permeability rheologically-stiff plug, implying that the overpressure will be a function of the dissolved water concentration. Source magmas with high water content that generate plugs with a large microlite population and high-viscosity liquid component are likely to sustain large overpressures at shallow depths. The plugging of volcanic conduits turns the volcanic edifice into a pressure vessel. Any failure of that vessel is likely to result in explosive behaviour as water-supersaturated magma regains equilibrium. See Sparks (1997) for discussion of specific volcanic events resulting from such a process.

Vesicularity, density and permeability

The final vesicularity of experimental flows was between 80–95%, over which range axial permeability increased by at least an order of magnitude (Fig. 9a). Blower (2001) and Takeuchi *et al.* (2008) reviewed the relationship between bubble volume fraction and permeability of vesiculated magmas based on the experimental, numerical and field approaches of themselves and others. The relationship is non-linear and, at any particular vesicularity, a wide range of permeabilities is possible based on detailed local factors such as bubble size and shape (Blower 2001), the presence of crystals (Clarke *et al.* 2007) and the prevailing fluid-dynamic behaviour (Saar & Manga 1999). However, in general terms, the permeabilities of samples of vesiculated natural magmas appear to be about 10^{-12} ($\pm$ at least 1 decade) m^2 at *c.* 80% bubble fraction. Experimentally, axial permeability of whole gum rosin flows was some 10–100 times larger at 80% bubble fraction than that of vesicular magma samples, but similar to that of 'whole flow' estimates for vesiculated corn syrup (Namiki & Manga 2008). At 95% vesicularity, the experimental permeability was similarly elevated above extrapolations of magma permeabilities. This could indicate that, in comparison to natural magmas, the experimental bubble size was larger, the bubbles were (more) highly elongated, and/or, bubble coalescence was more rapid on

the experimental timescale (Blower 2001). Alternatively, the high experimental 'whole flow' permeability could indicate the remnants of flow-generated preferential gas-escape pathways (measurements were taken post-experiment) along the lines of those found by Taddeucci *et al.* (2006), or the influence of highly elongated bubbles enhancing axial permeability (Namiki & Manga 2008). Gas-escape pathways in magmas may elude detection not only because of size-scaling of the samples (too small), but also because of time-scaling, i.e. pathways may be transient (post-rupture healing) and permeability may change in the transition from magma to pyroclast.

In the model of Melnik & Sparks (1999) (Fig. 9b), degassing was only by means of vertical permeable flow through the magma column and was, therefore, relevant to our impermeable-walled experiments. Experimentally, vesicularity (Fig. 9a) declines gradually with height in the source region, then more rapidly in the flow cap/head or 'plug'. In the model of Melnik & Sparks (1999), such densification could theoretically occur, with gas volume fractions reducing within the top few tens of metres of the conduit when magma permeability is within a plausible range and a 'plug' has formed. However, at SHV the plug is considered a few hundreds of metres in vertical extent, a decade more than the model of Melnik & Sparks (1999). The high pressure-gradient in the experimental cap/head (Fig. 13a) acts to locally compress the gum rosin foam. Consequently, gas escape pathways close and permeability reduces as gas volume fraction declines. The relatively gas-tight nature of the experimental cap/head supports the finding that the main gas escape pathway was most likely along regions of active shear at the margin of the extruding flow, tapping deeper, more permeable fluid. The presence of this gas escape mechanism may explain why vesicularity declined more extensively in the experiment than suggested in the model of Melnik & Sparks (1999) and also why the magma plug is more extensive than a few tens of metres in depth.

Vulcanian explosive activity at SHV removed the magma plug and underlying material to depths approaching 2 km (Clarke *et al.* 2007). Analysis of natural samples from these eruptions (Clarke *et al.* 2007) suggests that magma density increases sharply (i.e. vesicularity and permeability decline) at pressures between *c.* 5–10 MPa, which equates to a high-density region (plug) *c.* 250–700 m deep. Figure 8 of Clarke *et al.* (2007) (Fig. 9c) indicates that a much greater proportion of the plug undergoes density increase than the model of Melnik & Sparks (1999) would suggest, and is closer to experimental observations. 1-D modelling (radial heterogeneity excluded) of conduit flow

(Fig. 8 of Clarke *et al.* (2007), see Fig. 9c) demonstrates that vertical degassing through a permeable plug does not explain the density increase found from field evidence. However, allowing radial degassing (but not including development of radial flow structure) by incorporating permeable conduit walls does result in the generation of a high-density plug from the point at which magma permeability becomes sufficiently high to allow open-system degassing. The extent of the plug may be sensitive to the magma input rate (Diller *et al.* 2006), where under static conditions a thin high-density cap is predicted to form over a high-vesicularity magma column. Experimentally, there was no input of fresh fluid, but the flow did not reach final equilibrium on experimental timescales and the development of a 'transitional magma plug' (Diller *et al.* 2006, Fig. 1) in the experiments is consistent with models having a finite input of fresh magma.

Using experimental and textural evidence, Kennedy *et al.* (2005) proposed strong radial heterogeneity of vesicularity within the SHV conduit, with dense, degassed material at the conduit margin and more gas-rich magma in the conduit centre. This structure was observed in the present experiments, with degassed gum rosin forming a shell up which lower viscosity fluid ascended. Vertical shear within the established experimental extrusive flow did not take place at the wall of the glass tube, but within the self-constructed gum rosin shell, i.e. the *conduit lining* described by Kennedy *et al.* (2005). Such radial heterogeneity is likely to be sustained if gas can readily escape from the margins of the flow, either through the wall/country rock or along a fracture system. Experiments demonstrate that both vertical and radial densification develops as a consequence of non-explosive flow of a degassing fluid whose viscosity increases as gas is lost; therefore the processes identified by both Kennedy *et al.* (2005) and Clarke *et al.* (2007) are here experimentally supported and may happen simultaneously to generate considerable axial and radial heterogeneity in parameters such as vesicularity, rheology and permeability.

Experimentally, the tube walls are impermeable, but the presence of open pathways generated by flow processes serves the same purpose as permeable walls, and a sizeable dense plug forms (Fig. 9a). It may also be possible that plug flow in the conduit of SHV (Collier & Neuberg 2006) and other volcanoes generates and sustains preferential pathways for gas escape and provides a significant radial component to magma degassing. The experiments call into question the need to invoke permeable country rocks in the conduit wall. It is pertinent to question how long such permeability

would remain open as particles and liquid intrude and block the rock pore space and fractures in similar fashion to those created by shear-fracture of magma (Tuffen & Dingwell 2005), but which are continually reformed by flow/fracture processes (Tuffen *et al.* 2003). It is interesting to note that Lane *et al.* (2001) demonstrated experimentally that the development of radial structure, most obvious in the form of inverse annular flow, precedes the onset of fragmentation in an expanding and degassing foam (but not with expansion alone (Namiki & Manga 2005)). All this suggests that the margins of a conduit-confined magma flow play a major role in determining physical behaviour under both explosive and effusive conditions, and is significant in switching between these two regimes (e.g. Melnik *et al.* 2005). This re-emphasizes the importance of considering radial flow structures in experimental, theoretical and field observations of magma flow in conduits (Massol & Jaupart 1999, Massol *et al.* 2001).

There is field evidence for the presence of fracture networks at the margins of flow within volcanic conduits (e.g. Stasiuk *et al.* 1996; Polacci *et al.* 2001; Gonnermann & Manga 2003; Tuffen *et al.* 2003), and the nature of plug flow is likely to create and sustain these over a significant vertical extent during periods of motion. Analogue experiments (Taddeucci *et al.* 2006) have demonstrated the development of fracture networks during expansion of degassing, viscoelastic silicone putty. The quasi-1-D model of Gonnermann & Manga (2003) showed that the development of high permeability at the margins of degassing magma flow through shear-induced fragmentation of the viscoelastic phase was a theoretical possibility. Estimating the onset of shear-induced fragmentation in gum-rosin flows using the threshold of Gonnermann & Manga (2003), an elastic modulus of $2-4$ GPa (Lane *et al.* 2001; Bagdassarov & Pinkerton 2004), and flow velocity of 1 mm/s (Fig. 11), yields a relaxed liquid viscosity approaching 10^9 Pa s and a thickness of about 2 mm for the permeable fractured zone at the flow margin. Such viscosities are high, but plausible for heavily degassed gum rosin solution (Phillips *et al.* 1995), and the estimate of 2 mm thickness for the fragmentation zone is neither unrealistically high nor low in relation to experimental dimensions, and therefore also plausible. An 'effective' viscosity of about 10^4 Pa s was suggested for the flow head, considerably less than the 10^9 Pa s emerging from the model of Gonnermann & Manga (2003). However, the estimate of 10^4 Pa s, which was too low to be consistent with the volatile concentration, may represent friction in the slip plane of a plug rather than a true viscosity. The experiments of Taddeucci *et al.* (2006) demonstrated that fracture, rather than wholesale fragmentation, is a likely outcome of margin shear rate exceeding the elastic threshold. Figure 4 of Taddeucci *et al.* (2006) lucidly illustrates that the onset of margin fracture promoted plug flow and consequent rapid expansion as wall drag was removed. In gum rosin flows, the slip plane of plug flow is hidden from view by the dense, degassed shell; however, whether fracture or fragmentation, the outcome of margin decrepitation is the formation of preferential gas-escape pathways and potential onset of plug flow. Vesicular materials are compressible, which raises the possibility of incremental rather than wholesale margin decrepitation; a process consistent with plug densification.

Ida (2007) develops a model for lateral gas flow in ascending magma, and postulates that gas migrates into fracture networks that provide means of gas loss. The consequence of Ida's (2007) model is a criterion for switching between effusive and explosive behaviour dependent on permeability and viscosity. The experiments presented here qualitatively support the criterion shown in Figure 10 of Ida (2007), in that explosive behaviour is demonstrated when average liquid viscosities are low (volatile content high), but this changes to effusive behaviour as average viscosity increases (volatile content low, Fig. 2). The experiments presented here appear to demonstrate a self-generating mechanism of promoting open system degassing, thereby suppressing explosive behaviour, and support the approach of Gonnermann & Manga (2003), Taddeucci *et al.* (2006) and Ida (2007). Viscous heating during vigorous magma flow could decrease viscosity in regions of very high shear (Polacci *et al.* 2001), increasing the shear rate required for margin decrepitation. However, in volcanic systems with low effusion rates, heat loss to country rock may act to increase liquid viscosity at the flow margin, suggesting that both the experiments presented here and the model of Gonnermann & Manga (2003) will underestimate the onset of margin decrepitation.

The hypothesis of shear-induced decrepitation generating preferential degassing pathways at the flow margin is also supported by the fortuitous real-time observations of circumferential degassing at Santiaguito Volcano by Bluth & Rose (2004). Here, gas and ash were erupted from around a magma plug. This suggests the gas was sourced from beneath the plug (and this could possibly be tested by gas analysis, Burton *et al.* 2007), in similarity to experimental observations, with the ash being generated as a consequence of margin decrepitation and plug motion. This behaviour may be promoted by the low-permeability plug that could otherwise erupt more explosively if gas were not able to escape from beneath. This opens up the possibility of a control or feedback mechanism

where pressure driving the flow dictates the physical extent of any preferential degassing pathways, and rates of degassing due to the preferential pathways change the flow-driving pressure; analogous to the weighted vent plug on a pressure cooker. The onset of margin decrepitation also suggests a positive feedback mechanism by encouraging volatile loss and reducing the shear rate at which decrepitation occurs, creating stable flow behaviour over a range of volatile concentrations.

Tilt cycles

During December 1996, Voight *et al.* (1998) measured cyclic tilt variation of $1-2$ μrad with a period of $6-8$ hr at SHV. Following installation of new tilt meters, tilt cycles of $10-25$ μrad and period $12-18$ hr were measured with deflation being more rapid than inflation. The tilt cycles were attributed to pressure changes some few hundred metres below the lava dome, and tilt peaks were seen to coincide with emissions of ash. Figure 14 illustrates the qualitative similarity between pressure-generated tilt cycles from SHV (Fig. 14d) and experimental (Fig. 14c) pressure cycles. The tilt cycles at SHV correlated with cycles in seismic amplitude, and SO_2 emission rates followed tilt-cycle amplitude (Watson *et al.* 2000). Seismic-amplitude cycles of similar period were also observed following the climactic eruption of Mt. Pinatubo in June 1991 (Denlinger & Hoblitt 1999). The seismic amplitude peaks at both volcanoes correspond to periods of rapid magma extrusion. More recently, Iverson *et al.* (2006) analyse the regular occurrence (30 to 300 s repeat period) of drumbeat earthquakes at Mt. St. Helens, although Waite *et al.* (2008) provide an alternative analysis. Chouet *et al.* (2005) seismically imaged conduit sections at Popocatépetl undergoing pressure cycles with a similar repeat period. Tilt cycles associated with vulcanian activity were also generated at Colima Volcano (Zobin *et al.* 2007).

Voight *et al.* (1999) describe the growth of the SHV dome as '...cyclic patterns of ground deformation... ...overprinted on a background of continuous lava extrusion.' This also describes the nature of the gum rosin experiments very well, with continuous flow being driven by an overpressure with superimposed smaller-scale pressure oscillations. Voight *et al.* (1999) model the deformation with a pressurized line source buried in an elastic half space. The depth of the top of the pressurized source was calculated to be 270 m below the base of the dome using this method, and they determine an upper bound on the amplitude of the pressure cycle of 60 MPa. This degree of pressurization is thought to be an overestimate since it is about twice the pressure estimated from the ballistic launch velocities (around 180 m s^{-1}) in the 17 September 1996 explosion (Robertson *et al.* 1998), which yielded 10 to 27 MPa. Experiments suggest that the pressure changes associated with the tilt cycles are likely to be small in comparison to the total superstatic pressure; a factor of 20 was found experimentally.

The model of Sparks (1997) identifies that a rheologically stiff magma plug is theoretically capable of sustaining tens of MPa of overpressure. Pressure continues to increase under the plug due to degassing-induced rheology change until the yield stress of the plug is overcome, allowing flow or explosion in some cases. A fresh batch of magma then ascends and the cycle repeats. A similar process was described by Stix *et al.* (1997) to explain the series of vulcanian explosions occurring at Galeras between 1992 and 1993.

Denlinger & Hoblitt (1999) modelled magma flow as a compressible Newtonian fluid with constant viscosity. Cyclic behaviour is introduced into the model by imposing a slip condition in the shallow parts of the conduit. Slip is thought to occur at shallow depths because of decreasing wall rock shear strength with decreasing pressure (Paterson 1978). When the magma flow rate exceeds a certain threshold value, flow stresses overcome the shear strength of the conduit wall at shallow levels; this causes an increase in the exit flow rate that results in pressure drop in the conduit. In turn, this pressure drop reduces the exit flow rate until the flow stresses again fall to a level that allows re-instatement of the no-slip condition. Melnik & Sparks (1999) proposed similar cyclic behaviour for an elastic conduit connected to a rigid chamber. The process of decoupling the magma from the wall rock is similar to the sliding flow phenomena exhibited experimentally before the onset of extrusive flow, the difference being that there are no wall rocks to break, but wall friction to overcome. However, in experimental extrusive flow during which pressure cycles are observed, margin decrepitation initiates the slip, where the margin is the boundary between the stationary shell of degassed gum rosin that lines the tube and the core of flowing foam. Margin decrepitation and wall rock failure under shear may be considered equivalent in terms of the effect on flow, with one a function of fluid rheology and the other a function of conduit mechanics.

Denlinger & Hoblitt (1999) do not take into account the probable non-linear viscosity of the magma. Vesiculation, microlite crystallization and degassing processes all combining to make the magma non-Newtonian and vertically (e.g. Sparks 1997) and radially (e.g. Gonnermann & Manga 2003) heterogeneous. However, magma heterogeneity could increase the likelihood of transition to a slip flow condition near the surface because, in this region of decreasing wall rock shear strength, the increasing magma viscosity would serve to intensify wall shear stress for a given velocity gradient at the wall. This mechanism is analogous to the transitions between sliding flow and extrusion flow observed in the experiments; however, the large number of unknowns make the timescales for this mechanism difficult to evaluate.

Diffusion pumping

Volcanic application of the experimental diffusion-pumping model requires well-characterized events; we choose the Soufrière Hills Volcano, Montserrat. The following parameters are required: the level of superstatic pressure related to the tilt cycles; the average radius of the bubbles in the pressurizing magma; the average melt shell radius for the bubbles in the pressurizing magma; the diffusion coefficient of water in the magma; the temperature of the magma; and the timescale of pressurization. Green *et al.* (2006) used plausible conduit dimensions, combined with the flow dynamics and surface topography, to explain the tilt amplitudes in terms of shear-induced pressure gradients in the region of $3.5 \times 10^4 \text{ Pa/m}$ (note that pressure gradients of 25 kPa/m were found in the experimental flow body). Green *et al.* (2006) estimate a tilt-cycle pressure change in the 0.5–1.5 MPa range, showing that viscous drag on the conduit wall is an effective tilt mechanism. Interestingly, the total pressure therefore appears to be in the region of 20 MPa, and the tilt-cycle pressure change in the region of 1 MPa, a ratio very similar to that found experimentally.

The magma temperature at the SHV was estimated at 820–880 °C using geothermometric methods (Barclay *et al.* 1998, avg. 850 °C or 1123 K). The diffusivity of water in the melt is given an arbitrarily reasonable value of $10^{-12} \text{ m}^2 \text{ s}^{-1}$ (Zhang & Behrens 2000). Melnik & Sparks (1999) estimate magma vesicularity as a function of depth at the Soufrière Hills volcano. The model allows for magma permeability, but the effect of permeability is small at the depth of interest. Their data estimate a value of vesicularity of about 0.3 at a depth of 500 m below the top of the dome, which is approximately the estimated depth of the pressurization source at the SHV (Voight *et al.* 1999). The melt shell volume was estimated from the melt shell radius and bubble radius used as initial values in the rhyolitic bubble growth model of Blower *et al.* (2001b), i.e. 10^{-3} m and 10^{-5} m respectively. This gives a melt shell volume *c.* $4 \times 10^{-9} \text{ m}^3$. The melt shell radius is linearly related to the bubble radius and can be estimated as the bubble radius divided by the cube root of gas volume fraction (Lensky et al. 2001). The estimated melt shell volume and porosity yield a bubble radius and melt shell radius in the region of 7.5×10^{-4} m and 1.1×10^{-3} m respectively at the pressurization source depth. These values for bubble and melt shell radii are somewhat arbitrary, but will suffice to illustrate the pressure cycle mechanism.

Applying the experimental diffusion-based pressure cycle model, the degree of water supersaturation of the magma (ΔC) required to produce the pressure cycles observed at the SHV is estimated by

$$\Delta C = \frac{\Delta P r_b^2 (S_b - r_b)}{3 S_b D k T \Delta}. \qquad (6)$$

This yields a value for the supersaturation of magmatic water needed to drive the tilt cycles by diffusion pumping of 6×10^{25} m^{-3}, which corresponds to about 0.07% w/w water. By considering the pressure change required to alter water solubility by this amount, the reasonableness of this degree of supersaturation may be assessed. Assuming pressure is at least lithostatic under the magma plug (>10 MPa) largely avoids the non-linear section of a Henry's law solubility model (Zhang 1999). With overpressure likely to reach 6–8 MPa (Melnik & Sparks 1999), or 3–15 MPa (Diller et al. 2006), an absolute pressure of 20 MPa seems reasonable, and is supported by petrological (Harford et al. 2003) and eruption (Robertson et al. 1998) evidence. At this pressure, a water concentration change of 0.07% w/w implies a pressure change of about 1.4 MPa, which given the approximations made, is close to the 1 MPa pressure change used in applying the diffusion pumping model, and consistent with Green et al. (2006). Chouet et al. (2006) give a more rigorous theoretical exploration of diffusion pumping, but the generation of plausible parameter values on application of the experimental model to the tilt cycles at SHV indicates this mechanism is just as likely for the pressurization section of SHV tilt cycles as it is at Popocatépetl (Chouet et al. 2005; 2006).

Figure 14d illustrates the effect that the dome collapse at about 12:45 on 25 June 1997 had on tilt cycles (Voight et al. 1998). Amplitude increased by about a factor of 1.6 and cycle period reduced by about 1.7. Also note that the pre-collapse 'noise' created by rock falls (Neuberg pers. comm.) disappears once the dome has collapsed. The post-collapse decrease in cycle period is largely explained by a shorter tilt-increase period. The experiments indicate that more rapid pressurization

could result from an increase in volatile supersaturation. Using equation 6, ΔP increases by 1.6 and Δt decreases by 1.7, with other variables not likely to change as much during dome collapse. This suggests a 2.7-fold increase in supersaturation to about 0.19% w/w water to account for the changed tilt cycle behaviour caused by dome collapse. This degree of supersaturation suggests a pressure reduction of about 3.7 MPa. The collapse event reduced dome height by about 100 m (Green & Neuberg 2006). Assuming an average dome density of 3000 kg m^{-3} in this region (Fig. 9c), a reduction in static pressure of about 3.0 MPa within the conduit is expected. Given the approximations and assumptions made, we consider 3.7 MPa and 3.0 MPa to be in good agreement and support the diffusion-pumped mechanism. This analysis indicates that tilt cycles may potentially be used to estimate both absolute values of, and changes in, volatile supersaturation.

Dome collapse results in partial removal of the rheologically stiffened plug. Magma flow rate must then increase as the pressure gradient equilibrates to the new rheology. The reservoir of overpressured vesicular magma imprisoned by the plug is then likely to drive the increased flow rate for some time. The increased flow rate is likely to cause greater tilt amplitude through wall shear stress (Green et al. 2006), implying that directly scaling ΔP changes to tilt amplitude gives a maximum value. It is possible that more volatile-rich magma is drawn into the base of the tilt-cycle mechanism by the removal of degassed material from the top of the conduit. The increase in volatile content is likely to give a larger diffusivity and increase the rate of diffusive pressure increase. The depressurization caused by dome collapse may also result in the nucleation of new small bubbles, enhancing the surface area over which diffusion can increase pressure. In the diffusion-pumped mechanism, these processes all act to decrease the tilt cycle period following dome collapse; therefore the changes that occurred over the dome collapse event are consistent with diffusive control of pressure rises driving tilt cycles.

Given the limited and dynamic nature of the tilt-cycle record from SHV, there is no direct evidence of systematic increase in the tilt cycle period as observed for the pressure cycle period in the experiments. This suggests that the relevant diffusion coefficient remained approximately constant at SHV. This would be consistent with differences in the reservoir of volatile available. This was small experimentally and the gas escape during each depressurization was a measurable proportion of the total diethyl ether available. However, in comparison to the reservoir of water available in the SHV conduit, the quantity escaping during each tilt decline was too small to significantly decrease the diffusivity.

Green & Neuberg (2006) propose that fracturing, allowing gas loss, could account for the observed relationship between seismic and tilt cycles at SHV. Correlation of seismic families with tilt cycles (Green & Neuberg 2006; Neuberg et al. 2006) indicated that seismicity increased greatly as rate-of-tilt went from rising to falling, i.e. driving pressure was starting to decline as gas escape pathways formed. This mechanism also explains the experimental observation that gas escape rates were increasing for some time before the maximum base pressure was reached (Fig. 15c). The permeability of the zone of margin decrepitation is not rapidly switched, therefore, but changes more gradually over time. This is consistent with a compressible plug. Maximum experimental gas escape rates occurred during the peak rate of decline in base pressure (Fig. 15c), which correlates with the observation of Watson et al. (2000) that SO_2 emissions from SHV tended to be greatest during tilt waning. Maximum gas escape indicates maximum permeability of the margin decrepitation region, probably representing the time over which plug motion was occurring and the fracture network was most extensive and connected to the flow front. Seismicity appears to decline once the tilt maximum has been reached (Neuberg et al. 2006, Fig. 3), i.e. rate of gas escape exceeds rate of replenishment. Seismicity then all but ceases as the rate-of-tilt changes from decreasing to increasing, i.e. the driving pressure is building up again as gas escape pathways are sealed. Experimentally, gas escape declines to background levels at about the base pressure minimum. This would suggest that seismicity is due to formation of cracks and gas escaping through these cracks, but that once gas flow rates fall from their maximum, seismicity rapidly declines. It is worth noting that seismic sources associated with tilt cycles at SHV appear fixed at c. 500 m below sea level (Neuberg et al. 2006). It is not clear why this should be so, and it is possible that the source location is being defined by the geometry of the conduit at this depth (Chouet, comment at workshop 'The Physics of Fluid Oscillations in Volcanic Systems', Lancaster University, 2006).

Explosive behaviour

The nature of experimental fragmenting flows, where inverse annular flow is a pre-requisite for sustained explosive behaviour (Lane et al. 2001), suggests that any transition from dome growth to explosive activity is likely to result from failure of the plug-flow to Poiseuille-flow transition. The onset of the declining stage of a tilt cycle is,

therefore, a high-risk period for explosive behaviour to emerge; Voight et al. (1999) observed vulcanian explosions occurring at this point of the tilt cycle. The deflation accompanying the explosive behaviour appears more rapid than in the absence of explosions, suggesting that the degassing mode has changed. The rate of the tilt decline due to a vulcanian explosion in the final measured cycle (4 Aug, 1997, 16:40 hr) exceeds the sampling rate of the tilt meter. This very rapid tilt decline may represent a detachment of the flow from the conduit lining at the onset of explosive activity, thereby reducing the contribution of wall shear to the tilt signal (Green et al. 2006) on the timescale of the change in flow pattern. Such behaviour is consistent with the experimental observation of inverse annular flow accompanying decrepitation (Lane et al. 2001).

The plug-flow to Poiseuille-flow transition may fail because the low-permeability plug becomes extruded from the conduit when average flow rate is faster than plug growth rate. The degree of permeability of the bubbly fluid underlying the plug may be significant here, because the transition from plug flow to Poiseuille flow relies on efficient gas escape to reduce driving pressure, rather than extrusion of the plug. If permeability declines, then the plug may be extruded to the point that overpressured material can expand explosively; the analogy being the uncorking of an 'overpressured' bottle. Tilt cycles at SHV occurred during dome growth, which evolved into a series of vulcanian explosions with mean separation times similar to the tilt cycle period (Druitt et al. 2002). The sequence of the vulcanian explosions suggests that the basic flow mechanism had not changed, but the mode of degassing certainly had. We postulate that the transition from dome growth with decoupled degassing, to vulcanian explosion with partly coupled degassing, resulted from a decrease in the permeability of magma underlying the plug; Figure 17 schematically illustrates this process. If gas cannot escape so readily, then the plug motion will be more extensive. As wall friction declines to small values, the plug flow changes to inverse annular flow, with fragmentation ensuing. Both states then return to Poiseuille flow following pressure decrease and the cycle repeats.

The use of tilt-cycle data, and the longer timescale tilt record on which it is superimposed, potentially provide a means of forecasting volcanic activity, at least on short timescales. Tilt cycles at SHV occurred on a rising tilt background preceding the dome collapse of 25 June, 1997. The increasing tilt background is likely to represent an increase in both conduit overpressure and effusion rate on which the tilt cycles are superimposed, and increasing pressure is likely to promote dome collapse

(Voight & Elsworth 2000). The increase in tilt may represent an increase in pressure at some depth being communicated to the upper conduit.

Tilt cycles occurred on a declining tilt background just before and during the explosive behaviour of 4 August, 1997, prior to the unfortunate destruction of the tilt meter. Declining tilt background suggests that conduit overpressure is decreasing. One cause of this could be the progressive removal of the rheologically stiff plug that would allow the axial pressure gradient in the conduit to become more linear. The onset of explosive activity accompanying tilt decline supports this mechanism. If the cause of plug removal were a decrease in the permeability of magma underlying the plug, then a reduction in decoupled degassing rates may well accompany the declining tilt signal, and partly coupled degassing (explosive activity) becomes the dominant gas-escape mode, with diffusive pumping still acting to increase pressure.

Motion of a magma plug

There appear to be significant parallels between the flow mechanisms occurring in the experiments documented here and those taking place during dome building at SHV, and possibly other volcanoes. A rheologically stiff plug impedes flow and localizes the pressure gradient causing significant superstatic pressure to exist below the plug. The plug moves by viscous flow, but as the pressure increases below the plug, increasing the driving force on the plug base and hence the plug rise rate, the viscoelastic transition is approached within regions of high strain rate at the flow edges (e.g. Gonnermann & Manga 2003). Margin decrepitation then creates gas-escape pathways along the plug boundary to the surface (or into the country rock), and viscous flow gives way to plug flow. Volcanically, the fracturing and high-pressure gas escape cause increased seismic activity. Once gas can readily escape from the permeable foam beneath the plug, the driving pressure rapidly declines and the plug rise rate slows. This allows the re-establishment of viscous control on the flow and any fractures quickly fill with debris and/or heal by liquid flow. The rapid pressure drop creates a state of water supersaturation, and the resulting diffusion of water vapour onto bubbles increases the pressure until margin decrepitation is again initiated. Essentially, the process oscillates around the viscous-elastic transition at the flow margin.

In order to constrain the reasonableness of this mechanism, we now explore the motion of a high viscosity degassed plug of magma in a volcanic conduit using a simplified one-dimensional model.

In this model, friction acting in a thin layer between the solid plug and the conduit wall resists the motion of the plug; we model this here as a viscous force. We have assumed a circular cross-section for both plug and conduit leading to Couette flow solutions for the viscous drag, but in general, the model solutions are not sensitive to this choice of geometry. The model formulation is one-dimensional in the vertical direction and is illustrated schematically in Figure 18a. The rheologically stiff plug overlies a permeable region of assumed constant volume into which volatile gases can diffuse, which results in an increase in pressure in this region. The gas pressure exerts a force on the base of the plug that is equal in magnitude to the pressure multiplied by the base area of the plug (PA). This force acts vertically upwards and is opposed by the weight of the plug (Mg). The viscous resistance to flow in the annular region between the plug and the conduit wall is velocity-dependent and can be written

$$F_v = \mu \frac{A_f}{t_f} \dot{x}, \qquad (7)$$

where μ is the fluid viscosity, A_f is the area of the fluid film in contact with the conduit wall; t_f is the liquid film thickness and $\dot{x}$ is the velocity of the plug (Schlichting 1979).

As a first approximation, we assume that the plug acts as an 'on−off valve', such that when a critical pressure P_c is reached in the permeable region below the plug, motion of the plug is initiated and the pressure is released by gas venting. We envisage that pressurization occurs due to gas diffusion from the melt, so the pressure beneath the plug would increase as the square root of time. However, the tilt cycles reflecting conduit pressurization at SHV (and the experimental pressure rise) show an approximately linear increase of pressure with time (Voight *et al.* 1999), therefore we use this time dependence in these illustrative calculations. For pressures less than P_c, we can write the pressure below the plug in the form

$$P = k_1 Q_{in} t, \qquad (8)$$

where Q_{in} is the volumetric flux of pressurizing gas and k_1 is a coefficient that describes the pressurization per unit volume of the permeable region to which gas is being added by diffusion. When the pressure increases to a value that equals P_c, viscous flow switches to plug flow and gas can be released. If we describe the gas release using a volumetric release flux, Q_{out}, defined by analogy

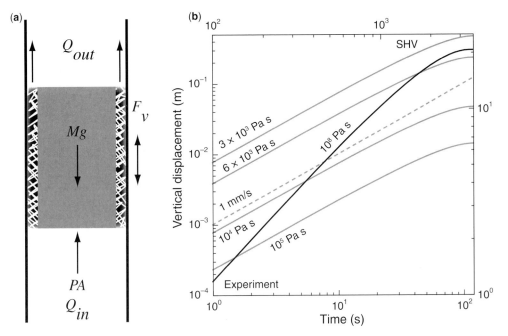

Fig. 18. Conceptual diagram (**a**) for formulation of 1-D plug-motion model indicates the input and output of gas and the three forces acting to control plug acceleration. In (**b**), model results are shown for experiment conditions (grey lines) with comparison to measured flow velocity of approximately 1 mm/s. Model result for SHV (black line) with a film viscosity of 10^8 Pa s.

with the pressurizing gas flux (cf Equation 8), the pressure underneath the plug can be written

$$P = P_c - k_2(Q_{out} - Q_{in})t, \qquad (9)$$

where $Q_{out} \gg Q_{in}$ and k_2 is a coefficient that describes the depressurization per unit volume of the permeable region which is determined from observations (in this case, the tilt cycle decay-typically 1 hr-as a proxy for the decompression rate at SHV, and experimental depressurization time (typically 10 s)). Note that the value P_c chosen includes the static pressure appropriate for the initiation of motion, but the dynamics of that motion depend on the weight of the plug and the viscous resistance.

We can write an equation of motion for the plug as:

$$M\ddot{x} = A(P_c - k_2)(Q_{out} - Q_{in})t) - Mg$$
$$- \mu \frac{A_f}{t_f} \dot{x}. \qquad (10)$$

The terms on the right-hand side represent the force on the base of the plug generated by gas pressurization, the weight of the plug and the viscous resistance respectively. Note that this equation has a similar form to that derived by Iverson *et al.* (2006) for the motion of a rheologically stiffened

plug. Equation 10 was solved numerically using a fourth-order Runge-Kutta scheme (in this study, we used the mathematical software package Maple^{TM}) with initial conditions $x = 0$, $\dot{x} = 0$ at $t = 0$. Parameters used to evaluate equation 10 for SHV and experiments are shown in Table 3.

Application of this model to experiments is shown in Figure 18b for a range of viscous friction for the annular film. The model results are sensitive to the liquid viscosity, and in the parameter space appropriate for the experimental conditions, the viscous resistance term dominates, indicating that propagation of the plug essentially takes the form of a viscously damped expansion. For a flow-front velocity of 1 mm/s (Fig. 11), the annular film has a friction similar to that of a liquid of viscosity 10^4 Pa s. This is many orders of magnitude less than the relaxed viscosity (c. 10^9 Pa s) emerging from the model of Gonnermann & Manga (2003), and illustrates the lubricating effect of the failure of bubble walls during margin decrepitation. Experimentally, the decompression phase of a pressure cycle typically lasted 10 s; however, Figure 18 suggests that plug flow can continue for up to about 100 s. This results from the model assumption that plug flow continues until the upward velocity falls to zero, when the annular film seals and true viscous-flow resumes. Because of the low density of the experimental

Table 3. *Parameters used for the evaluation of equation 10*

Parameter	SHV, Montserrat	Experiment
Tube/conduit radius	15 m	0.02 m
Plug dimension	500 m	0.1 m
	(Clarke *et al.* 2007)	
Plug density	2500 kg m^{-3}	200 kg m^{-3}
Film viscosity	10^8 Pa s	Range
	(Sparks 1997)	
Film thickness	0.1 m	2 mm
Critical pressure P_c	20 MPa	20 kPa
Pressure drop	1 MPa in 1 hour	1 kPa in 10 s
Pressure rise	1 MPa in 10 hours	1 kPa in 100 s

plug, this assumption results in loss of most of the driving pressure before plug velocity reaches zero. Note that Mg results in a static pressure (c. 200 Pa) only about 20% that of the pressure cycle height (c. 1 kPa) and that the ratio of PA/Mg is of order 100 (Fig. 18a). The measured 1 kPa pressure loss is constrained to take place over 10 s, when model plug velocity is still significant. This suggests that the viscous resistance, dominant under the experimental conditions, has velocity dependence such that plug motion can cease abruptly below some critical non-zero plug velocity. This is typically the case for natural systems controlled by viscous resistance, or granular particle motion, which exhibit 'stick-slip' friction (e.g. Iverson *et al.* 2006). For the present experiments, we postulate that this mechanism is the re-establishment of significant viscous behaviour of material in the brittle-fragmented flow margin.

The model was also applied to parameters relevant to SHV (Table 3, Fig. 18b). Here, the plug velocity fell close to zero during the constrained 1 MPa pressure drop over one hour. The viscous term did not dominate under these conditions, with the Mg term creating a static pressure (c. 10 MPa), at least an order of magnitude larger than the volcanic pressure-cycle height (c. 1 MPa) and the ratio of PA/Mg was of order 2 (Fig. 18a). The smaller relative contribution made by the F_v term, and more dominant Mg inertial term, suggests that the role of velocity-dependent viscous friction is less important in the volcanic case than experimentally. Figure 18b indicates a plug displacement in the region of 20 m during the waning of a tilt cycle. Spine growth was a common feature of dome growth at SHV (Smithsonian Institution 2002), and we postulate that the spines grew as a result of the plug-flow phase during tilt cycles. Observations of spine growth over timescales of one hour are not available, but growth rates of 20 m/day have been reported at SHV (Smithsonian Institution 2002). However, the timescale of one tilt cycle approximates to one day and field observation of spine growth is, therefore, consistent with the fluid-dynamic interpretation of tilt cycles presented here.

Conclusions

Flows of fluids driven by an exsolving gas species, and exhibiting rapidly increasing liquid viscosity on degassing, show similar behaviour at both laboratory and volcanic scales. In time, a rheology gradient forms a plug and allows the development of a non-linear axial pressure gradient leading to overpressure beneath the plug. An oscillation may emerge between Poiseuille flow and plug flow that results in pressure cycles in experiments and tilt cycles in volcanoes, where the oscillatory pressure change is a small fraction of the prevailing overpressure.

The physical mechanisms driving flow oscillation are diffusive pumping and margin decrepitation. During Poiseuille flow, pressure rise is driven by diffusion of exsolving gas into bubbles within a closed system sealed by a densified, low-permeability plug. Maximum strain rate, which is focused at the flow margin, increases to the point where bubble walls fracture resulting in margin decrepitation. This both initiates plug flow and opens fracture pathways, with plug flow maintaining a network of fractures at the flow margin. This results in open-system degassing of permeable foam situated below the low-permeability plug. Gas escapes more rapidly than diffusion can replenish, and pressure falls. The plug decelerates, strain-rate at the margin decreases and fractures heal by blockage and viscous flow; thus shutting off gas-escape pathways. Poiseuille flow re-establishes and diffusive pumping increases pressure into the next cycle.

Barry Voight is gratefully acknowledged for access to tilt cycle data. GAR thanks the NERC for PhD funding. SJL thanks the Royal Society for an Equipment Grant. We are very grateful to Amanda Clarke and Jacopo Taddeucci for highly constructive criticism that greatly improved the manuscript.

References

BAGDASSAROV, N. & PINKERTON, H. 2004. Transient phenomena in vesicular lava flows based on laboratory experiments with analogue materials. *Journal of Volcanology and Geothermal Research*, **132**, 115–136.

BARCLAY, J., RUTHERFORD, M. J., CARROLL, M. R., MURPHY, M. D., DEVINE, J. D., GARDNER, J. & SPARKS, R. S. J. 1998. Experimental phase equilibria constraints on pre-eruptive storage conditions of the Soufriere Hills magma. *Geophysical Research Letters*, **25**, 3437–3440.

BLOWER, J. D. 2001. Factors controlling permeability-porosity relationships in magma. *Bulletin of Volcanology*, **63**, 497–504.

BLOWER, J. D., KEATING, J. P., MADER, H. M. & PHILLIPS, J. C. 2001a. Inferring volcanic degassing processes from vesicle size distributions. *Geophysical Research Letters*, **28**, 347–350.

BLOWER, J. D., MADER, H. M. & WILSON, S. D. R. 2001b. Coupling of viscous and diffusive controls on bubble growth during explosive volcanic eruptions. *Earth and Planetary Science Letters*, **193**, 47–56.

BLUTH, G. J. S. & ROSE, W. I. 2004. Observations of eruptive activity at Santiaguito volcano, Guatemala. *Journal of Volcanology and Geothermal Research*, **136**, 297–302.

BURTON, M., ALLARD, P., MURÉ, F. & LA SPINA, A. 2007. Magmatic gas composition reveals the source depth of slug-driven Strombolian explosive activity. *Science*, **317**, 227–230.

CHOUET, B., DAWSON, P. & ARCINIEGA-CEBALLOS, A. 2005. Source mechanism of Vulcanian degassing at Popocatépetl volcano, Mexico, determined from waveform inversions of very long period signals. *Journal of Geophysical Research*, **110**, B07301, doi:10.1029/2004JB003524.

CHOUET, B., DAWSON, P. & NAKANO, M. 2006. Dynamics of diffusive bubble growth and pressure recovery in a bubbly rhyolitic melt embedded in an elastic solid. *Journal of Geophysical Research*, **111**, B07310, doi:10.1029/2005JB004174.

CLARKE, A. B., STEPHENS, S., TEASDALE, R., SPARKS, R. S. J. & DILLER, K. 2007. Petrologic constraints on the decompression history of magma prior to Vulcanian explosions at the Soufrière Hills volcano, Montserrat. *Journal of Volcanology and Geothermal Research*, **161**, 261–274.

COBBOLD, P. R. & JACKSON, M. P. A. 1992. Gum rosin (colophony): a suitable material for thermomechanical modeling of the lithosphere. *Techonophysics*, **210**, 255–271.

COLLIER, L. & NEUBERG, J. 2006. Incorporating seismic observations into 2D conduit flow modelling. *Journal of Volcanology and Geothermal Research*, **152**, 331–346, doi:10.1016/j.jvolgeores.2005.11.009.

CRANK, J. 1975. *The Mathematics of Diffusion* (2nd edn). Oxford University Press.

DENLINGER, R. P. & HOBLITT, R. P. (1999). Cyclic eruptive behaviour of silicic volcanoes. *Geology*, **27**, 459–462.

DILLER, K., CLARKE, A. B., VOIGHT, B. & NERI, A. 2006. Mechanisms of conduit plug formation: implications for vulcanian explosions. *Geophysical Research Letters*, **33**, L20302, doi:10.1029/2006GL027391.

DRUITT, T. H., YOUNG, S. R., BAPTIE, B. J. ET AL. 2002. Episodes of cyclic vulcanian explosive activity with fountain collapse at Soufrière Hills Volcano, Montserrat. *In*: DRUITT, T. H. & KOKELAAR, B. P. (eds) *The Eruption of Soufrière Hills Volcano, Montserrat, from 1995 to 1999*. Geological Society of London Memoir, **21**, 281–306.

EICHELBERGER, J. C., CARRIGAN, C. R., WESTRICH, H. R. & PRICE, R. H. 1986. Non-explosive silicic volcanism. *Nature*, **323**, 589–602.

GARDINER, B. S., DLUGOGORSKI, B. Z. & JAMESON, G. J. (2000). The steady shear of three-dimensional wet polydisperse foams. *Journal of Non-Newtonian Fluid Mechanics*, **92**, 151–166.

GONNERMANN, H. M. & MANGA, M. 2003. Explosive volcanism may not be the inevitable consequence of magma fragmentation. *Nature*, **426**, 432–435.

GREEN, D. N. & NEUBERG, J. 2006. Waveform classification of volcanic low-frequency earthquake swarms and its implication at Soufrière Hills Volcano, Montserrat. *Journal of Volcanology and Geothermal Research*, **153**, 51–63, doi:10.1016/j.jvolgeores.2005.08.003.

GREEN, D. N., NEUBERG, J. & CAYOL, V. 2006. Shear stress along a conduit wall as a plausible source of tilt at Soufrière Hills volcano, Montserrat. *Geophysical Research Letters*, **33**, L10306, doi:10.1029/2006GL-25890.

HANSELMANN, W. & WINDHAB, E. 1996. Foam flow in pipes. *Applied Rheology*, **6**, 253–260.

HARFORD, C. L., SPARKS, R. S. J. & FALLICK, A. E. 2003. Degassing at the Soufrière Hills Volcano, Montserrat, recorded in matrix glass compositions. *Journal of Petrology*, **44**, 1503–1523.

HESS, K. U. & DINGWELL, D. B. 1996. Viscosities of hydrous leucogranite melts: a non-Arrhenian model. *American Mineralogist*, **81**, 1297–1300.

IDA, Y. 2007. Driving force of lateral permeable gas flow in magma and the criterion of explosive and effusive eruptions. *Journal of Volcanology and Geothermal Research*, **162**, 172–184.

IVERSON, R. M., DZURISIN, D., GARDNER, C. A. ET AL. 2006. Dynamics of seismogenic volcanic extrusion at Mount St Helens in 2004–05. *Nature*, **444**, 439–443, doi:10.1038/nature05322.

JAUPART, C. 1998. Gas loss from magmas through conduit walls during eruptions. *In*: GILBERT, J. S. & SPARKS, R. S. J. (eds) *The Physics of Explosive Volcanic Eruptions*. Geological Society, London, Special Publications, **145**, 73–90.

JAUPART, C. & ALLÈGRE, C. J. 1991. Gas content, eruption rate and instabilities of eruption regime in silicic volcanoes. *Earth and Planetary Science Letters*, **102**, 413–429.

JENKINS, S. F., MAGILL, C. R. & MCANENEY, K. J. 2007. Multi-stage volcanic events: a statistical investigation. *Journal of Volcanology and Geothermal Research*, **161**, 275–288.

KENNEDY, B., SPIELER, O., SCHEU, B., KUEPPERS, U., TADDEUCCI, J. & DINGWELL, D. B. 2005. Conduit implosion during Vulcanian eruptions. *Geology*, **33**, 581–584, doi:10.1130/G21488.1.

LANE, S. J., CHOUET, B. A., PHILLIPS, J. C., DAWSON, P., RYAN, G. A. & HURST, E. 2001. Experimental observations of pressure oscillations and flow regimes in an analogue volcanic system. *Journal of Geophysical Research*, **106**, 6461–6476.

LENSKY, N. G., LYAKHOVSKY, V. & NAVON, O. 2001. Radial variations of melt viscosity around growing bubbles and gas overpressure in vesiculating magmas. *Earth and Planetary Science Letters*, **186**, 1–6.

MADER, H. M., PHILLIPS, J. C., SPARKS, R. S. J. & STURTEVANT, B. 1996. Dynamics of explosive degassing of magma: observations of fragmenting two-phase flows. *Journal of Geophysical Research*, **101**, 5547–5560.

MASSOL, H. & JAUPART, C. 1999. The generation of gas overpressure in volcanic eruptions. *Earth and Planetary Science Letters*, **166**, 57–70.

MASSOL, H., JAUPART, C. & PEPPER, D. W. 2001. Ascent and decompression of viscous magma in a volcanic conduit. *Journal of Geophysical Research*, **106**, 16,223–16,240.

MELNIK, O. & SPARKS, R. S. J. 1999. Nonlinear dynamics of lava dome extrusion. *Nature*, **402**, 37–41.

MELNIK, O., BARMIN, A. A. & SPARKS, R. S. J. 2005. Dynamics of magma flow inside volcanic conduits with bubble overpressure buildup and gas loss through permeable magma. *Journal of Volcanology and Geothermal Research*, **143**, 53–68, doi:10.1016/j.jvolgeores.2004.09.010.

MOURTADA-BONNEFOI, C. C. & MADER, H. M. 2001. On the development of highly-viscous skins of liquid around bubbles during magmatic degassing. *Geophysical Research Letters*, **28**, 1647–1650.

MOURTADA-BONNEFOI, C. C. & MADER, H. M. 2004. Experimental observations of the effects of crystals and pre-existing bubbles on the dynamics and fragmentation of vesiculating flows. *Journal of Volcanology and Geothermal Research*, **129**, 83–97, doi:10.1016/S0377-0273(03)00233-6.

NAMIKI, A. & MANGA, M. 2005. Response of a bubble bearing viscoelastic fluid to rapid decompression: implications for explosive volcanic eruptions. *Earth and Planetary Science Letters*, **236**, 269–284.

NAMIKI, A. & MANGA, M. 2006. Influence of decompression rate on the expansion velocity and expansion style of bubbly fluids. *Journal of Geophysical Research*, **111**, B11208.

NAMIKI, A. & MANGA, M. 2008. Transition between fragmentation and permeable outgassing of low viscosity magmas. *Journal of Volcanology and Geothermal Research*, **169**, 48–60.

NAVON, O., CHEKHMIR, A. & LYAKHOVSKY, V. 1998. Bubble growth in highly viscous melts: theory, experiments, and autoexplosivity of dome lavas. *Earth and Planetary Science Letters*, **160**, 763–776.

NEUBERG, J., TUFFEN, H., COLLIER, L., GREEN, D., POWELL, T. & DINGWELL, D. 2006. The trigger mechanism of low-frequency earthquakes on Montserrat. *Journal of Volcanology and Geothermal Research*, **153**, 37–50, doi:10.1016/j.jvolgeores.2005.08.008.

NEWHALL, C. G. & MELSON, W. G. 1983. Explosive activity associated with the growth of volcanic domes, *Journal of Volcanology and Geothermal Research*, **17**, 111–131.

PAPALE, P. 1999. Strain-induced magma fragmentation in explosive eruptions. *Nature*, **397**, 425–428.

PATERSON, M. S. (1978). *Experimental Rock Deformation — The Brittle Field*. Berlin, Germany, Springer-Verlag.

PHILLIPS, J. C., LANE, S. J., LEJEUNE, A. M. & HILTON, M. 1995. Gum rosin-acetone system as an analogue to the degassing behaviour of hydrated magmas. *Bulletin of Volcanology*, **57**, 263–268.

POLACCI, M., PAPALE, P. & ROSI, M. 2001. Textural heterogeneities in pumices from the climactic eruption of Mount Pinatubo, 15 June 1991, and implications for magma ascent dynamics. *Bulletin of Volcanology*, **63**, 83–97.

PROUSSEVITCH, A. A., SAHAGIAN, D. L. & ANDERSON, A. T. 1993. Dynamics of diffusive bubble growth in magmas: isothermal case. *Journal of Geophysical Research*, **98**, 22283–22307.

PROUSSEVITCH, A. A. & SAHAGIAN, D. L. 1998. Dynamics and energetics of bubble growth in magmas: Analytical formulation and numerical modeling. *Journal of Geophysical Research*, **103**, 18223–18251.

ROBERTSON, R., COLE, P., SPARKS, R. S. J. *ET AL.* 1998. The explosive eruption of Soufrière Hills Volcano, Montserrat, West Indies, 17 September, 1996. *Geophysical Research Letters*, **25**, 3429–3432.

RYAN, G. A. 2002. *The flow of rapidly decompressed gum rosin di-ethyl ether and implications for volcanic eruption mechanisms*. PhD thesis, University of Lancaster, UK.

SAAR, M. O. & MANGA, M. 1999. Permeability-porosity relationship in vesicular basalts. *Geophysics Research Letters*, **26**, 111–114.

SCHLICHTING, H. 1979. *Boundary-Layer Theory*, 7th edn. McGraw-Hill, NY.

SMITHSONIAN INSTITUTION. 2002. Monthly report for Soufrière Hills Volcano (Montserrat). *Bulletin of the Global Volcanism Network*, **27**, 06.

SPARKS, R. S. J. 1997. Causes and consequences of pressurisation in lava dome eruptions. *Earth and Planetary Science Letters*, **150**, 177–189.

SPARKS, R. S. J. & PINKERTON, H. (1978). Effect of degassing on rheology of basaltic magma. *Nature (London)*, **276**, 385–386.

STASIUK, M. V., BARCLAY, J., CARROLL, M. R., JAUPART, C., RATTE, J. C., SPARKS, R. S. J. & TAIT, S. R. 1996. Degassing during magma ascent in the Mule Creek vent (USA). *Bulletin of Volcanology*, **58**, 117–130.

STASIUK, M. V., JAUPART, C. & SPARKS, R. S. J. 1993. Influence of cooling on lava-flow dynamics. *Geology*, **21**, 335–338.

STIX, J., TORRES, R., NARVAEZ, L., CORTES, G. P., RAIGOSA, J., GOMEZ, D. & CASTONGUAY, R. 1997. A model of Vulcanian eruptions at Galeras volcano, Colombia. *Journal of Volcanology and Geothermal Research*, **77**, 285–303.

TADDEUCCI, J., SPIELER, O., ICHIHARA, M., DINGWELL, D. B. & SCARLATO, P. 2006. Flow and fracturing of viscoelastic media under diffusion-driven bubble growth: an analogue experiment for eruptive volcanic conduits. *Earth and Planetary Science Letters*, **243**, 771–785.

TAKEUCHI, S., NAKASHIMA, S. & TOMIYA, A. 2008. Permeability measurements of natural and experimental volcanic materials with a simple permeameter: toward an understanding of magmatic degassing processes. *Journal of Volcanology and Geothermal Research*, doi:10.1016/j.jvolgeores.2008.05.010.

TORAMARU, A. 1995. Numerical study of nucleation and growth of bubbles in viscous magmas. *Journal of Geophysical Research*, **100**, 1913–1931.

TUFFEN, H. & DINGWELL, D. 2005. Fault textures in volcanic conduits: evidence for seismic trigger mechanisms during silicic eruptions. *Bulletin of Volcanology*, **67**, 370–387, doi:10.1007/s00445–004-0383–5.

TUFFEN, H., DINGWELL, D. B. & PINKERTON, H. 2003. Repeated fracture and healing of silicic magma generate flow banding and earthquakes? *Geology*, **31**, 1089–1092.

VOIGHT, B. & ELSWORTH, D. 2000. Instability and collapse of hazardous gas-pressurized lava domes. *Geophysical Research Letters*, **27**, 1–4.

VOIGHT, B., SPARKS, R. S. J., MILLER, A. D. *ET AL.* 1999. Magma flow instability and cyclic activity at Soufrière Hills Volcano, Montserrat, British West Indies. *Science*, **283**, 1138–1141, doi:10.1126/science.283.5405.1138.

VOIGHT, B., HOBLITT, R. P., CLARKE, A. B., LOCKHART, A. B., MOLLER, A. B., LYNCH, L. & MCMAHON, J. 1998. Remarkable cyclic ground deformation monitored in real-time on Montserrat, and its use in eruption forecasting. *Geophysical Research Letters*, **25**, 3405–3408.

WAITE, G. P., CHOUET, B. A. & DAWSON, P. B. 2008. Eruption dynamics at Mount St. Helens imaged from broadband seismic waveforms: Interaction of the shallow magmatic and hydrothermal systems. *Journal of Geophysical Research*, **113**, B02305, doi:10.1029/2007JB005259.

WATSON, I. M., OPPENHEIMER, C., VOIGHT, B. *ET AL.* 2000. The relationship between degassing and ground deformation at Soufrière Hills Volcano, Montserrat. *Journal of Volcanology and Geothermal Research*, **98**, 117–126.

ZHANG, Y. X. 1999. H_2O in rhyolitic glasses and melts: measurement, speciation, solubility, and diffusion. *Reviews of Geophysics*, **37**, 493–516.

ZHANG, Y. X. & BEHRENS, H. 2000. H_2O diffusion in rhyolitic melts and glasses. *Chemical Geology*, **169**, 243–262.

ZOBIN, V. M., SANTIAGO-JIMÉNEZ, H., RAMÍREZ-RUIZ, J. J., REYES-DÁVILA, G. A., BRETÓN-GONZÁLEZ, M. & NAVARRO-OCHOA, C. 2007. Quantification of volcanic explosions from tilt records: Volcán de Colima, México. *Journal of Volcanology and Geothermal Research*, **166**, 117–124.

Index